Livestock Health and Production

Livestock Health and Production

Editor: Johann Casini

RCALLISTO
REFERENCE
www.callistoreference.com

Callisto Reference,
118-35 Queens Blvd., Suite 400,
Forest Hills, NY 11375, USA

Visit us on the World Wide Web at:
www.callistoreference.com

ISBN: 978-1-64116-201-2 (Hardback)

Trademark Notice: Registered trademark of products or corporate names are used only for explanation and identification without intent to infringe.

Cataloging-in-Publication Data

Livestock health and production / edited by Johann Casini.
    p. cm.
Includes bibliographical references and index.
ISBN 978-1-64116-201-2
1. Animal health. 2. Livestock. 3. Livestock--Diseases. 4. Animal culture.
5. Animal industry. I. Casini, Johann.
SF745 .L58 2019
636.089--dc23

Table of Contents

Preface..IX

Chapter 1 *In vitro* Ruminal Acid Load and Methane Emission Responses to Supplemented
Lactating Dairy Cow Diets with Inorganic Compounds Varying in
Buffering Capacities..1
M. Jafarpour Boroujeni, M. Danesh Mesgaran, A. R. Vakili and A. A. Naserian

Chapter 2 Joint Analysis of the *DGAT1, OPN* and *PPARGC1A* Genes Effects on Variation
of Milk Production and Composition in Holstein Cattle Population..............................8
H. Kharrati Koopaee, M. Pasandideh, M. Dadpasand,
A. Esmailizadeh Koshkoiyeh and M. R. Mohammad Abadi

Chapter 3 Reproductive Performance Evaluation of Holstein Friesian and their Crosses
with Boran Cattle Breeds in Ardaita Agricultural Technical Vocational Education
Training College Dairy Farm, Oromia Region, Ethiopia..15
D. W. Mengistu, K. A. Wondimagegn and M. H. Demisash

Chapter 4 Predictive Ability of Statistical Genomic Prediction Methods when Underlying
Genetic Architecture of Trait Is Purely Additive...25
M. Momen, A. Ayatollahi Mehrgardi, A. Sheikhy, A. K. Esmailizadeh and
M. Assadi Foozi

Chapter 5 Performance Traits of Buffalo under Extensive and Semi-Intensive Bathan
System..33
M. M. Momin, M. K. I. Khan and O. F. Miazi

Chapter 6 Comparison of Different Methods of Oestrus Synchronization on Reproductive
Performance of Farahani Sheep in Iran...42
A. Mirshamsollahi

Chapter 7 Effect of Different Progesterone Protocol and Low Doses of Equine Chorionic
Gonadotropin (eCG) on Oestrus Synchronization in Arabian Ewes............................47
F. Khalilavi, M. Mamouei, S. Tabatabaei and M. Chaji

Chapter 8 Investigation of *GDF9* and *BMP15* Polymorphisms in Mehraban Sheep to Find
the Missenses as Impact on Protein...53
A. Ahmadi, F. Afraz, R. Talebi, A. Farahavar and S. M. F. Vahidi

Chapter 9 The Expression of Myogenin and Myostatin Genes in Baluchi Sheep.........................63
K. Forutan, M. Amin Afshar, K. Zargari, M. Chamani and N. Emam Jome Kashan

Chapter 10 Use of Microsatellite Polymorphisms in *Ovar-DRB1* Gene for Identifying
Genetic Resistance in Fat-Tailed Ghezel Sheep to Gastrointestinal Nematodes...................69
R. Hajializadeh Valilou, S. A. Rafat, M. Firouzamandi and M. Ebrahimi

Chapter 11   **Inbreeding and Inbreeding Depression on Body Weight in Iranian Shal Sheep**............................77
Z. Patiabadi, S. Varkoohi and S. Savar-Sofla

Chapter 12   **Comparison of Artificial Neural Network and Multiple Regression Analysis for Prediction of Fat Tail Weight of Sheep**............................84
M. A. Norouzian and M. Vakili Alavijeh

Chapter 13   **Performance Hematology and Correlation between Economical Traits under the Effects of Dietary Lysine and Methionine in Broilers**............................89
M. Bouyeh and O. K. Gevorgyan

Chapter 14   **AMELX and AMELY Structure and Application for Sex Determination of Iranian Maral deer (Cervus elaphus maral)**............................96
T. Farahvash, R. Vaez Torshizi, A. A. Masoudi, H. R. Rezaei and M. Tavallaei

Chapter 15   **Economic Value and Produced Milk Quality in Holstein Lactating Cows in Organic System**............................102
M. Sharifi, R. Pahlavan and A. Aghaei

Chapter 16   **Association between MTNR1A and CYP19 Genes Polymorphisms and Economic Traits in Kurdi Sheep**............................111
Z. Davari Varanlou, S. Hassani, M. Ahani Azari, F. Samadi, S. Zakizadeh and A. R. Khan Ahmadi

Chapter 17   **Validation of Reference Genes for Real Time PCR Normalization in Milk Somatic Cells of Holstein Dairy Cattle**............................117
M. Muhaghegh-Dolatabady, H. Hossainy-Dolatabady, E. Heidari Arjlo and R. Mahmoudi

Chapter 18   **Transcriptome Sequencing of Guilan Native Cow in Comparison with bosTau4 Reference Genome**............................123
M. Moridi, S. H. Hosseini Moghaddam and S. Z. Mirhoseini

Chapter 19   **A Research on Association between SCD1 and OLR1 Genes and Milk Production Traits in Iranian Holstein Dairy Cattle**............................130
M. Hosseinpour Mashhadi

Chapter 20   **Ruminal Methane Emission, Microbial Population and Fermentation Characteristics in Sheep as Affected by Malva sylvestris Leaf Extract: in vitro Study**............................136
S. Khamoshi, F. Kafilzadeh, H. Jahani-Azizabadi and V. Naseri

Chapter 21   **Effects of Barley Grain Particle Size on Ruminal Fermentation and Carcass Characteristics of Male Lambs Fed High Urea Diet**............................142
S. R. Ebrahimi-Mahmoudabad and M. Taghinejad-Roudbaneh

Chapter 22   **Genetic Diversity and Molecular Phylogeny of Iranian Sheep based on Cytochrome b Gene Sequences**............................148
S. Savar So la, H. R. Seyedabadi, A. Javanrouh Aliabad and R. Seyed Shari i

Chapter 23   **Effects of Specific Gravity and Particle Size of Passage Marker on Particulate Rumen Turnover in Holestine Dairy Cattle**............................153
A. Teimouri Yansari

Chapter 24   **Evaluation of Total Antioxidant, Total Calcium, Selenium, Insulin, Free Triiodothyronine and Free Thyroxine Levels in Cows with Ketosis**......................162
S. Kozat and N. Yüksek

Chapter 25   **Effect of Different Levels of Milk Thistle (*Silybum Marianum*) in Diets Containing Cereal Grains with Different Ruminal Degradation Rate on Rumen Bacteria of Khuzestan Buffalo**......................169
Z. Nikzad, M. Chaji, K. Mirzadeh, T. Mohammadabadi and M. Sari

Chapter 26   **Effects of Sodium Bentonite on Blood Parameters, Feed Digestibility and Rumen Fermentation Parameters of Male Balouchi Sheep Fed Diet Contaminated by Diazinon, an Organophosphate Pesticide**......................178
M. H. Aazami, A. M. Tahmasbi, V. Forouhar and A. A. Naserian

Chapter 27   **Estimation of Economic Values for Fertility, Stillbirth and Milk Production Traits in Iranian Holstein Dairy Cows**......................186
H. Ghiasi, A. Pakdel, A. Nejati-Javaremi, O. González-Recio, M. J. Carabaño, R. Alenda and A. Sadeghi-Se idmazgi

Chapter 28   **Anti-oxidative Effects of Ethanol Extract of *Origanum vulgare* on Kinetics, Microscopic and Oxidative Parameters of Cryopreserved Holstein Bull Spermatozoa**......................191
H. Daghigh Kia, R. Farhadi, I. Ashrafi and M. Mehdipour

Chapter 29   **Effect of Total and Differential Somatic Cell Counts, Lactation Stage and Lactation Number on Lipolysis and Physicochemical Composition in Camel (*Camelus dromedaries*) and Cow Milk**......................198
H. Hamed, A. F. El Feki and A. Gargouri

Permissions

List of Contributors

Index

Preface

In my initial years as a student, I used to run to the library at every possible instance to grab a book and learn something new. Books were my primary source of knowledge and I would not have come such a long way without all that I learnt from them. Thus, when I was approached to edit this book; I became understandably nostalgic. It was an absolute honor to be considered worthy of guiding the current generation as well as those to come. I put all my knowledge and hard work into making this book most beneficial for its readers.

Animals like cattle, pigs, chickens, goats, etc. that are raised for the purpose of producing labor and commodities like meat, milk, eggs, fur, leather, etc. are known as livestock. The maintenance, breeding and slaughter of livestock is studied under animal husbandry. Livestock health maintenance is one of the most essential aspects of livestock management. This includes the diagnosis, prevention and cure of diseases like scrapie, foot-and-mouth disease, swine fever, etc. affecting livestock. It also develops management strategies for internal and external parasites. Intensive animal farming is a modern farming technique in which animals are raised in high-density feedlots, broiler houses, etc. This book provides comprehensive insights into the field of livestock health and production. The topics covered herein deal with the core aspects of this discipline. Researchers and students in this field will be assisted by this book.

I wish to thank my publisher for supporting me at every step. I would also like to thank all the authors who have contributed their researches in this book. I hope this book will be a valuable contribution to the progress of the field.

Editor

In vitro Ruminal Acid Load and Methane Emission Responses to Supplemented Lactating Dairy Cow Diets with Inorganic Compounds Varying in Buffering Capacities

M. Jafarpour Daraujani[1], M. Danesh Mesgaran[1*], A.R. Vakili[1] and A.A. Naserian[1]

[1] Department of Animal Science, Faculty of Agriculture, Ferdowsi University of Mashhad, Mashhad, Iran

*Correspondence E-mail: danesh@um.ac.ir

ABSTRACT

Using 24 hours *in vitro* cultures of rumen microorganisms, this study investigates the effect of buffering capacity of 2 inorganic compounds (M1=119.43 and M2=116.50 meq/L) on the *in vitro* rumen acidogenic value (AV), medium pH, dry matter disappearance (INVDMD) and methane emission of lactating dairy cow diets containing various forage to concentrate ratios as 40:60 ($FC_{40:60}$) and 30:70 ($FC_{30:70}$) in a completely randomized design. Inorganic compounds were included in the experimental diets at the rate of 0.0, 10 or 20 g/kg DM. Diet with higher amount of concentrate caused a decline in medium pH, an increase in both AV and IVDMD. The acidogenic value of $FC_{40:60}$ containing M2 at 20, M1 at 10 and 20 g/kg DM and FC 30:70 plus M1 and M2 at 20 g/kg DM was the lowest. The lowest level of CH_4 emission (mL/0.20 g DM) was observed in $FC_{30:70}$ plus M1 at the rate of 10 g/kg DM, while the highest level belongs to $FC_{40:60}$ plus M1 at 10 g/kg DM and $FC_{30:70}$ containing M1 at 20 and M2 at 10 g/kg DM. It has been concluded that the higher buffering capacity of a lactating diet might reduce the rumen acid load and increased IVDMD, while a diet with higher amount of concentrate causes to decline rumen methane emission.

KEY WORDS acidogenic value, buffering capacity, dairy cows, methane, rumen.

INTRODUCTION

The problem primarily met in dairy cow feeding is to provide an energetically high-density ration without jeopardizing ruminal ecosystem, animal welfare and production performances (Zebeli *et al.* 2008). Enhancing energy supply through increased use of concentrates or rapidly fermentable fiber can swallow the rumen into acidosis. Subacute ruminal acidosis (SARA) is a common and economically important problem in well managed dairy herds. Failure to maintain a consistent rumen pH in high yielding dairy cows may result in metabolic disorders and reduced production performance (Tajik and Nazifi, 2011).

The rumen pH will fall when organic acids that are produced during fermentation by rumen microbes accumulate and rumen buffering is not sufficient to prevent the increase in acidity (Plaizier *et al.* 2008). Chemical buffers in diets for ruminants to provide rumen pH in a range that is optimal for the activity of cellulose-degrading organisms (pH=6-7) are used. The need for buffering agents in dairy cattle diets depends on the secretion of salivary buffers, the buffering capacity of feed and feed acidogenic value. Wadhwa *et al.* (2001) have extended a simple laboratory based technique for evaluating ruminal acid load from feedstuffs based on the dissolution of Ca from $CaCO_3$. The acidogenic value (AV) of the feedstuffs different with the nonfiber carbohydrate (NFC) content, protein and fiber concentrations. The highest AV was for starch rich feeds, forages were intermediate and protein sources had the lowest AV. Rustomo *et al.* (2006) reported that fiber sources had intermediate AV and protein sources had the lowest AV, wheat straw and alfalfa hay, as a fiber sources in ration, had lower AV than alfalfa pellet or corn silage. Additionally, the AV of feed ingredients were positively correlated to changes in rumen fluid pH after incubation, suggesting that high AV feeds were expected to increase the risk of rumen acidosis in dairy cows than low AV feeds.

Ruminal methanogenesis represents an alternative mechanism of reducing equivalent disposal for carbohydrate-fermenting bacteria, but interspecies hydrogen transfer is only exergonic at very low partial pressures of hydrogen (Wolin, 1975).

If the methanogens are inhibited, hydrogen accumulates, the hydrogenases are inhibited and the carbohydrate-fermenting bacteria utilize other mechanisms of reducing equivalent disposal (e.g. the dehydrogenases of propionate production) (Gottschalk, 1986).

Sauvant and Giger-Reverdin (2007) realize the relationship between methane production and proportion of concentrate in the diet to be curvilinear, with methane losses of 6-7% of gross energy (GE) being constant at 30-40% concentrate levels in the diet and then decreasing to 2-3% of GE with a concentrate proportion of 80-90%. The objective of the present experiment was to determine the effect of buffering capacity (BC) of various mixed inorganic compounds on *in vitro* rumen acidogenic value and methane emission from diets containing various forage to concentrate ratios.

MATERIALS AND METHODS

Diets, chemical composition and inorganic buffering compounds

The experimental diets were designed to provide two different forage to concentrate ratios as 40:60 ($FC_{40:60}$) and 30:70 ($FC_{30:70}$), respectively. Ration ingredients and nutrient compositions are presented in Table 1. Inorganic mixtures were made of different compositions of $NaHCO_3$, Na_2CO_3, $CaCO_3$, $MgCO_3$, MgO and bentonite Na as 500, 200, 90, 100, 100 and 10 g/kg (M1) or 300, 200, 200, 150, 100 and 50 g/kg (M2), respectively. This inorganic mixtures had the highest buffering capacity and obtained from Acros brand. A modification of the procedure of Evans and Ali (1967) was used to measure the buffering capacity. Approximately, 1 g DM sample of individual buffering compound or the mixtures of M1 and M2 was suspended in 100 mL distilled water and stirred continuously with a magnetic stir bar.

Titrations were performed by addition of HCl (83.3 mL/L) or NaOH (40 g/L) (Merck brand) in variable increments until pH was decreased to 4 or increased to 9. Buffering capacity and initial pH of the individual and the compositions used in the present experimental diets are shown in Tables 2 and 3, respectively.

Table 1 Ingredient and chemical compositions of the experimental diets

Diet (forage:concentrate)	$FC_{40:60}$	$FC_{30:70}$
Ingredient (% of DM)		
Alfalfa hay	19.72	13.89
Corn silage	19.72	13.89
Barley grain	18.33	21.11
Corn grain	16.67	20.00
Wheat bran	9.44	12.22
Soybean meal	12.22	13.89
Canola meal	3.33	2.78
Wheat straw	0.56	2.22
Chemical composition (mg/g of DM) of rations		
Crude protein	177.60	170.80
Neutral detergent fiber (NDF)	371.00	338.10
Acid detergent fiber (ADF)	208.40	171.40
Ash	52.90	49.50

Each composition (M1 or M2) were added to the diets of $FC_{40:60}$ and $FC_{30:70}$ at the rate of 0.0, 10 and 20 g/kg DM. Ingredients used in the diets were ground through a mill with a 1-mm sieve, then dried using air-forced oven (48 h, 65 °C). Nitrogen content of each ingredient was determined using Kjeldahl method (Kjeltec 2300 Autoanalyzer Foss Tecator AB, Hoganas, Sweden) and CP was calculated as N × 6.25.

Neutral detergent fiber (NDF) and acid detergent fiber (ADF) were determined according to Van Soest *et al.* (1991). Samples were also analyzed for ash by igniting the samples in muffle furnace at 525 °C for 8 h.

In vitro acidogenic value

In vitro technique used in this experiment adapted from Wadhwa *et al.* (2001). Appropriately, one-gram (DM) of each diet was placed into a 125-mL incubation bottle, then M1 or M2 was added as the experimental protocol, then bottles were held at 39 °C in a water-bath. The samples were incubated in a 3 run and quadruplicate with 30 mL of buffered rumen liquor comprising 60% buffer and 40% rumen liquor.

The buffer (5.880 g/L $NaHCO_3$; 5.580 g/L Na_2HPO_4; 0.282 g/L NaCl; 0.342 g/L KCl; 0.028 g/L $CaCl_2.2H_2O$ and 0.036 g/L $MgCl_2$) (Merck brand) was made up at 20% the strength of the Tilley and Terry (1963). Rumen fluid was collected from two rumen fistulated dairy cows fed corn silage, alfalfa hay and concentrates 25, 25 and 50%, DM, respectively; at 3 h after the morning feeding. Cysteine hydrochloride monohydrate (Merck brand) (0.025% wt/vol) was added just prior to the incubations.

Table 2 Buffering capacities and initial pH of the individual inorganic buffering inorganic compounds

Parameters	Buffering inorganic components					
	$NaHCO_3$	Na_2CO_3	$CaCO_3$	$MgCO_3$	MgO	Bentonite Na
Initial pH	8.31	11.18	8.97	10.45	10.56	9.58
Buffering capacity (meq/L)	114.93	171.13	152.56	171.86	184.00	1.96

The bottles were closed with gas release valves and shaken continuously. A set of bottles without feed sample was also incubated similarly which served as blank. After 24 of the incubation, the bottles were transferred to an ice bath to stop fermentation, and then opened to measure medium pH using a pH meter (Metrohm pH meter, model 691). Bottle contents were filtered and a 2 mL sample of the supernatant from each bottle was taken to analyze residual acidity (acidogenic value).

Table 3 Buffering capacities and initial pH of different composition of inorganic compounds

Parameters	Inorganic compounds[*]	
	M1	M2
Initial pH	9.67	9.33
Buffering capacity (meq/L)	119.43	116.50

[*] M1 and M2 were made of from $NaHCO_3$, Na_2CO_3, $CaCO_3$, $MgCO_3$, MgO and bentonite Na as 500, 200, 90, 100, 100 and 10 g/kg or 300, 200, 200, 150, 100 and 50 g/kg, respectively.

The filtrated residual was oven dried (75 °C for 48 h), weighted and used to calculate *in vitro* dry matter disappearance (IVDMD). The supernatant of each bottle was transferred into a 2 mL centrifuge tube containing excess amount of $CaCO_3$ powder (Merck brand) (50 mg). The mixture was shaken manually for 5 s and then centrifuged at 4000 × g for 10 min.

The supernatant Ca concentration was then immediately determined using an autoanalyzer (A15 Biosystem). A measurement of dissolution of Ca from insoluble $CaCO_3$ powder makes it possible to assess residual acidity after fermentation of feeds.

In vitro rumen methane emission

An *in vitro* incubations were carried out as proposed by Menke and Steingass (1988). Appropriately, 200 mg (DM basis) of each experimental diet was placed into a 125 mL incubation bottle, then inorganic mixtures of M1 or M2 was introduced as rate of 0.0, 10 or 20 g/kg DM. A set of bottles without feed sample was also incubated similarly which served as blank. The bottles were incubated with 30 mL of buffered rumen fluid (artificial saliva to the rumen liquid in ratio of 2:1) and held at 39 °C in a water-bath.

The artificial saliva was made up of 475 mL/L distilled water, 240 mL/L buffer solution (ammonium bicarbonate 4 g/L and sodium bicarbonate 35 g/L, Merck brand), 240 mL/L macromineral solution (5.7 g anhydrous Na_2HPO_4, 6.2 g anhydrous KH_2PO_4 and 0.6 g $MgSO_4.7H_2O$ per liter,

Merck brand), 0.12 mL/L micro-mineral solution (13.2 g $CaCl_2.2H_2O$, 10.0 g $MnCl_2.4H_2O$, 1 g $CoCl_2.6H_2O$ and 8.0 g $FeCl_3.6H_2O$ per 100 mL, Merck brand), 1.22 mL/L Resazurin aqueous (Merck brand) (1 mg/1 mL). The medium was then reduced by addition of reducing agent (47.5 mL distilled water, 2 mL 1 N NaOH and 336 mg $Na_2S.9H_2O$, Merck brand) per liter of medium. Rumen fluid was collected as described previously.

The incubation was carried out in two set (run) and in triplicate. After 24 of incubation, the bottles were transferred to an ice bath to stop fermentation, then total gas was recorded by digital pressure indicator (model SEDPGB0015PG5) and methane emission was determined using a biological gas recorder (SR2-BIO).

Calculations and statistical analysis

The buffering capacity (BC, meq/L) was calculated by the following formula (Evans and Ali, 1967):

$$BC= [(milliliters\ of\ 1\ N\ HCl) + (milliliters\ of\ 1\ N\ NaOH)] \times 10^3/30$$

In vitro acidogenic value (mg Ca/g DM) of each sample was calculated as the product of Ca concentration (mg/mL, from the analysis) and fluid volume (30 mL) divided by the sample weight (1 g). *In vitro* dry matter disappearance was calculated as follows (Jahani Azizabadi *et al.* 2011):

$$IVDMD\ (\%)= [(A-(B-C)) / A] \times 100$$

Where:
A: dry weight of sample.
B: dry weight of residue after incubation.
C: dry weight of blank.

Data were analyzed as a completely randomized design to compare the diets and buffering composition in each experimental diet with replications using Dunnett's test (P<0.05) procedure in SAS (2002).

RESULTS AND DISCUSSION

The effect of inorganic mixtures supplementation on the *in vitro* medium pH, AV and IVDMD after 24 hours incubation within the experimental diets containing 40:60 or 30:70 forage to concentrate ratios are shown in Table 4 and Table 5, respectively.

Table 4 *In vitro* ruminal medium pH, acidogenic value [AV, (mg Ca/g DM)] and dry matter disappearance (IVDMD, %) of a dairy cow diet containing different forage: concentrate ratio as 40:60 which supplemented with different composition of inorganic buffering compounds (M1 and M2), after 24 h incubation

| Ration | Inorganic chemical compounds applied | | Parameters | | |
	Inorganic compounds[1]	Concentration (g/kg DM ration)	pH	AV	IVDMD
$FC_{40:60}$[2]	-	0.0	5.50	6.60	55.50
$FC_{40:60}$	M1	10	5.59*	6.43	57.50
$FC_{40:60}$	M1	20	5.60*	6.93	64.50
$FC_{40:60}$	M2	10	5.54	7.59	54.25
$FC_{40:60}$	M2	20	5.63*	7.09	54.00
SEM	-	-	0.022	1.034	5.205
P-value	-	-	0.016	0.254	0.185

[1] M1 and M2 were made of from $NaHCO_3$, Na_2CO_3, $CaCO_3$, $MgCO_3$, MgO and bentonite Na as 500, 200, 90, 100, 100 and 10 g/kg or 300, 200, 200, 150, 100 and 50 g/kg, respectively.
[2] $FC_{40:60}$: a dairy cow diet containing 40% forage and 60% concentrate.
SEM: standard error of the means.
The means within the same column with at least one common letter, do not have significant difference (P>0.05).

Table 5 *In vitro* ruminal medium pH, acidogenic value [AV, (mg Ca/g DM)] and dry matter disappearance (IVDMD, %) of a dairy cow diet containing different forage: concentrate ratio as 30:70 which supplemented with different composition of inorganic buffering compounds (M1 and M2), after 24 h incubation

| Ration | Inorganic chemical compounds applied | | Parameter | | |
	inorganic compound[1]	Concentration (g/kg DM ration)	pH	AV	IVDMD
$FC_{30:70}$[2]	-	0.0	5.44	8.25	67.25
$FC_{30:70}$	M1	10	5.49	10.23	68.00
$FC_{30:70}$	M1	20	5.54*	7.09	69.00
$FC_{30:70}$	M2	10	5.52	8.58	66.00
$FC_{30:70}$	M2	20	5.54*	10.72	67.00
SEM	-	-	0.038	0.809	5.580
P-value	-	-	0.050	0.100	0.797

[1] M1 and M2 were made of from $NaHCO_3$, Na_2CO_3, $CaCO_3$, $MgCO_3$, MgO and bentonite Na as 500, 200, 90, 100, 100 and 10 g/kg or 300, 200, 200, 150, 100 and 50 g/kg, respectively.
[2] $FC_{30:70}$: a dairy cow diet containing 30% forage and 70% concentrate.
SEM: standard error of the means.
The means within the same column with at least one common letter, do not have significant difference (P>0.05).

Increased concentrate from 60 to 70%, due to increasing the level of rapidly fermentable carbohydrates caused a decrease in medium pH and a significant (P< 0.05) increase in AV. By increasing the amount of acid produced as a result of fermentation of carbohydrates, pH levels on the medium was reduced. High concentrate diets contain high amounts of non structural carbohydrates which are quickly fermented by ruminal microbes, resulting in a greater decline in ruminal pH (Kalscheur *et al.* 1997).

Results of previous studies have shown a decline in ruminal pH when more rapidly fermentable carbohydrates were included in the diet (Krause *et al.* 2002b). Danesh Mesgaran *et al.* (2009) reported that there is a positive correlation between non fibrous carbohydrates in diet and acidogenic value, so that by increasing the non fibrous carbohydrates in diet, the AV enhanced. As described by Rustomo *et al.* (2006), energy feeds and fiber sources have the highest and intermediate AV, respectively.
Results indicated that the adding of M1 at both rate and M2 at 20 mg/kg to $FC_{40:60}$ caused a significant (P<0.05) increase in medium pH. Inorganic buffers are capable in preventing pH reduction in the medium through neutralizing acids produced by bacterial activities.

The highest significant (P<0.05) level of medium pH was belonged to $FC_{30:70}$ plus M1 and M2 which were added at the rate of 20 g/kg DM. Inorganic buffers enhance ruminal environmental conditions by modulating acidity of the ruminal contents, preventing severe drops in pH (Le Ruyet and Tucker, 1992).

Tripathi *et al.* (2004) have also reported that $NaHCO_3$ supplementation caused a linear enhancement in ruminal fluid pH. In experiment of West *et al.* (1987), the addition of various buffers to the diet also resulted in a significant increase in rumen pH. Santra *et al.* (2003) reported that dietary buffers prevent the reduction of rumen pH when animal fed a high levels of concentrate. With regard to AV (Tables 4 and 5), none of the M1 and M2 supplementation had significant (P>0.05) effect compared with the non-supplemented diet, which might express to this cause that the inorganic mixtures could not significantly affected the buffering capacity and ultimately the acid load created in the medium.

Dietary buffers are widely used to improve the harmful effects of acidity in high concentrate diets (Coppock *et al.* 1986), but the response of buffer is variable and sometimes unpredictable.

Erdman (1988) have also reported that the buffering agents that possess a pK_a above the typical ruminal fluid pH will act as alkalinizing agents rather than simply as buffers to increase the resistance of the rumen to a change in pH. Present results indicated that the adding of the inorganic mixtures to the experimental diets, had no significant effect on IVDMD (Tables 4 and 5), which, probably indicated no effect of the inorganic mixtures on fermentation at the medium. Bodas *et al.* (2009) used sodium bicarbonate in the diet of lambs and did not obtain difference in dry matter digestibility. However, Mould and Qrskov (1983) reported that the addition of buffer due to maintenance of ruminal pH above the critical level might improve the DM digestibility.

The effect of inorganic mixtures supplementation on the *in vitro* total gas, CH_4 and CO_2 emission from diets containing different forage to concentrate ratios as 40:60 and 30:70 are shown in Table 6 and Table 7, respectively. Increased concentrate from 60 to 70%, caused a significant decrease in both total gas and CO_2 emission, which is probably due to the negative impact of rapid fermentation of carbohydrates on microorganisms. Rumen pH is one of the most critical determinants for rumen function as cell-

ulolytic bacteria fail to grow below pH 6.0, while a slight increase in ruminal pH favors the activity of these bacteria (Santra *et al.* 2003). Within both the experimental diets, the supplementation with M1 and M2 alter total gas, methane and carbon dioxide produced in the medium significantly (P<0.05). The lowest levels of total gas were observed in $FC_{40:60}$ containing M1 at 20 and M2 at both 10 and 20 g/kg DM. Methane emission was significantly (P<0.05) higher when M1 was added to $FC_{40:60}$ at the rate of 10 g/kg DM and $FC_{30:70}$ at the rate of 20 g/kg DM compared with the non-supplemented diets.

Present results indicated that $FC_{40:60}$ containing M1 at 20 g/kg DM, M2 at 10 and 20 g/kg DM and $FC_{30:70}$ plus M1 at 10 g/kg DM had the lowest level CO_2 emission (P<0.05).

The pattern of the responses was influenced by the kind of inorganic composition and the concentration applied. Inorganic mixture M1 to M2, had the highest buffering capacity and by increasing the amount of M1 in the $FC_{40:60}$, decreased total gas, methane and carbon dioxide production, while, with the increasing the amount of M2 in the diet, increased total gas, methane and carbon dioxide production. But the results in the $FC_{30:70}$ against the results of the $FC_{40:60}$.

Table 6 *In vitro* total gas (mL/0.20 g DM), CH_4 (mL/0.20 g DM) and CO_2 (mL/0.20 g DM) of a dairy cow diet containing different forage: concentrate ratio as 40:60 which supplemented with different composition of inorganic buffering compounds (M1 and M2), after 24 h incubation

Ration	Inorganic chemical compounds applied		Parameter		
	Inorganic compound[1]	Concentration (g/kg DM ration)	Total gas	CH_4	CO_2
$FC_{40:60}$[2]	-	0.0	41.17	3.54	37.05
$FC_{40:60}$	M1	10	43.57*	3.84*	39.08*
$FC_{40:60}$	M1	20	38.77*	3.48	34.89*
$FC_{40:60}$	M2	10	36.42*	3.27	32.77*
$FC_{40:60}$	M2	20	37.67*	3.39	33.90*
SEM	-	-	0.360	0.061	0.327
P-value	-	-	0.0001	0.0008	0.0001

[1] M1 and M2 were made of from $NaHCO_3$, Na_2CO_3, $CaCO_3$, $MgCO_3$, MgO and bentonite Na as 500, 200, 90, 100, 100 and 10 g/kg or 300, 200, 200, 150, 100 and 50 g/kg, respectively.
[2] $FC_{40:60}$: a dairy cow diet containing 40% forage and 60% concentrate.
SEM: standard error of the means.
The means within the same column with at least one common letter, do not have significant difference (P>0.05).

Table 7 *In vitro* total gas (mL/0.20 g DM), CH_4 (mL/0.20 g DM) and CO_2 (mL/0.20 g DM) of a dairy cow diet containing different forage: concentrate ratio as 30:70 which supplemented with different composition of inorganic buffering compounds (M1 and M2), after 24 h incubation

Ration	Inorganic chemical compounds applied		Parameter		
	Inorganic compound[1]	Concentration (g/kg DM ration)	Total gas	CH_4	CO_2
$FC_{30:70}$[2]	-	0.0	39.17	3.52	35.44
$FC_{30:70}$	M1	10	36.67*	3.30*	33.00*
$FC_{30:70}$	M1	20	41.17*	3.70*	37.05
$FC_{30:70}$	M2	10	42.17*	4.21*	37.53*
$FC_{30:70}$	M2	20	38.17	3.43	34.35
SEM	-	-	0.465	0.041	0.421
P-value	-	-	0.0001	0.0001	0.0001

[1] M1 and M2 were made of from $NaHCO_3$, Na_2CO_3, $CaCO_3$, $MgCO_3$, MgO and bentonite Na as 500, 200, 90, 100, 100 and 10 g/kg or 300, 200, 200, 150, 100 and 50 g/kg, respectively.
[2] $FC_{30:70}$: a dairy cow diet containing 30% forage and 70% concentrate.
SEM: standard error of the means.
The means within the same column with at least one common letter, do not have significant difference (P>0.05).

It seems that increase the buffering capacity environment by adding the inorganic mixtures, as well as increasing its concentration with effects on pH, rumen bacterial fermentation performance is affected.

Dietary buffers by increasing the buffering capacity of the medium might produce a situation by which a huge decrease in pH may prevent and thereby causing an increase in methane levels in the culture medium. It was reported that dietary buffer would prevent depression in rumen pH and improve rumen ecology associated with high concentrate feeding (Santra *et al.* 2003). The amount of decrease in pH after an increase in the fermentation rate will depend on the buffering capacity of the rumen fluid (Counotte *et al.* 1979). Supplementation of minerals in the diet of animals is known to increase the number of total ruminal bacteria especially the cellulolytic bacteria which contributed to better cellulose digestibility (Koul *et al.* 1998). However, diets high in cereals consequently reduce ruminal pH and cellulolytic activity (Franzolin and Dehority, 1996). It has long been recognized that the addition of cereal grains to ruminant diets causes a decrease in methane and an increase in propionate production (Czerkawski, 1986), but the cause of this fermentation shift was not clear.

CONCLUSION

It has been concluded that there is a chance to increase the buffering capacity of inorganic composition when different amount of the inorganic chemical compounds were used compared with the sodium bicarbonate. Adding inorganic compound had the highest buffering capacity, reducing the amount of the acid load and increased IVDMD, although the effect was not significant compared to the control that might be due to lack of impact of the inorganic compound on the buffering capacity of the medium. Increasing the concentration of mixture M1 in the diet, causes was reduced in the total gas, CH_4 and CO_2. But with regard to mixture M2, the opposite was observed may be due to the effect of buffering capacity created by this mixtures on pH medium and microbial activity.

ACKNOWLEDGEMENT

The authors wish to acknowledge the Ferdowsi University of Mashhad, Iran, for financial support.

REFERENCES

Bodas R., Frutos P., Giraldez F.J.G.H. and Lopez S. (2009). Effect of sodium bicarbonate supplementation on feed intake, digestibility, digesta kinetics, nitrogen balance and ruminal fermentation in young fattening lambs. *Spanish J. Agric. Res.* **7(2)**, 330-341.

Counotte G.H.M., van't Klooster A.T., van der Kuilen J. and Prins R.A. (1979). An analysis of the buffer system in the rumen of dairy cattle. *J. Anim. Sci.* **49**, 1536-1544.

Coppock C.E., Schelling G.T., Byers F.M., West J.M. and Labore J.M. (1986). A naturally occurring mineral as a buffer in the diet of lactating dairy cows. *J. Dairy Sci.* **69**, 111-118.

Czerkawski J.W. (1986). An Introduction to Rumen Studies. Pergamon Press, New York.

Danesh Mesgaran S., Heravi Moussavi A., Jahani-Azizabadi H., Vakili A.R., Tabatabaiee F. and Danesh Mesgaran M. (2009). The effect of grain sources on *in vitro* rumen acid load of close-up dray cow diets. Pp. 146-147 in Proc. 11th Int. Symp. Rumin. Physiol.Wageningen, Netherlands.

Erdman R.A. (1988). Dietary buffering requirements of the lactating dairy cow: a review. *J. Dairy Sci.* **71**, 3246-3252.

Evans J.L. and Ali R. (1967). Calcium utilization and feed efficiency in the growing rat as affected by dietary calcium, buffering capacity, lactose and EDTA. *J. Nutr.* **92**, 417-425.

Franzolin R. and Dehority B.A. (1996). Effect of prolonged concentrate feeding on ruminal protozoa concentration. *J. Anim. Sci.* **74**, 2803-2809.

Gottschalk G. (1986). Bacterial Metabolism. Springer-Verlag, New York.

Jahani Azizabadi H., Danesh Mesgaran M., Vakili A., Rezayazdi K. and Hashemi M. (2011). Effect of various medicinal plant essential oils obtained from semi-arid climate on rumen fermentation characteristics of a high forage diet using *in vitro* batch culture. *African J. Microbiol. Res.* **5**, 4812-4819.

Kalscheur K.F., Teter B.B., Piperova L.S. and Erdman R.A. (1997). Effect of dietary forage concentration and buffer addition on duodenal flow of trans-C18:1 fatty acids and milk fat production in dairy cows. *J. Dairy Sci.* **80**, 2104-2114.

Koul V., Kumar U., Sareen V.K. and Singh S. (1998). Effect of sodium bicarbonate supplementation on ruminal microbial populations and metabolism in buffalo calves. *Indian J. Anim. Sci.* **68**, 629-631.

Krause K.M., Combs D.K. and Beauchemin K.A. (2002b). Effects of particle size and grain fermentability in mid lactation cows. II. Ruminal pH and chewing activity. *J. Dairy Sci.* **85**, 1947-1957.

Le Ruyet P. and Tucker B. (1992). Ruminal buffers: temporal effects on buffering capacity and pH of ruminal fluid from cows fed a high concentrate diet. *J. Dairy Sci.* **75**, 1069-1077.

Menke K.H. and Steingass H. (1988). Estimation of the energetic feed value obtained from chemical analysis and *in vitro* gas production using rumen fluid. *Anim. Res. Dev.* **28**, 7-55.

Mould F.L. and Ørskov E.R. (1983). Manipulation of rumen fluid pH and its influence on cellulolysis *in sacco* dry matter degradation and rumen microflora of sheep offered either hay or concentrate. *Anim. Feed Sci. Technol.* **10**, 1-14.

Plaizier J.C., Krause D.O., Gozho G.N. and McBride B.W. (2008). Subacute ruminal acidosis in dairy cows: the physiological causes, incidence and consequences. *Vet. J.* **176**, 21-31.

Rustomo B., Cant J.P., Fan M.Z., Duffield T.F., Odongo N.E. and McBride B.W. (2006). Acidogenic value of feeds. I. The relationship between the acidogenic value of feeds and *in vitro* ruminal pH changes. *Canadian J. Anim. Sci.* **86**, 109-117.

Santra A., Chaturvedi O.H., Tripathi M.K., Kumar R. and Karim S.A. (2003). Effect of dietary sodium bicarbonate supplementation on fermentation characteristics and ciliate protozoal population in rumen of lambs. *Small Rumin. Res.* **47,** 203-212.

SAS Institute. (2002). SAS®/STAT Software, Release 9.1. SAS Institute, Inc., Cary, NC. USA.

Sauvant D. and Giger-Reverdin S. (2007). Empirical modeling meta-analysis of digestive interactions and CH_4 production in ruminants. Pp. 561-563 in Energy and Protein Metabolism and Nutrition. I. Ortigues-Marty, N. Miraux and W. Brand-Williams, Eds. Wageningen Academic, Wageningen, Netherlands.

Tajik J. and Nazifi S. (2011). Diagnosis of subacute ruminal acidosis: a review. *Asian J. Anim. Sci.* **5,** 80-90.

Tilley J.M.A. and Terry R.A. (1963). A two-stage technique for the *in vitro* digestion of forage crops. *J. British Grassland. Soc.* **18,** 104-111.

Tripathi M.K., Santra A., Chaturvedi O.H. and Karim S.A. (2004). Effect of sodium bicarbonate supplementation on ruminal fluid pH, feed intake, nutrient utilization and growth of lambs fed high concentrate diets. *Anim. Feed Sci. Technol.* **111,** 27-39.

Van Soest P.J., Robertson J.B. and Lewis B.A. (1991). Methods for dietary fibre, neutral detergent fibre and non-starch polysaccharides in relation to animal nutrition. *J. Dairy Sci.* **74,** 3583-3597.

Wadhwa D., Beck N.F.G., Borgida L.P., Dhanoa M.S. and Dewhurst R.J. (2001). Development of a simple *in vitro* assay for estimating net rumen acid load from diet ingredients. *J. Dairy Sci.* **84,** 1109-1117.

West J.W., Coppock C.E., Millam K.Z., Nave D.H. and Labore J.M. (1987). Potassium carbonate as a potassium source and dietary buffer for lactating Holstein cows during hot weather. *J. Dairy Sci.* **70,** 309-320.

Wolin M.J. (1975). Interactions between the bacterial species of the rumen. Pp. 134-148 in Digestion and Metabolism in the Ruminant. I.W. McDonald and A.C.I. Warner, Eds. University of New England Publishing Unit, Armidale, Australia.

Zebeli Q., Dijkstra J., Tafaj M., Steigass H., Ametaj B.N. and Drochner W. (2008). Modeling the adequacy of dietary fiber in dairy cows based on the response of ruminal pH and milk fat production to composition of the diet. *J. Dairy Sci.* **91,** 2046-2066.

Joint Analysis of the *DGAT1*, *OPN* and *PPARGC1A* Genes Effects on Variation of Milk Production and Composition in Holstein Cattle Population

H. Kharrati Koopaee[1], M. Pasandideh[2*], M. Dadpasand[3], A. Esmailizadeh Koshkoiyeh[4] and M.R. Mohammad Abadi[4]

[1] Institute of Biotechnology, Shiraz University, Shiraz, Iran
[2] Department of Genetics and Animal Breeding, Faculty of Animal Science and Fishery, Sari Agricultural Sciences and Natural Resources University, Sari, Iran
[3] Department of Animal Science, Faculty of Agriculture, Shiraz University, Shiraz, Iran
[4] Department of Animal Science, Faculty of Agriculture, Shahid Bahonar University of Kerman, Kerman, Iran

*Correspondence E-mail: majidpasandideh@gmail.com

ABSTRACT

The aim of this study was to investigate effects of *DGAT1*, *OPN* and *PPARGC1A* candidate genes on milk production traits in Iranian Holstein cattle. Several papers have studied single nucleotide polymorphisms (SNPs) and their association with economic traits in dairy cows, but the combined effect of these genes has not been examined in Iranian Holstein cattle population. Blood samples were collected from 398 registered Holstein cows. Total DNA was extracted using the salting out protocol. The PCR-RFLP technique was used for SNPs genotyping. The largest genotype frequency was estimated as 0.65 for *PPARGC1A* (c.1892)^{CT} and the least frequency was estimated as 0.09 for *DGAT1^{KK}* genotype. The allele frequencies were in the range 0.36 to 0.64 for *PPARGC1A* (c.3359) A and C alleles, respectively. The allelic substitution effects were estimated using a multiple regression model. The effects of allelic substitution for *DGAT1^{K}* and *PPARGC1A* (c.1892)^{T} were significant on estimated breeding values for fat percentage (EBV$_{FP}$) (P<0.01). In addition, the results of multivariate analysis indicated the significant effect of *DGAT1* and *PPARGC1A* (c.1892) on EBV$_{FP}$ (P<0.05). However, there were no association between *OPN* and *PPARGC1A* (c.3359) polymorphisms and studied traits.

KEY WORDS *DGAT1*, Holstein, *OPN*, *PPARGC1A*.

INTRODUCTION

Candidate gene approach and whole genome scans are two main strategies for QTL identification (Andersson, 2001). The candidate gene approach studies the relationship between the traits and known genes that may be associated with the physiological pathways underlying the trait (Liu *et al.* 2008). This approach has been successful to some extent. For example, several studies have identified QTLs for milk composition on chromosomes 6 and 14 (Riquet *et al.*

1999; Farnir *et al.* 2002; Olsen *et al.* 2005). The economic traits are polygenic. It means that they are controlled by many loci. Several studies indicated genetic variation in milk production traits cannot be explained by few candidate genes (Kaupe *et al.* 2007). Therefore, the effects of all candidate loci should be explored together in the same statistical model diacyl glycerol acyltransferase 1 (*DGAT1*) is located near the centromeric region of *Bos taurus* autosome 14 (BTA14). The first evidence for the effect of *DGAT1* variation on milk yield and composition in Holstein cattle was reported by Grisart *et al.* (2002). *DGAT1* is considered as the key enzyme in controlling the synthesis rate of triglycerides in adipocytes. A non-conservative K232A substitution (conservation of alanine to lysine) in *DGAT1* was associated with milk production and compositions in Holstein cattle (Thaller *et al.* 2003). Spelman *et al.* (2002) and Banos *et al.* (2008) reported that the K232A substitution in exon 8 of the *DGAT1* gene was associated with increasing of milk fat yield and decreasing of milk production and protein yield. Some studies showed that there are significant associations between *DGAT1* and milk, fat yield and protein yield. Bovine chromosome six (BTA6) harbors at least six QTLs influencing milk production traits of dairy cattle.

The osteopontin (*OPN*) and peroxisome proliferator activated receptor gamma co-activator 1 Alpha (*PPARGC1A*) are about 6 Mb apart, which is about 12 cM for this region of chromosome 6 (Olsen *et al.* 2005). *OPN* is a strong functional candidate for milk production and it is a highly phosphorylated glycoprotein (Leonard *et al.* 2005).

Schnabel *et al.* (2005) reported an association between *OPN* and milk protein percentage in the North American Holstein population. *PPARGC1A* has main role in fat and glucose metabolism and plays a critical role in the activation of nuclear hormone receptors and transcription factors regulating energy homeostasis (Liang and Ward, 2006; Kowalewska-Luczak *et al.* 2010). Structure of *PPARGC1A* gene is made from 13 exons and expressed at different levels in a great number of tissues (Liang and Ward, 2006). Khatib *et al.* (2007) showed significant associations between *PPARGC1A* (*c.3359*) gene, milk yield, milk protein percentage, and somatic cell score in the North American Holstein population. The aim of this study was to investigate the joint effects of *OPN*, *PPARGC1A* and *DGAT1* candidate genes on milk production traits in Iranian Holstein cattle population.

MATERIALS AND METHODS

Animals and traits

Totally 398 blood samples were collected from Holstein-Friesian cows of Iran, which were distributed in ten dairy herds in two provinces of Iran. The cows were under official milk recording of Animal Breeding Center (Karaj-Iran).

Finally 372 records for estimated breeding values for milk production adjusted for 305 days (EBV_M), fat yield (EBV_F) (kg) and fat percentage (EBV_{FP}) were obtained from the Animal Breeding Center for analyzing association between genotypes and economics traits. The EBVs were estimated by random regression test day model.

DNA extraction, PCR amplification and SNPs genotyping

DNA extractions were performed using standard salting out protocol (Miller *et al.* 1988). PCR reactions were performed using standard PCR (Thermo cycler, Biometra, Germany). More details about primers are shown in Table 1 (Kaupe *et al.* 2004; Weikard *et al.* 2005; Khatib *et al.* 2007).

PCR reaction for *DGAT1* (GenBank: EU077528), *OPN* (GenBank: NW_255516) and *PPARGC1A* (GenBank: AY321517) loci were performed in a 25 μL volume using 100 ng genomic DNA, PCR buffer (1X), 1.5 m*M* MgCl₂, 0.2 m*M* dNTPs, 0.6 pmol of each primer and Taq polymerase enzyme (2U). All accession numbers are available in NCBI site. For *DGAT1* gene, the addition of DMSO to the PCR reactions allowed an equal amplification of both alleles. The annealing temperature for *DGAT1*, *OPN* and *PPARGC1A* are considered as 60, 53 and 55 centigrade and finally 411, 290, 195 and 357 base pairs fragments were amplified for *DGAT1*, *OPN*, *PPARGC1A* (*c.1892*) and *PPARGC1A* (*c.3359*). The PCR products (5 μL) were digested using 2 units of the restriction enzymes (FERMENTAS, Lithuania) and separated on a 2% agarose gel. The gels were stained with ethidium bromide and visualized under UV light. Finally, SNPs were genotyped by PCR-RFLP technique. Table 2 shows more detail about restriction enzyme conditions.

Gene and genotype frequencies

The population genetic parameters including gene and genotypic frequencies, Hardy-Weinberg equilibrium (Chi-square test), Indices of genetic diversity in population (Nei (H) and Shannon (I)) were estimated using the PopGene software version 3.1d (Nei, 1977).

Joint analysis of *DGAT1*, *OPN* and *PPARGC1A* variants

The effects of genotypes were tested using the GLM procedure (Pillais trace test) of SPSS (2010) implementing the following fixed model:

$$y_{ijkmn} = \mu + P_i + A_j + D_k + O_m + e_{ijkmn}$$

Where:
y: observation for each trait.
μ: overall mean.
Pi, A_j, D_k and O_m: fixed effects of genotypes of *PPARGC1A* (*c.1892*), *PPARGC1A* (*c.3359*), *DGAT1* and *OPN* genes.
e_{ijkmn}: residual effect.

The mean comparisons were performed using the Tukey test for significant genotypes.

Table 1 Primer sequence for PCR reaction

Loci	Forward and revers primer
DGAT1	F 5'-GCACCATCCTCTTCCTCAAG-3' R 5'-GGAAGCGCTTTCGGATG-3
OPN	F 5'-GCAAATCAGAAGTGTGATAGAC-3' R 5'-CCAAGCCAAACGTATGAGTT-3
PPARGC1A (c.3359)	F 5'-GCGAGCACGGTGTTACATTACTAAGGAGAGTTGGCTAG-3' R 5'-GTTGTGTTGCACTCAATGGAC-3'
PPARGC1A (c.1892)	F 5'-CATAGCCGGCGCCCCAGGTAAGATGCACGTTGGC-3' R 5'-CTGGTACTCCTCGTAGCTGTC-3'

Table 2 Restriction enzymes and the digestion conditions

Locus	Position	Enzyme	Digestion temperature (°C)	Digestion time
DGAT1	K232A	*CfrI*	37	3 h
OPN	c.8514	*BsrI*	65	5 h
PPARGC1A	c.3359	*NheI*	37	3 h
PPARGC1A	c.1892	*HaeIII*	37	5 h

The effects of allele substitutions on milk production traits were tested using the following multiple linear regression models (Knott *et al.* 1996):

$$y_{ijkl} = \mu + b_i x_i + b_j x_j + b_k x_k + b_l x_l + e_{ijkl}$$

Where:

y: observation for EBV_M, EBV_F and EBV_{FP} traits.

μ: overall mean.

b_i, b_j, b_k, b_l: regression coefficients representing the allelic substitutions for ($DGAT^K$, OPN^T, $PPARGC1A$ (c.3359)^A, $PPARGC1A$ (c.1892)^T.

x_i, x_j, x_k, x_l: indicator variables for genotypes of *DGAT1*, *OPN*, *PPARGC1A* (c.3359), *PPARGC1A* (c.1892) loci.

e_{ijkl}: residual effect.

RESULTS AND DISCUSSION

The most extreme genotypes frequencies were estimated as 0.65 and 0.09 for *PPARGC1A* (c.1892)^{CT} and *DGAT1^{KK}* loci, respectively. Similar results were obtained about genotype frequencies in Holstein cattle population by Khatib *et al.* (2007), Thaller *et al.* (2003) and Komisarek and Dorynek (2009). In addition, the most and the least allele frequencies were calculated as 0.64 and 0.36 for A and C alleles of *PPARGC1A* (c.3359).

The joint testing of all of the candidate loci in the population under study indicated significant deviation from the Hardy-Weinberg equilibrium (P<0.05), which seems expectable due to the long time selection for milk production. The genetic diversity indices showed that the population has desirable genetic variation. More detail about values and frequencies are shown in Table 3. Figure 1 gives more information about the results of digestion. Similar results were obtained using the multivariate analysis and between subject test Table 4.

The results indicated that the *PPARGC1A* (c.1892) and *DGAT1* polymorphisms had significant association with EBV_{FP}. The summaries of statistical analysis are illustrated in Tables 5 and 6. The results obtained are supported from other studies. Weikard *et al.* (2005) reported significant association between SNP in intron 9 of the *PPARGC1A* (c.1892) gene and fat yield, which means that the *PPARGC1A* gene might be involved in genetic variation underlying the QTL for milk fat synthesis on BTA6. Schennink *et al.* (2009) showed that two SNPs in *PPARGC1A* (c.3359 and c.1892) had significant effects on fat yield. However, we found that the effect of *PPARGC1A* (c.3359) and *OPN* polymorphism were not significant, but different results reported by Zhang *et al.* (1998) and Mosig *et al.* (2001), who identified candidate gene affecting milk production traits close to *OPN* location.

Table 3 Summery of frequencies, H-W equilibrium and genetic variation indices

Locus	Allele frequency		Genotype frequency			H-W	Index	
DGAT1	K	A	KK	KA	AA	χ^2	Shannon	Nei
	0.37	0.63	0.09	0.56	0.35	17.71**	0.66	0.46
OPN	C	T	CC	CT	TT			
	0.47	0.53	0.19	0.57	0.24	7.50**	0.69	0.49
PPARGC1A (c.3359)	A	C	AA	CA	CC			
	0.64	0.36	0.38	0.52	0.10	6.01*	0.65	0.46
PPARGC1A (c.1892)	C	T	CC	CT	CC			
	0.56	0.44	0.23	0.65	0.12	8.51**	0.68	0.49

* (P<0.05) and ** (P<0.01).

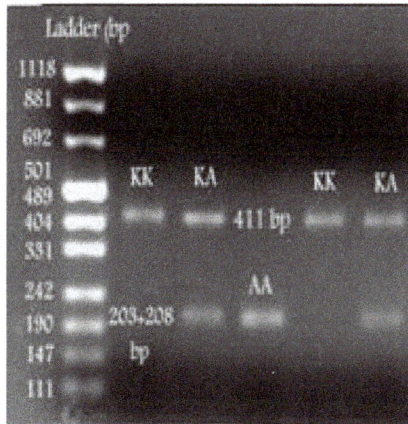

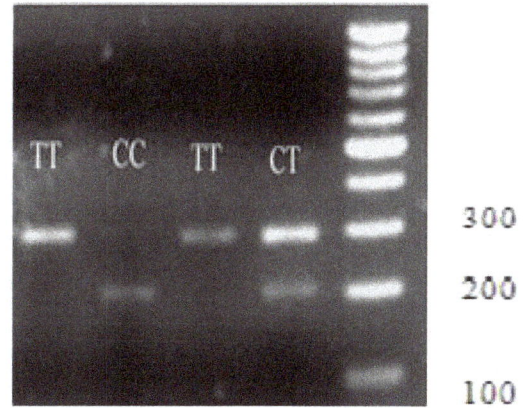

A

Uncut fragment 411 bp (KK)

Fragments 411, 208 and 203 bp (KA)

Fragments 208 and 203 bp (AA)

B

Uncut fragment 290 bp (TT)

Fragments 290, 200 and 90 bp (CT)

Fragments 200 and 90 bp (CC)

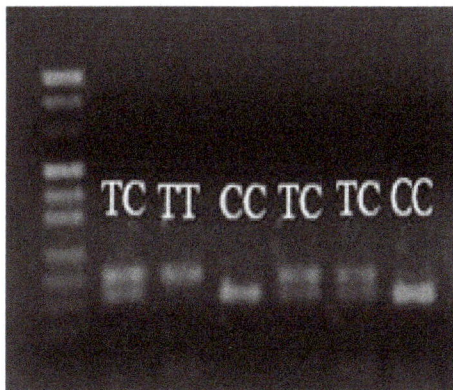

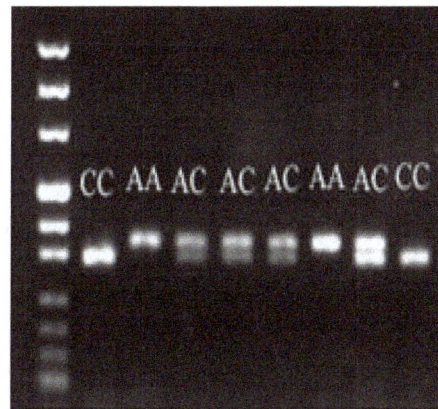

C

Uncut fragment 195 bp (TT)

Fragments 195, 163 and 32 bp (TC)

Fragments 163 and 32 bp (CC)

D

Uncut fragment 357 bp (AA)

Fragments 357, 319 and 38 bp (AC)

Fragments 319 and 38 bp (CC)

Figure 1 Electrophoretic separation of *DGAT1* (A), *OPN* (B), *PPARGC1A* (c.1892) (C) and *PPARGC1A* (c.3359) (D) genes PCR products
Figures a, c and d have a common ladder (PUC mix 8)
Figure b: ladder (gene ruler DNA ladder)

Table 4 The results of multivariate analysis (F-statistics) for EBV_{FP}

Locus	*DGAT1*	*OPN*	*PPARGC1A* (c.3359)	*PPARGC1A* (c.1892)
F-value	3.19[*]	1.52	1.93	2.41[*]

[*] ($P<0.05$).

Table 5 The analysis results of between subjects effects (F-values)

Locus/trait	DGAT1	OPN	PPARGC1A (c.3359)	PPARGC1A (c.1892)
EBV$_M$	2.12	1.18	0.44	1.89
EBV$_F$	1.40	1.20	0.14	0.10
EBV$_{FP}$	8.45[**]	1.98	2.07	4.79[**]

** (P<0.01).

Table 6 The results of mean comparisons for significant genes

DGAT1			PPARGC1A (c.1892)		
KK	KA	AA	CC	CT	TT
0.05±0.02[b]	0.04±0.01[b]	-0.01±0.001[a]	-0.008±0.001[a]	0.03±0.01[a]	0.09±0.02[b]

The means within the same column with at least one common letter, do not have significant difference (P>0.05).

Table 7 The effect of allele substitution for candidate genes

Locus/trait	DGAT^K	OPN^T	PPARGC1A (c.3359)^A	PPARGC1A (c.1892)^T
EBV$_M$	-76.75±42.54	37.25±108.05	-40.77±43.22	92.29±45.91
EBV$_F$	+0.75±1.32	1.97±1.26	-0.08±1.41	0.45±1.50
EBV$_{FP}$	+0.04±0.01[**]	0.01±0.01	0.02±0.01	-0.37±0.01[**]

** (P<0.01).

In addition, Leonard et al. (2005) showed significant association of OPN gene with milk protein percentage. Therefore, we suggest the further studies need to clarify the association between OPN, PPARGC1A (c.3359) and milk production traits in other populations. There are several possible reasons for different results of studies, including differences in allele frequency, the statistical models used to undertake the association analysis and genetic background of the animals in the study (Berry et al. 2010) and environmental circumstances where the animals were producing. The association between DGAT1 gene and EBV$_{FP}$ was significant which may be due to critical role of DGAT1 gene in the synthesis rate of triglycerides (Grisart et al. 2002). Kadlecova et al. (2014) reported significant association between DGAT1 genotypes and fat percentage in primiparous Holstein cows. Anton et al. (2012) indicated the significant effects of the DGAT1 K232A polymorphism on milk yield, fat and protein percentage, as well. In addition, Fontanesi et al. (2015) illustrated that DGAT1 polymorphism was highly associated with fat yield and fat percentage in Reggiania dairy cows (local breed in north of Italy). The results of mean comparisons illustrated that the genotypes of DGAT1^{KK} and PPARGC1A (c.1892)^{TT} had highest EBV$_{FP}$. More results of mean comparisons are shown in Table 6. Table 7 gives the additive effects (allele substitutions) of the alleles. The result indicated that and DGAT^K allele increased the EBV$_{FP}$ by +0.04 ± 0.01. Some standard errors were estimated more than their allele substitution effect. It can be due to low number of data, which are used in this study.

The results of allele substitutions were confirmed by other studies. Winter et al. (2002) and Strzałkowska et al. (2005) showed that the DGAT1^K allele has a positive effect on milk fat content in different cattle breeds.

Naslund et al. (2008) reported that DGAT1^K variant was associated with an increase in milk fat and protein percentages but decrease milk yield compared with the DGAT^A variant. Similar results showed that the DGAT^K allele exceeds of DGAT^A allele, by (+0.34) percentage unit in fat (Grisart et al. 2002). The DGAT1^K allele increases milk fat yield, whereas the DGAT1^A allele increases both milk and protein yield (Kaupe et al. 2007; Thaller et al. 2003).

According to our finding PPARGC1A (c.1892)^T allele decreased the EBV$_{FP}$ by -0.37 ± 0.01. In addition, an association of the PPARGC1A (c.1892)^T allele with higher fat yield has been suggested in German Holsteins (Weikard et al. 2005). Alim et al. (2012) indicated that PPARGC1A (c.1892)^T allele increased protein yield and protein concentration but there was no association between PPARGC1A (c.1892)^T allele and fat yield (%, kg).

CONCLUSION

Milk and its products are regarded as the most important nutritional resource, meeting the energy requirements and offering high quality protein and various vitamins and minerals. Earlier, most genetic improving programs of agriculturally important livestock population have been carried out through complete phenotypic and pedigree information. However, applying molecular genetic information in breeding stock may lead to a better understanding of quantitative traits. Briefly, the results show that there is significant association between PPARGC1A, DGAT1 and EBV$_{FP}$ trait. Generally, detection and estimation of associations of identified genes and genetic markers with economic traits are the basis of a successful application of marker-assisted selection (MAS) in breeding programs. The MAS strategies can be used for pre-selection of young bulls prior to progeny test.

ACKNOWLEDGEMENT

This study was financially supported by Agricultural Biotechnology Research Institute of Iran.

REFERENCES

Alim M.A., Fan Y., Xie Y., Wu X., Sun D., Zhang Y., Zhang S., Zhang Y., Zhang Q. and Lin L. (2012). Single nucleotide polymorphism (SNP) in *PPARGC1A* gene associates milk production traits in Chinese Holstein cattle. *Pakistan Vet. J.* **32,** 609-612.

Andersson L. (2001). Genetic dissection of phenotypic diversity in farm animals. *Nat. Genet.* **2,** 130-138.

Anton I., Kovacs K., Hollo G., Farkas V., Szabo F., Egerszegi I., Ratky J., Zsolnai A. and Brüssow K.P. (2012). Effect of *DGAT1*, leptin and *TG* gene polymorphisms on some milk production traits in different dairy cattle breeds in Hungary. *Arch. Tierzucht.* **55(4),** 307-314.

Banos G., Woolliams J.A., Woodward B.W., Forbes A.B. and Coffey M.P. (2008). Impact of single nucleotide polymorphisms in leptin, leptin receptor, growth hormone receptor, and diacylglycerol acyltransferase (*DGAT1*) gene loci on milk production, feed and body energy traits of UK dairy cows. *J. Dairy Sci.* **91,** 3190-3200.

Berry D.P., Howard D., Boyle P.O., Water S., Kearney J.F. and McCabe M. (2010). Associations between the K232A polymorphism in the diacylglycerol-O-transferase 1 (*DGAT1*) gene and performance in Irish Holstein-Friesian dairy cattle. *Irish J. Agric. Food Res.* **49,** 1-9.

Farnir F., Grisart B., Coppieters W., Riquet J., Berzi P., Cambisano N., Karim L., Mni M., Simon P. and Wagenaar D. (2002). Simultaneous mining of linkage and LD to fine map QTL in outbred half-sib pedigrees: revisiting the location of a QTL with major effect on milk production on bovine chromosome 14. *Genetics.* **161,** 275-287.

Fontanesi L., Scotti L., Samore A.B., Bagnato A. and Russo V. (2015). Association of 20 candidate gene markers with milk production and composition traits in sires of reggiana breed, a local dairy cattle population. *Livest. Sci.* **176,** 14-21.

Grisart B., Coppieters W., Farnir F., Karim L., Ford C., Berzi P., Cambisano N., Mni M., Reid S., Simon P., Spelman R., Georges M. and Snell R. (2002). Positional candidate cloning of a QTL in dairy cattle: identification of a missense mutation in the bovine *DGAT1* gene with major effect on milk yield and composition. *Genome. Res.* **12,** 222-231.

Kadlecova V., Nemeckova D., Katerina Jecminkova K. and Stadni K.L. (2014). Association of bovine *DGAT1* and leptin genes polymorphism with milk production traits and energy balance indicators in primiparous Holstein cows. *Mljekarstvo.* **64(1),** 19-26.

Kaupe B., Winter A., Fries R. and Erhardt G. (2004). *DGAT1* polymorphism in *Bos indicus* and *Bos taursus* cattle breeds. *J. Dairy Res.* **71,** 182-187.

Kaupe B., Brandt H., Prinzenberg E.M. and Erhardt G. (2007). Joint analysis of the influence of *CYP11B1* and *DGAT1* genetic variation on milk production, somatic cell score, conformation, reproduction and productive lifespan in German Holstein cattle. *J. Anim. Sci.* **85,** 11-21.

Khatib H., Zaitoun I., Wiebelhaus-Finger J., Chang Y.M. and Rosa G.J.M. (2007). The association of bovine *PPARGC1A* and *OPN* genes with milk composition in two independent Holstein cattle populations. *J. Dairy Sci.* **90,** 2966-2970.

Knott S.A., Elsen J.M. and Haley C.S. (1996). Methods for multiple-marker mapping of quantitative trait loci in half-sib populations. *Theor. Appl. Genet.* **93,** 71-80.

Komisarek J. and Dorynek Z. (2009). Effect of *ABCG2*, *PPARGC1A*, *OLR1* and *SCD1* gene polymorphism on estimated breeding values for functional and production traits in Polish Holstein-Friesian bulls. *J. Appl. Genet.* **50,** 125-132.

Kowalewska-Luczak I., Kulig H. and Kmiec M. (2010). Associations between the bovine *PPARGC1A* gene and milk production traits. *Czech J. Anim. Sci.* **55,** 195-199.

Leonard S., Khatib H., Schutzkus V., Chang Y.M. and Maltecca C. (2005). Effects of the osteopontin gene variants on milk production traits in dairy cattle. *J. Dairy Sci.* **88,** 4083-4086.

Liang H. and Ward W. (2006). PGC-1α: a key regulator of energy metabolism. *Adv. Physiol. Educ.* **30,** 145-151.

Liu X., Zhang H., li H., li N., Zhang Y., Zhang Q., Wang S., Wang Q. and Wang H. (2008). Fine mapping quantitative trait loci for body weight and abdominal fat traits: effects of marker density and sample size. *Poult. Sci.* **87,** 1314-1319.

Miller S.A., Dykes D.D. and Polesky H.F. (1988). A simple salting-out procedure for extracting DNA from human nucleated cells. *Nucleic Acids Res.* **16,** 1215-1219.

Mosig M.O., Lipkin E., Khutoreskaya G., Tchourzyna E., Soller M. and Friedmann A. (2001). A whole genome scan for quantitative trait loci affecting milk protein percentage in Israeli-Holstein cattle, by means of selective milk DNA pooling in a daughter design, using an adjusted false discovery rate criterion. *Genetics.* **157,** 1683-1698.

Naslund J., Fikse W.F., Pielberg G.R. and Lunden A. (2008). Frequency and effect of the bovine acyl-CoA: diacylglycerolAcyltransferase 1 (*DGAT1*) K232A polymorphism in Swedish dairy cattle. *J. Dairy Sci.* **91,** 2127-2134.

Nei M. (1977). F-statistics and analysis of gene diversity in subdivided populations. *Ann. Hum. Genet.* **41,** 225-233.

Olsen H.G., Lien S., Gautier M., Nilsen H., Roseth A., Berg P.R., Sundsaasen K.K., Svendsen M. and Meuwissen T.H. (2005). Mapping of a milk production quantitative trait locus to a 420-kb region on bovine chromosome 6. *Genetics.* **169,** 275-283.

Riquet J., Coppieters W., Cambisano N., Arranz J.J., Berzi P., Davis S.K., Grisart B., Farnir F., Karim L., Mni M., Simon P., Taylor J.F., Vanmanshoven P., Wagenaar D., Womack J.E. and Georges M. (1999). Fine-mapping of quantitative trait loci by identity by descent in outbred populations: application to milk production in dairy cattle. *Proc. Nat. Acad. Sci. USA.* **16,**

9252-9257.

Schennink A., Bovenhuis H., Leon-Kloosterziel K.M., Arendonk A.M. and Visker M.H.P. (2009). Effect of polymorphisms in the *FASN, OLR1, PPARGC1A, PRL* and *STAT5A* genes on bovine milk-fat composition. *Anim. Genet.* **40,** 909-916.

Schnabel R.D., Kim J.J., Ashwell M.S., Sonstegard T.S., Van Tassell C.P., Connor E.E. and Taylor J.F. (2005). Fine-mapping milk production quantitative trait loci on BTA6: analysis of the bovine osteopontin gene. *Proc. Nat. Acad. Sci. USA.* **102,** 6896-6901.

Spelman R.J., Ford C.A., Mcelhinney P., Gregory G.C. and Snell R.G. (2002). Characterization of the *DGAT1* gene in the New Zealand dairy population. *J. Dairy Sci.* **85,** 3514-3517.

SPSS Inc. (2010). Statistical Package for Social Sciences Study. SPSS for Windows, Version 20. Chicago SPSS Inc.

Strzałkowska N., Siadkowska E., Słoniewski K., Krzyżewski J. and Zwierzchowski L. (2005). Effect of the *DGAT1* gene polymorphism on milk production traits in Black-and-White (Friesian) cows. *Anim. Sci. Pap. Rep.* **23,** 189-197.

Thaller G., Kramer W., Winter A., Kaupe B., Erhardt G. and Fries R. (2003). Effects of *DGAT1* variants on milk production traits in German cattle breeds. *J. Anim. Sci.* **81,** 1911-1918.

Weikard R., Kuhn C., Goldammer T., Freyer G. and Schwerin M. (2005). The bovine *PPARGC1A* gene: molecular characterization and association of an SNP with variation of milk fat synthesis. *Physiol. Genomics.* **21,** 1-13.

Winter A., Krämer W., Werner F.A., Kollers S., Kata S., Durstewitz G., Buitkamp J., Womack J.E., Thaller G. and Fries R. (2002). Association of a lysine-232/alanine polymorphism in a bovine gene encoding acyl-CoA: diacylglycerol acyltransferase (*DGAT1*) with variation at a quantitative trait locus for milk fat content. *Proc. Nat. Acad.*

Zhang Q., Boichard D., Hoeschele I., Ernst C., Eggen A., Murkve B., Pfistergenskow M., Witte L.A., Grignola F.E., Uimari P., Thaller G. and Bishop M.D. (1998). Mapping QTL for milk production and health of dairy cattle in a large outbred pedigree. *Genetics.* **149,** 1959-1973.

Reproductive Performance Evaluation of Holstein Friesian and Their Crosses with Boran Cattle Breeds in Ardaita Agricultural Technical Vocational Education Training College Dairy Farm, Oromia Region, Ethiopia

D.W. Mengistu[1*], K.A. Wondimagegn[2] and M.H. Demisash[2]

[1] Department of Animal Science, Selale University, Selale, Ethiopia
[2] Department of Animal Production and Technology, Bahirdar University, Bahirdar, Ethiopia

*Correspondence E-mail: destawworku@gmail.com

ABSTRACT

The study was conducted at Ardaita Agricultural Technical Vocational Education Training (ATVET) college dairy farm, to evaluate the reproductive performance of Holstein Friesian and its crosses with Boran cattle breeds. Data collected between 2000 and 2015 on reproductive traits (n=2632) were studied and analyzed using general linear model procedure. The overall estimated means for age at first service (AFS), age at first calving (AFC), calving interval (CI), days open (DO) and number of services per conception (NSC) were: 31.33 ± 0.44 months, 41.08 ± 0.44 months, 405.50 ± 3.32 days, 134.84 ± 3.51 days and 1.36 ± 0.03, respectively. Except age at first service, which is influenced by level of Holestien Friesian percentage, season of calving and level of Holestien Friesian percentage was not significant (P>0.05) on all reproductive traits. The traits calving interval and days open significantly (P<0.001) influenced by year of calving and parity. Season of birth (P<0.05) and year of birth (P<0.001) significantly influenced age at first service and age at first calving. Service per conception was significantly influenced by year of calving (P<0.001) only. Except number of service per conception, the result obtained for age at first service, age at first calving, days open and calving interval of Holstein Friesian and its crosses with Boran cows in the study area were below the standards set for commercial dairy farms. Therefore, consideration should be given to the farm to improve those genetic and non-genetic factors affected performance.

KEY WORDS Ardaita, crossbred, Holstein Friesian, reproductive performance.

INTRODUCTION

Ethiopia has the largest livestock population in Africa. The cattle population in Ethiopia is estimated to be 54 million heads, with about 55% females (CSA, 2013). Livestock production constitutes to be an important sub-sector of the agricultural production in Ethiopia, contributing 45% of the total Agricultural Gross Domestic Product (IGAD, 2010). With rapid population and income growth, and increasing urbanization, the demand for livestock and livestock products is growing, presenting huge opportunities for the sector. Even though livestock sector has a significant contribution to the Ethiopian economy but production per animal is extremely low (Kumar and Tkui, 2014). Poor reproductive performance of dairy cows in Ethiopia could be related to genetic, environmental/management factors or both. In order to improve the low productivity of local cattle, selection of the most promising breeds and crossbreeding of these indigenous breed with high producing exotic cattle has been considered as a practical solution (Bekele, 2002).

In this regard, selection among indigenous breeds is a huge task and will take too long to arrive at an acceptable production level. Therefore crossbreeding has long been practiced in Ethiopia to combine the high yield of the European dairy breeds with the toughness of indigenous breeds. Besides their genotype, the performance of dairy animals is also affected by many environmental factors. Un favorable environmental factors may suppress the animal's true genetic ability and create a bias in the selection of animals (Lateef et al. 2008). It has been well documented that, to increase production, improving environmental condition and management practices coupled with improving genetic potential of dairy animals is an effective approach (Lateef, 2007).

Appropriate periodical evaluation of factors affecting reproductive performance of animals is very important for future planning and management. Previous studies have been conducted to evaluate the productive and reproductive performances of dairy cows at institutional large scale dairy farms (Goshu et al. 2007; Fekadu et al. 2010; Abera et al. 2010; Tadesse et al. 2010). However, information is limited about the productive and reproductive performance of dairy cows in different dairy farms, particularly in Ethiopia (Lobago, 2007). Thus there is a need to have data of animals under different farming conditions. The few data from studies on the performance of Holstein Friesian and its crosses with indigenous cattle in Ethiopia are not enough. Moreover, there are not studies that show the reproductive performance of Holstein Friesian and HF × Boran crossbred dairy cows.

Ardaita dairy farm is among the oldest state farms in Ethiopia, where Holstein Friesian and HF × Boran crossbred dairy cows kept. This study was designed to measure and compare various parameters of reproductive performance of Holestien Friesian and HF × Boran crossbred dairy cows.

MATERIALS AND METHODS

Description of the study area
The study was conducted at Ardaita ATVET College dairy farm, which is situated 305 kilo meters far away to southeast of Addis Ababa. The farm is surrounded with farm lands at is at an altitude of 2410-2610 meter above sea level with 4°17'20" north latitude and 37°11'30" east longitude in south eastern Ethiopia of Oromia National Regional State in the high land of Arsi zone. The mean annual rain fall is 1200 mm and maximum and minimum temperature ranges from 20 °C and 5 °C, respectively.

The area has three distinct seasons. A short rainy season which extends from March to June, a long rainy season, which extends from July to October and a dry season that extends from November to February (NMSA, 2010). The dominant soil types are black clay soils with sand–silt clay having a pH of 7.9.

Establishment of the farm and breed groups
The Ardaita dairy farm was established in 1984 by purchasing 50 first crosses of HF × Boran dairy cows. The farm then started milk production with a herd of Holstein Friesian and HF × Boran crosses.

Animal management
Animals were kept under an intensive feeding and production systems. Herds were divided in groups based on sex, age, time of calving and lactation. Animals were liberally stall fed with green fodders and roughages, concentrates were also fed to the animals according to the need (Table 1). Heifers and dry cows were mainly fed on green fodder and other roughages throughout the year. Semen of pure bred Holestien Friesian Bulls from the National Artificial Insemination Center was used for insemination. The insemination was carried out by AI technicians. Detection of estrus was carried out twice a day, early in the morning and late in the afternoon. During the rainy season of the day, cows were grazed on pastures from 2:00-3:00 am local time. Later on the day animals were tied and fed with dry and green fodder, concentrates and mineral licks while being in the shade. Animals were fed according to calculated requirements with concentrate feeds and mineral licks during late pregnancy and lactation (Table 2). Non-lactating cows were only given hay. Lactating cows were received 1 kg concentrates per 2.5 kg of milk produced before each milking. Hay produced from various types of annual and perennial plants of graminaceous and leguminous species were used for feeding animals (Table 3).

Pregnant cows were managed separately during the last trimester. Calving was in well-constructed calving pens. Newborn calves were taken away from their dams shortly after birth and were given colostrums for the first five days age. Fresh milk was offered twice a day in a bucket till the age of 6 months. They were kept in individual pens. Lactating cows were hand-milked twice daily, early in the morning (4:00-5:00 p.m.) and late in the afternoon (3:00-4:00 p.m.).

Animals were regular vaccinated against anthrax, pasteurellosis, blackleg, foot and mouth disease, lumpy skin disease, and contagious bovine pleura pneumonia. Internal and external parasitic infestation were dewormed and sprayed regularly.

Table 1 Ingredients used to make concentrates in Ardaita dairy farm

Classes of animals	Grain (%)	Noug seed cake (%)	Wheat bran (%)	Wheat short (%)	Salt (%)	Lime (%)
Calves	23	25	20	20	1	1
Heifers	20	30	28	20	1	1
Lactating	20	30	27	20	1	1

Table 2 Recommended levels of concentrate feed in Ardaita dairy farm

Classes of animals	Concentrate fed
Heifer	3 kg
Pregnant	3 kg at the last trimester
Lactating	1 kg/2.5 liter of milk produced
Dry cows	2 kg

Table 3 Legumes and grasses used to make hay in Ardaita dairy farm

Legume species	Grass species
Vigna unguiclata	*Sorgum sudanese*
Dolichos lablab	*Cynodon dactylon*
Cajanus cajan	*Chloris gayana*
Dismodium intortum	*Cenchurs ciliaris*
Dismodium unicinatum	*Punisetum purperium*
Medicago sativa	-

Data collection

The data concerning reproductive performance of the farm was collected for the period from 2000 to 2015 G.C. Parameters of reproductive performance were calculated. These include age at first service (AFS), age at first calving (AFC), days open (DO), calving interval (CI) and number of services per conception (NSC). The compiled record cards were checked for completeness and unclear and incomplete data were cleaned out.

Data analysis

The data were entered into Microsoft excel spread sheet and the reproductive traits (AFS, AFC, DO, CI and NSC) were analyzed by the general linear model (GLM) procedures using of the SAS (SAS, 2008). Cases that aborted or had still birth were removed from the data bases. The model included fixed effects of level of HF percentage (75% HF, 87.5% HF and 100% HF), season, parity and year (2000-2015 G.C). The year were divided into 3 seasons based on rain fall distribution; a short rainy season which extends from March to June, a long rainy season, which extends from July to October and a dry season that extends from November to February. Although the maximum parity in the original data was 10. There were too few animals with a parity ≥ 7 for further meaning full analysis. Therefore, all parities above 7 were pooled together in parity ≥ 7. For DO, CI and NSC, the number of observation of animals that calved during 2002 and earlier were too small, therefore all animals that calved prior to 2002 were pooled with those from 2002. Likewise the number of observation of animals that calved during 2014 and later were too small. Therefore, all animals that calved in 2014 and above were pooled together in 2014.

For AFS and AFC cows born during 2000 and 2001 were included. Preliminary analysis showed that interaction effects of the fixed factors were not significant and thus not included in the model. The following statistical models were used to analyze reproductive traits in the farms.

Model 1: age at first service (AFS) and age at first calving (AFC) were analyzed using the following model:

$$Y_{ijk} = \mu + B_i + S_j + Y_k + e_{ijk}$$

Where:

Y_{ijk}: n^{th} record of ith level of HF percentage.

j^{th}: season of birth and k^{th} year of birth.

μ: overall mean.

B_i: fixed effect of ith level of HF percentage (75% HF and 87.5 HF% and 100% HF).

S_j: fixed effect of jth season of birth (long rainy, short rainy and dry season).

Y_k: fixed effect of kth year of birth (2000-2015 G.C).

e_{ijk}: random error associated with each observation.

Model 2: calving interval (CI), days open (DO) and number of service per conception (NSC) were analyzed using the following model:

$$Y_{ijklm} = \mu + B_i + S_j + P_k + H_l + e_{ijkl,}$$

Where:

Y_{ijkl}: observation on NSC, DO and CI trait over n^{th} record of level of HF percentage.

j^{th}: season of calving.

l^{th}: year of calving.

k^{th}: parity.

μ: overall mean.

B_i: fixed effect of ith level of HF percentage (75% HF, 87.5% HF and 100% HF).

S_j: fixed effect of jth season of calving (long rainy, short rainy and dry season).

P_k: fixed effect of kth parity of Dam (1…7).

H_l: fixed effect of lth year of calving (2000-2015 G.C).

e_{ijkm}: random residual error.

RESULTS AND DISCUSSION

Age at first service (AFS)

The overall least square mean and standard error of AFS of Holstein Frisian and HF × Boran crossbred dairy cows was estimated to be 31.33 ± 0.44 months with coefficient of variation 17.11%. The mean AFS found in this study was higher than 23.2 months for local × Holstein Friesian as reported by Ibrahim *et al.* (2011), 24.30 ± 8.01 months for Zebu × Holstein-Friesian dairy cows in Jimma town, Ethiopia (Belay *et al.* 2012), 25.6 months for crossbred dairy cows in eastern lowlands of Ethiopia (Mureda and Mekuriaw, 2007). Besides, the result was higher than the report of Yohannes Shiferaw *et al.* (2003) who found that AFS of 29.58 months for crossbred dairy cows in central highland of Ethiopia, Dinka (2012) reported that the mean AFS as 24.9 months for smallholder cross breed cows in Asella. This longer AFS of Holstein Friesian and HF × Boran crossbred dairy cows in Ardaita farm might be due to poor management of heifers than most reports in the tropics.

Results are given in Table 4. The level of HF percentage (P<0.05), season of birth and year of birth (P<0.001) significantly influenced AFS. Animals with 87.5% Holstein Friesian percentage were younger at age at first service, while purebred Holstein Friesian cows were older (Table 4). The precocity of animals with 87.5% Friesian blood likely benefited more from the last 12.5% of genes coming from the local breed in adapting to highland condition than cattle that were 75% HF or 100% HF.

Fekadu *et al.* (2010) reported similar findings regarding the effect of season of birth on AFS. But on purebred HF cows this was less clear (Lemma *et al.* 2010; Tadesse *et al.* 2010) under Ethiopian conditions. After the rainy season, grass on natural pastures develops rapidly and lasts through dry season, therefore AFS is affected by the chances for young cattle to benefit from the wealth of grass. In order to do this calves must be able to digest fiber.

Calves born in the dry season tended to be younger at first service and calving than those were born in short, and long rainy season of Holstein Friesian cows at Alage dairy farm (Fekadu *et al.* 2010). The significant effect of year of birth on AFS reported in this study is in agreement with the reports of Iffa *et al.* (2006); Tadesse *et al.* (2010); Fekadu *et al.* (2011) and Menale *et al.* (2011). In contrast, it disagreed with the report of Tadesse *et al.* (2006) in Ethiopia. The trend of AFS was inconsistence (Figure 1). The lowest value of AFS was recorded in cows that were born during 2003 and 2004, whereas the highest value recorded with the cows that were born during 2008 and 2010, respectively. Age at first service showed a decreasing trend over the years between 2000 and 2004, while increasing trend from the cows that were born during 2005-2010 (Figure 1). As a result AFS is affected by management and climatic fluctuation in the years.

Age at first calving (AFC)

The overall least square mean and standard error of AFC of pure Holstein Frisian and HF × Boran crossbred dairy cows was estimated to be 41.08 ± 0.44 months with coefficient of variation 13.49%. This is comparable with Ayalew and Assefa (2013) who reported AFC of 40.9 ± 6.6 months for crossbred dairy cows in north Shoa zone and 40.6 months for crossbred dairy heifers in different dairy production systems in central highlands of Ethiopia reported by Yohannes Shiferaw *et al.* (2003). It is higher than 38 months for Boran and its crosses with Friesian and Jersey breed in tropical highlands of Ethiopia as reported by Demeke *et al.* (2004b), 32.1 months for crossbred cows under smallholder condition in Ziway as reported by Yifat Denbarga *et al.* (2009) and 34.7 months for crossbred cows in Gonder, Ethiopia by Ibrahim *et al.* (2011). The difference of AFC reported in the present study with other findings might be due to difference in the level of Holestien percentage, and management variation during calf rearing period and problem of heat detection. A number of previous works indicated that management factor especially nutrition determines pre-pubertal growth rates and reproductive development (Negussie *et al.* 1998; Masama *et al.* 2003).

Results are given in Table 4. Season of birth (P<0.05) and year of birth (P<0.001) significantly influenced AFC, while the effect of HF blood percentage (P>0.05) was not significant.

Similar significant effect of season of birth on AFC reported by Chenyambuga and Mseleko (2009) in Tanzania, Demeke *et al.* (2004b) and Tadesse *et al.* (2006) in Ethiopia. But this was different from the findings of Lemma *et al.* (2010) and Tadesse *et al.* (2010) for purebred HF cows in Ethiopia.

Cows that was born during long rainy season attained AFC higher than those were born during dry and short rainy season, respectively.

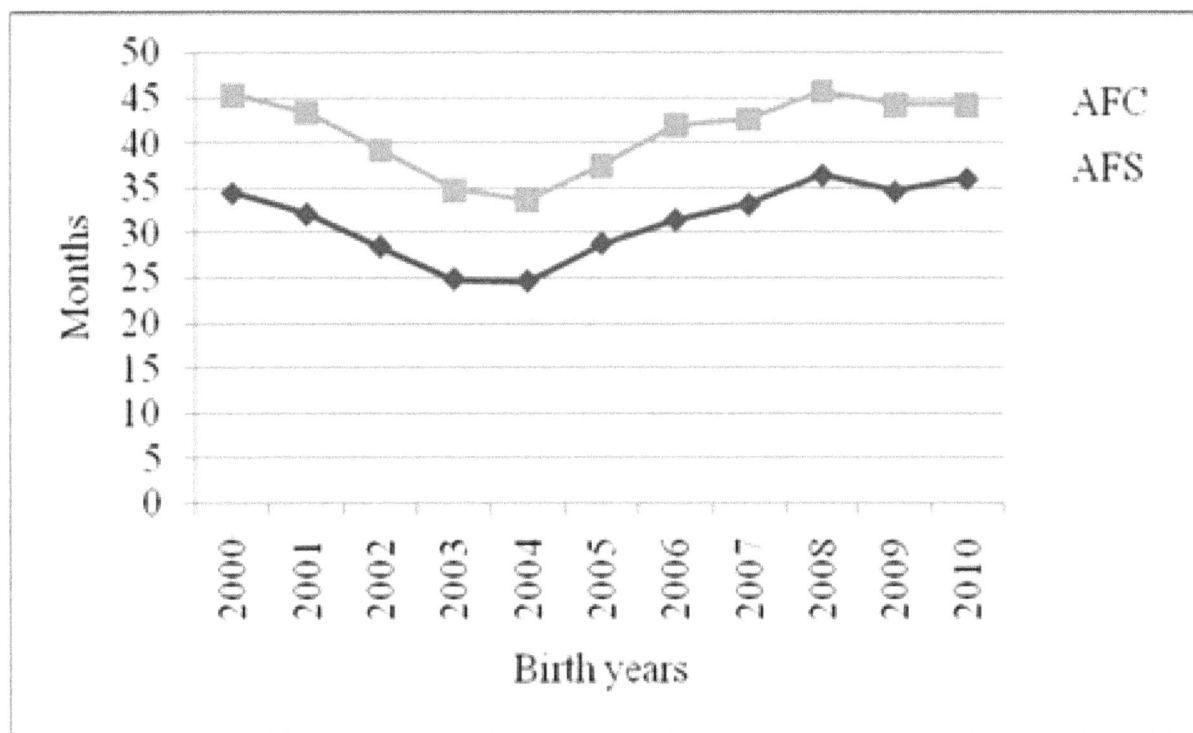

Figure 1 The trend of age at first service (AFS) and age at first calving (AFC) over year of birth at Ardaita College

Table 4 Least square means (LSM±SE) of age at first service (AFS) and age at first calving (AFC) of Holstein Friesian and HF × Boran crossbred dairy cows over the fixed effects of cross-breed percentage, season of birth and year of birth

Factor		AFS (months)	AFC (months)
	N	LSM ± SE	LSM ± SE
Overall	218	31.33±0.44	41.08±0.44
Coefficient of variation (CV) (%)		17.11	13.49
Season of birth		*	*
Long rainy	86	32.72±0.65^a	42.12±0.66^a
Short rainy	66	29.81±0.72^b	39.35±0.73^b
Dry season	66	31.46±0.77^{ab}	41.76±0.76^{ab}
Level of cross-breed		*	NS
¾ Friesian × ¼ Boran	65	31.40±0.95^{ab}	40.95 ±0.96
$^7/_8$ Friesian × $1/_8$ Boran	78	29.88±0.68^b	40.16±0.66
Holstein Friesian	75	32.71±0.69^a	42.13±0.70
Year of birth		***	***
2000	55	34.43±0.87^{ab}	45.25±0.86^a
2001	10	32.05±1.76^{ab}	43.35±1.81^{ab}
2002	16	28.43±1.42^{bc}	39.10±1.41^{ac}
2003	18	24.81±1.40^c	34.70±1.37^c
2004	24	24.59±1.15^c	33.52±1.16^c
2005	17	28.78±1.32^{bc}	37.44±1.36^c
2006	15	31.41±1.49^{ab}	41.92±1.49^{ab}
2007	20	33.13±1.26^{ab}	42.56±1.31^{ab}
2008	13	36.42±1.56^a	45.66±1.60^a
2009	18	34.57±1.37^{ab}	44.19±1.39^a
2010	12	35.99±1.72^a	44.15±1.64^{ab}

N: number of records.
SE: standard error.
The means within the same column with at least one common letter, do not have significant difference (P>0.05).
* (P<0.001) and *** (P<0.001).
NS: non significant.

Long rainy season is followed by dry season, which can be challenging due to stress condition and less chance of getting natural pastures. Therefore AFC is affected by the chance for young cattle to be benefited from the this season.

The significant effect of year of birth on AFC reported in this study agreed with the reports of Iffa *et al.* (2006) and Menale *et al.* (2011) and it disagreed with Tadesse *et al.* (2006) in Ethiopia. The trend of AFC over year of birth was inconsistent (Figure 1). A decreasing trend of AFC observed in cows that were born from 2000-2004, but increasing trend of AFC observed in cows that were born from 2005-2010 (Figure 1). This might be due to management fluctuation among years and the recommended amount of energy were not fed for calves.

Days open (DO)

The overall least square mean and standard error for DO of Holstein Frisian and HF × Boran crossbred dairy cows was estimated to be 134.84 ± 3.51 days with coefficient of variation 54.58%. This is nearly comparable with the report of Denberga *et al.* (2009) and Lyimo *et al.* (2004) who found that the mean value for DO of 130 days in the urban smallholder dairy farms around Ziway and for smallholder crossbred dairy cows in sub humid costal Tanzania, respectively. But this result was lower than the mean DO of 185 days for crossbred dairy cows in central highlands of Ethiopia (Yohannes Shiferaw *et al.* 2003) and much lower than Iffa *et al.* (2006) who reported 200.13 ± 25.55 days for cross breeds in highland of Ethiopia. However, the DO in the current study is longer than 104 and 86 days reported for the smallholder crossbred dairy cows in Zimbabwe (Masama *et al.* 2003). The difference in the DO of crossbred cows in the current study might be attributed to the existing differences in management which have accounted for the observed differences on DO (Masama *et al.* 2003; Yohannes Shiferaw *et al.* 2003; Lyimo *et al.* 2004).

Results are given in Table 5. Year of calving and parity had significant effect (P<0.001) on DO, but the level of HF percentage and season of calving had no significant effect (P>0.05). Similar significant effect of year of calving was reported by Tadesse *et al.* (2010) for HF cows in Ethiopia. The highest and lowest value of DO observed during 2002 (173.53±9.90 days) and 2005 (103.45±10.34 days), respectively. Although there was no consistence trend, mean DO declined from 2002-2006, slightly increased in between 2007 and 2008 and remains constant with minimal decreasing trend from 2009-2014 (Figure 2). Generally, days open directly affect CI, which plays an important role in the profitability of dairy farm. Therefore, more emphasis should be given for the inconsistence trend of DO over year of calving.

Tadesse *et al.* (2010) reported similar finding regarding the effect of parity on DO. In contrast, reported non-significant effect of parity on DO of crossbreed dairy cows in highlands of Ethiopia. Cows in the 1st parity had significantly longer DO and remain constant in the other parities. The result reflected in this study agreement with Giday (2001) who reported that the longest DO in young cows, which might be due to lower energy balance as they are not able to consume more for their own growth, production, reproduction and maintenance, thus lower energy balance delays the onset of postpartum heat.

Calving interval (CI)

The overall least square mean and standard error for CI of pure Holstein Friesian and HF × Boran crossbred dairy cows was estimated to be 405.50 ± 3.32 days with coefficient of variation 16.9%. The result is comparable with 408 days for crossbred dairy cows of urban smallholder dairy farms reported by Denberga *et al.* (2009). However, it was lower than the CI of 459 ± 2.4 days for HF crosses in central highlands of Ethiopia (Tadesse, 2001) and 552 days for crossbred dairy cows in central highlands of Ethiopia (Yohannes Shiferaw *et al.* 2003).

However, the CI in the present study is longer than the generally accepted calving interval of 365 days expected on a commercial dairy farm. This longer CI is the result of longer DO obtained which could be related to poor management of the existing farm including poor breeding management. Belay *et al.* (2012) suggested that relatively longer CI might be indicative of poor nutritional status, poor breeding management, lack of own bull and artificial insemination service, longer days open, diseases and poor management practices.

Results are summarized in Table 5. Year of calving and parity had significant (P<0.001) effect, but the level of HF percentage and season of calving (P>0.05) had no significant effect. The observed significant effect of year of calving is in line with the findings of various reports (Tadesse, 2006; Tadesse *et al.* 2010; Menale *et al.* 2011) in Ethiopia. The highest calving interval was recorded during 2002, whereas during 2005, 2006 and 2014, respectively, recorded lowest calving interval. Calving interval for cows that were calved from 2002-2006 decreased and remains constant from 2007-2012 and decreased then after. In general, the trend of CI over year of calving was inconsistence (Figure 2).

This inconsistent trend might be attributed to change in climatic condition, negligence of management practice, because these years were the period of regime change and hence directed to financial scarcity as the farm was funded by government across year of calving.

Table 5 Least square means (LSM±SE) of calving interval (CI), days open (DO) and number of service per conception (NSC) of Holstein Friesian and HF × Boran crossbred dairy cows over the fixed effects of level of cross-breed percentage, season of calving, year of calving and parity

Factor	CI (days)		DO (days)		NSC	
	N	LSM ± SE	N	LSM ± SE	N	LSM ± SE
Overall	700	405.50±3.32	691	134.84±3.51	809	1.36±0.03
Level of cross-breed		NS		NS		NS
¾ Friesian × ¼ Boran	275	397.13±5.19	255	128.99±5.99	284	1.42±0.04
⁷/₈ Friesian × ¹/₈ Boran	295	406.33±4.55	290	135.96±4.94	338	1.37±0.04
Holstein Friesian	130	413.04±7.31	146	139.58±7.91	187	1.30±0.06
Season of calving		NS		NS		NS
Long rainy	244	396.87±5.06	237	124.78±5.40	278	1.41±0.04
Short rainy	217	412.89±5.08	219	140.30±5.52	277	1.33±0.04
Dry season	239	406.74±5.01	235	139.45±5.56	254	1.35±0.04
Year of calving		***		***		***
2002	93	437.08±8.83[a]	94	173.53±9.90[a]	97	1.64±0.08[a]
2003	29	409.95±13.45[cab]	27	129.48±15.40[cab]	29	1.20±0.12[bcd]
2004	43	389.61±11.61[cb]	42	109.07±12.08[cb]	16	1.31±0.10[th]
2005	65	382.52±9.46[c]	66	103.45±10.34[c]	70	1.58±0.08[ab]
2006	67	376.78±9.22[c]	63	111.18±10.43[cb]	74	1.50±0.08[ab]
2007	67	427.37±9.03[cab]	63	154.31±10.16[cab]	81	1.32±0.07[cd]
2008	57	429.35±9.49[ab]	55	162.58±10.61[ab]	60	1.27±0.08[cd]
2009	59	413.10±9.20[cab]	60	136.45±10.11[cab]	70	1.31±0.08[cd]
2010	53	406.30±9.72[cab]	50	129.21±11.05[cab]	63	1.12±0.08[e]
2011	51	415.06±10.03[cab]	34	135.37±13.49[cab]	56	1.21±0.09[cd]
2012	54	428.28±9.65[cab]	48	154.55±11.37[cab]	52	1.27±0.09[cd]
2013	48	398.36±10.39[cab]	48	129.47±11.56[cab]	56	1.29±0.09[cd]
2014	14	357.74±19.04[c]	41	124.32±12.42[cab]	55	1.48±0.09[ab]
Parity		***		***		NS
1	161	441.22±5.89[a]	162	178.01±6.40[a]	200	1.42±0.05
2	145	410.11±6.13[b]	145	133.57±6.64[b]	155	1.34±0.05
3	116	394.98±6.74[b]	114	117.49±7.45[b]	132	1.35±0.06
4	86	399.89±7.80[b]	86	129.12±8.68[b]	105	1.29±0.06
5	69	395.38±8.78[b]	67	126.29±9.82[b]	73	1.33±0.08
6	54	401.96±10.07[b]	51	130.24±11.24[b]	58	1.33±0.09
7⁺	69	394.97±9.25[b]	66	129.18±10.20[b]	86	1.47±0.07

N: number of records.
SE: standard error.
The means within the same column with at least one common letter, do not have significant difference (P>0.05).
*** (P<0.001).
NS: non significant.

The significant effect of parity on CI of Holstein Friesian and HF × Boran crossbred dairy cows is in agreement with the report of several authors (Tadesse and Dessie, 2003; Goshu et al. 2007; Mekuriaw et al. 2009; Tadesse et al. 2010; Menale et al. 2011). On contrary to the current finding, Mulindwa et al. (2006) in Uganda found that parity did not significantly (P>0.05) affect CI. Cows with first parity recorded highest CI. The evidence from longer DO recorded in first parity of animal was the result of longer CI in 1st parity of this study.

Number of service per conception (NSC)
The overall mean and standard error of NSC of Holstein Friesian and HF × Boran crossbred dairy cows found in this study was estimated to be 1.36 ± 0.03 with coefficient of variation 44.9%.

This is nearly comparable with Moges (2012) who reported that services required for each conception was 1.3, 1.5 in urban and peri urban areas in Gonder, respectively. It was higher than 1.29 (Ibrahim et al. 2011) for local × Holstein in Gondar, Ethiopia.

However, the result observed was lower than 1.72 for Boran and its crosses at Holetta research center (Tessema et al. 2003) and 1.720 ± 0.06 for smallholder crossbreed cows in Ziway (Goshu et al. 2007). The NSC found in the study area suggests comparatively better result in Ethiopian conditions.

This might be due to semen quality, use of reproductive technology, heat detection ability and individual skill difference of AI technicians. Furthermore, favorable environmental condition in the study area makes more animals to have first service.

The results are summarized in Table 5. Number of service per conception was significantly affected by year of calving (P<0.001) only, but level of HF percentage, season of calving and parity had no significant effect (P>0.05). The non-significant effect of season of calving, blood level and parity in the current study disagreed with Asimwe and Kifaro (2007) and Ahmed et al. (2007). But Goshu et al. (2007); Tadesse et al. (2010) and Hammoud et al. (2010) reported non- significant effect of season on HF cows in Ethiopia and Egypt, respectively.

Non-significant effect of parity on NSC was reported by Ibrahim et al. (2011), while Goshu et al. (2007) and Yifat Denbarga et al. (2009) found significant effect of parity on NSC.

Number of service per conception of animals were highest during the year 2002 and 2005, whereas lowest in animals that were calved in 2010. Generally NSC observed in the current study showed a progressively decreasing trend from 2002-2010 except 2003, which is significantly decreased and again increased from 2011-2014 (Figure 3).

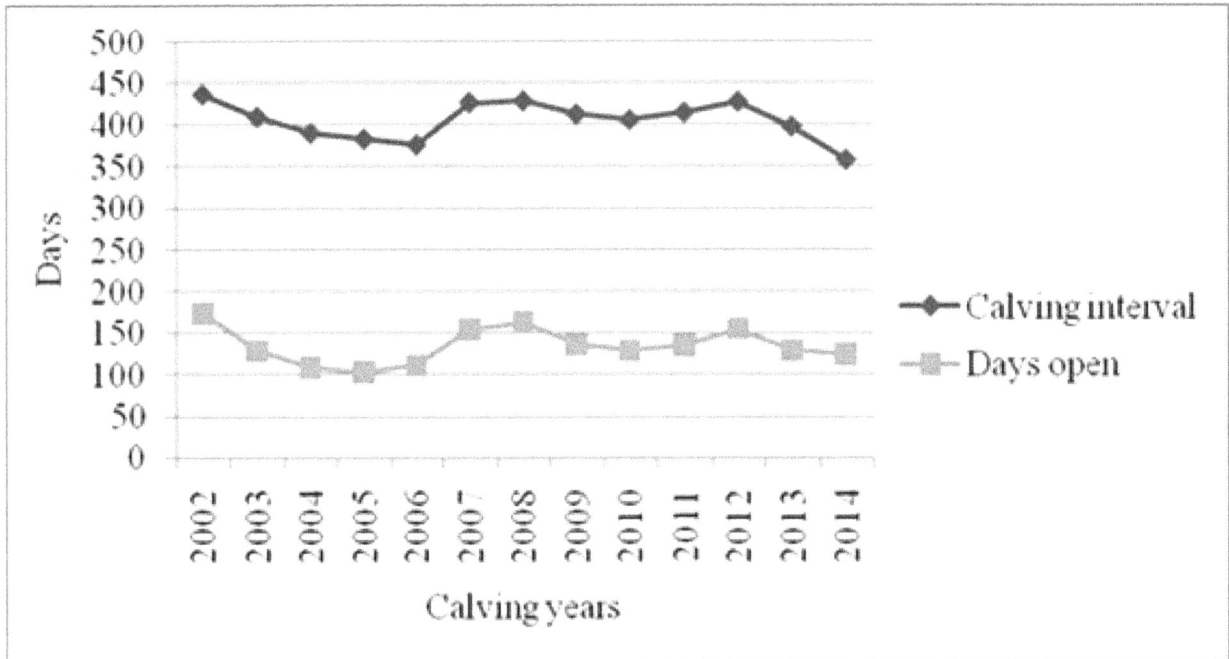

Figure 2 The trend of calving interval and days open over year of calving at Ardaita College

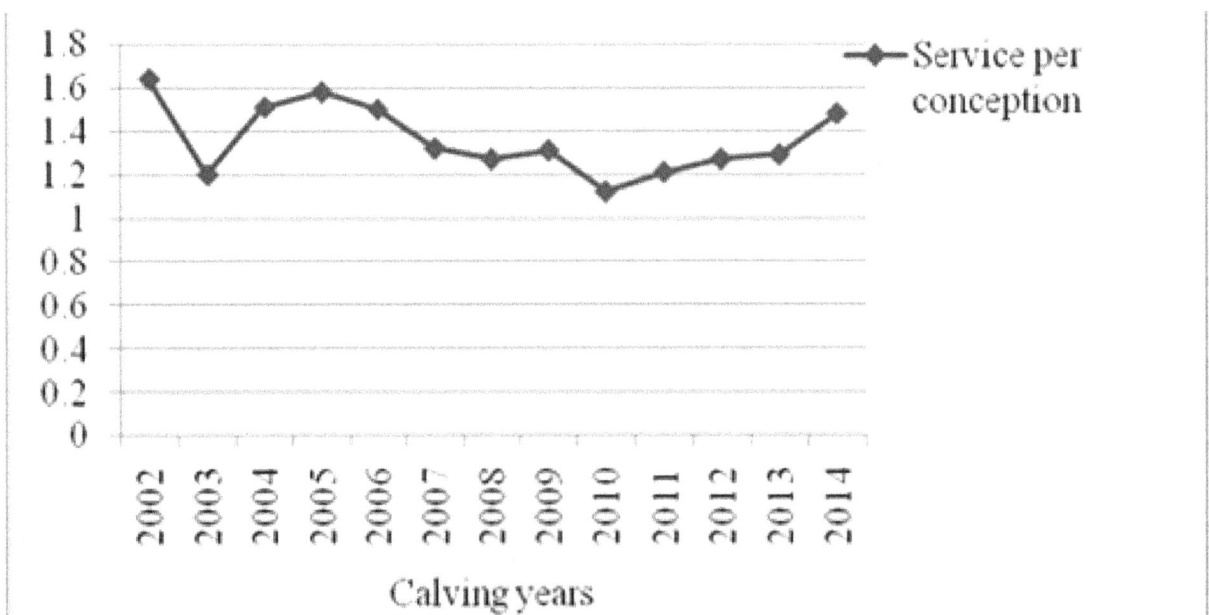

Figure 3 The trend of number of service per conception over year of calving at Ardaita College

The decreasing trend over the years could reflect change of management regime like adequate feeding, difference in inseminators, proper heat detection, artificial insemination (AI) service and semen quality over the years.

CONCLUSION

The mean obtained for NSC was generally good as compared to other studies. AFS, AFC, CI and DO of Holstein Friesian and HF × Boran crossbred dairy cows in the study area were not satisfactory and below the standard expected from commercial dairy farm. Animals with 100% HF had longer AFS than 75% and 87.5% HF percentage, respectively. This could be due to inability of higher graded cows to withstand the prevailing environmental and management condition. Season of birth (P<0.05) and year of birth (P<0.001) significantly influenced AFS and AFC. The traits CI and DO significantly (P<0.001) influenced by year of calving and parity. Except AFS the effect of level of HF percentage was not significant (P>0.05) on all reproductive traits. NSC was influenced by year of calving only. Due to this level of management and breeding practice, management difference, variation of climatic condition over season of birth, year of birth and year of calving might be accounted for the lower value of reproductive traits in the study. The result therefore would provide very useful information and assist decision making particularly regarding how to improve the low reproductive performance for future production. The longer DO and CI in this study resulted to have very limited number of calve delivered in its life time. It is recommended that, adequate plane of nutrition before and after calving, good breeding management and keeping the health condition of animals in the farm will help to increase the onset of cycling activity in the breeding season and help to minimize the longer DO and CI. Since year of calving and year of births had shown to influence the performance of the existing breed, great attention should be given for the inconsistent management practice across the years.

ACKNOWLEDGEMENT

Above all the Authors would like to thank Ardaita ATVET College for their co-operation and offering facilities. Special thanks would be offered to Bahirdar University College of Agriculture and Environmental Science Especially Animal Production and technology instructors for their provision of various services during teaching learning process. The Authors also thank farm attendants, farm managers, AI technicians and veterinary professionals working in dairy farms who agreed to participate in the study.

REFERENCES

Abera H., Abegaz S. and Mekasha Y. (2010). Estimation of genetic parameters for growth traits of Horro cattle and their crosses with Holstein Friesian and Jersey at Bako Agricultural Research Center, Western Oromia, Ethiopia. MS Thesis. Haramaya Univ., Ethiopia.

Ahmed M.K., Ahmed B.T., Musa L.M.A and Peters K.J. (2007). Milk production and reproduction traits of different grades of Zebu × Friesian crossbreds under semi-arid conditions. *Arch. Tierz. Dummerstorf.* **50(3)**, 240-249.

Asimwe L. and Kifaro G.C. (2007). Effect of breed, season, year and parity on reproductive performance of dairy cattle under smallholder production system in Bukoba district, Tanzania. *Livest. Res. Rural Dev.* Available at: http://www.lrrd.org

Ayalew M. and Asefa B. (2013). Reproductive and lactation performances of dairy cows in Chacha town and nearby selected kebeles, North Shoa Zone, Amhara Region, Ethiopia. *World. J. Agri. Sci.* **1(1)**, 8-17.

Bekele T. (2002). Reproductive performances of zebu (Fogera) breed in the central highlands of Ethiopia. DVM Thesis. Addis Ababa Univ., Debre Zeit, Ethiopia.

Belay D., Yisehak K. and Janssens G.P.J. (2012). Productive and reproductive performance of Zebu × Holstein-Friesian crossbred dairy cows in Jimma Town, Oromia, Ethiopia. *Glob. Vet.* **8(1)**, 67-72.

Chenyambuga S.W. and Mseleko K.F. (2009). Reproductive and lactation performances of Ayrshire and Boran crossbred cattle kept in smallholder farms in Mufindi district, Tanzania. *Livest. Res. Rural Dev.* Available at: http://www.lrrd.org.

CSA. (2013). Agricultural Sample Survey 2012. Report on livestock and livestock characteristics. Addis Abeba, Ethiopia.

Demeke S., Neser F.W.C and Schoeman S.J. (2004b). Estimates of genetic parameters for Boran, Friesian and crosses of Friesian and Jersey with the Boran cattle in the tropical highlands of Ethiopia: milk production traits and cow weight. *J. Anim. Breed. Genet.* **121**, 163-175.

Denberga Y., Belihu K., Merga M., Lobago F., Gustafson H. and Kindahl H. (2009). Study on reproductive performance of crossbred dairy cattle under smallholder conditions in and around Ziway, Ethiopia. *Livest. Res. Rural Dev.* Available at: http://www.lrrd.org.

Dinka H. (2012). Reproductive performance of crossbred dairy cows under smallholder condition in Ethiopia. *Int. J. Livest. Prod.* **3(3)**, 25-28.

Fekadu A., Kassa T. and Belihu K. (2010). Reproductive performance of Holstein-Friesian dairy cows at Alage Dairy Farm, Rift Valley of Ethiopia. *Trop. Anim. Health Prod.* **43**, 581-586.

Giday Y. (2001). Assessment of calf crop productivity and total herd life of Fogera cows at Andassa Ranch in north western Ethiopia. MS Thesis. University of Alemaya, Ethiopia.

Goshu G., Belihu K. and Berihun A. (2007). Effect of parity, season and year on reproductive performance and herd life of

Friesian cows at Stella private dairy farm, Ethiopia. *Livest. Res. Rural Dev*. Available at: http://www.lrrd.org.

Hammoud M.H., Zarkouny S.Z. and Oudah E.Z.M. (2010). Effect of sire, age at first calving, season and year of calving and parity on reproductive performance of Friesian cows under semiarid conditions in Egypt. *Arch. Zootec*. **13(1)**, 60-82.

Ibrahim N., Abraha A. and Mulugeta S. (2011). Assessment of reproductive performance of crossbred cattle (Holstein Friesian×Zebu) in Gondar town. *Glob. Vet*. **6(6)**, 561-566.

Iffa K., Hegde B.P. and Kumsa T. (2006). Lifetime production and reproduction performances of *Bos taurus* × *Bos indicus* cross bred cows in the Central Highlands of Ethiopia. *Ethiopian J. Anim. Prod*. **6(2)**, 37-52.

IGAD. (2010). Inter Governmental Authority on Development. The contribution of livestock to economies of IGAD member states. IGAD Livestock Policy Initiative, Great Wolford, United Kingdom.

Kumar N. and Tkui K. (2014). Reproductive performance of crossbred dairy cows in Mekelle, Ethiopia. *Sci. J. Anim. Sci*. **3(2)**, 35-40.

Lateef M. (2007). Productive performance of Holstein Friesian and Jersey cattle in subtropical environment of the Punjab, Pakistan. Ph D. Thesis. University of Agriculture Faisalabad, Pakistan.

Lateef M.K.Z., Gondal M., Younas M., Sarwary M.I., Mustafa M. and Bashir M.K. (2008). Milk production potential of pure bred Holstein Friesian and Jersey cows in subtropical environment of Pakistan. *Pakistan Vet. J*. **28(1)**, 9-12.

Lemma H., Belihu K. and Sheferaw D. (2010). Study on the reproductive performance of Jersey cows at Wolaita Sodo dairy farm, southern Ethiopia. *Ethiopia J. Vet*. **4(1)**, 53-70.

Lobago F. (2007). Reproductive and lactation performance of dairy cattle in the Oromia Central Highlands of Ethiopia with special emphasis on the pregnancy period. Ph D Thesis. Swedish Univ., Uppsala, Sweden.

Lyimo C., Nukya R., Schoolman L. and Van Eerdenbutg F.J. (2004). Post-partum reproductive performance of crossbred dairy cattle on smallholder farms in sub humid coastal Tanzania. *J. Trop. Anim. Health Prod*. **36**, 269-279.

Masama E., Kusina K.T., Sibanda S. and Majoni C. (2003). Reproduction and lactation performance of cattle in a smallholder dairy system in Zimbabwe. *Trop. Anim. Health Prod*. **35**, 117-129.

Mekuriaw G., Ayalew W. and Hegde P.B. (2009). Growth and reproductive performance of Ogaden cattle at Haramaya university, Ethiopia. *Ethiopian J. Anim. Prod*. **9(1)**, 13-38.

Menale M., Mekuriaw Z., Mekuriaw G. and Taye G. (2011). Reproductive performances of Fogera cattle at Metekel cattle breeding and multiplication ranch, north west Ethiopia. *J. Anim. Feed Res*. **1(3)**, 99-106.

Moges N. (2012). Study on reproductive performance of crossbred dairy cows under small holder conditions in and around Gondar, north western Ethiopia. *J. Reprod. Infer*. **3(3)**, 38-41.

Mulindwa H.E., Ssewannyana E. and Kifaro G.C. (2006). Extracted milk yield and reproductive performance of Teso cattle and their crosses with Sahiwal and Boran at Serere, Uganda. *Uganda J. Agri. Sci*. **12(2)**, 36-45.

Mureda E. and Mekuriaw Z. (2007). Reproductive performance of crossbred dairy cows in eastern lowlands of Ethiopia. *Livest. Res. Rural Dev*. Available at: http://www.lrrd.org.

Negussie E., Brannang E., Banjaw K. and Rottmann O.J. (1998). Reproductive performance of dairy cattle at Asella livestock farm. Arsi, Ethiopia. I. Indigenous cows versus their F1 crosses. *J. Anim. Breed. Genet*. **115**, 267-280.

SAS Institute. (2008). SAS®/STAT Software, Release 9.2. SAS Institute, Inc., Cary, NC. USA.

Tadesse M. (2001). Estimation of crossbreeding parameters for milk production traits at Debre Zeit Agricultural Research Centre, Ethiopia. *Ethiopian J. Anim. Prod*. **1**, 45-54.

Tadesse M. and Dessie T. (2003). Milk production performance of zebu, Holstein Friesian and their crosses in Ethiopia. *Livest. Res. Rural Dev*. Available at: http://www.lrrd.org/lrrd15/3/Tade153.htm.

Tadesse M, Dessie T, Tessema G, Degefa T and Gojam Y. (2006). Study on age at first calving, calving interval and breeding efficiency of *Bos taurus*, *Bos indicus* and their crosses in the Highlands of Ethiopia. *Ethiopia J. Anim. Prod*. **6(2)**, 1-16.

Tadesse M., Thiengtham J., Pinyopummin A. and Prasanpanich S. (2010). Productive and reproductive performance of Holstein Friesian dairy cows in Ethiopia. *Livest. Res. Rural Dev*. Available at: http://www.lrrd.org.

Tadesse Y. (2006). Genetic and non-genetic analysis of fertility and production traits in Holetta and Adea Berga dairy herd. MS Thesis. Alemaya Univ., Ethiopia.

Tessema G., Gebrewold A. and Jayaparakash J. (2003). Study on reproductive efficiency of Boran and its crosses at Holetta research farm: effect of genotype, management and environment. *Ethiopia J. Anim. Prod*. **3(1)**, 89-107.

Yohannes Shiferaw Y., Tenhagen B.A., Bekana M. and Tesfu K. (2003). Reproductive performance of crossbred dairy cows in different production systems in the central high lands of Ethiopia. *J. Trop. Anim. Health Prod*. **35(6)**, 551-561.

Predictive Ability of Statistical Genomic Prediction Methods When Underlying Genetic Architecture of Trait is Purely Additive

M. Momen[1], A. Ayatollahi Mehrgardi[1*], A. Sheikhy[2], A.K. Esmailizadeh[1], and M. Assadi Foozi[1]

[1] Department of Animal Science, Faculty of Agriculture, Shahid Bahonar University of Kerman, Kerman, Iran
[2] Department of [...] Faculty of mathematics and Computer Science, Shahid Bahonar University of Kerman, Kerman, Iran

*Correspondence E-mail: a_ayatmehr@yahoo.com

ABSTRACT

A simulation study was conducted to address the issue of how purely additive (simple) genetic architecture might impact on the efficacy of parametric and non-parametric genomic prediction methods. For this purpose, we simulated a trait with narrow sense heritability $h^2 = 0.3$, with only additive genetic effects for 300 loci in order to compare the predictive ability of 14 more practically used genomic prediction models based on four criteria (mean squared error (MSE), Bias, $\gamma_{y,GEBV}$ and $\gamma_{GEBV,TBV}$). Results suggested that parametric genomic prediction models have greater superiority over non parametric genomic models under a simple purely additive genetic architecture. Our result also showed that, all parametric methods, other than ridge-regression BLUP (RR-BLUP), could explain most of phenotypic variation because they showed lower MSE, higher predictive correlation ($\gamma_{y,GEBV}$), the least amount of bias ($b_{y,GEBV}$) and the higher correlations between true breeding values and the estimated genomic breeding values ($\gamma_{TBV,GEBV}$). Random forest regression had the worst performance among non parametric methods. The simulation results suggested that there is a large difference between performances of non parametric methods in comparison with parametric methods when underlying architecture is purely additive. But this may not happen when dominance and epistatic genetic effects contributing to both additive and non-additive genetic variances.

KEY WORDS genomic statistical method, non parametric, parametric, predictive ability.

INTRODUCTION

Today, genome-wide marker data and whole genome single nucleotide polymorphism (SNP) chips are easy available and widely used for evaluation a wide range of economically important traits in plant and animal breeding programs (Desta and Ortiz, 2015; Do *et al.* 2015). Animal and plant breeders, are mainly interested in estimating genomic breeding values for these traits with including whole genome wide marker information that call as genomic selection (GS); (Daetwyler *et al.* 2013; Meuwissen *et al.* 2016). The genomic selection goal is that maximum capturing

variance that can be explained by the markers (Su *et al.* 2012). To achieve this, any statistical method implemented for the predictions must be able handle the large numbers of markers and evaluate marker effects across the entire genome (Gianola and Rosa, 2015). In hence, the models poses two source of challenges, one is curse of dimensionality and the other is unknown genetic architecture of the quantitative traits (Daetwyler *et al.* 2010; de Los Campos *et al.* 2013). Genetic architecture is a description of the structure of the genotype-phenotype relationship that includes the nature of the loci contributing to phenotypic variation (e.g., number of loci and their genomic location) and a description of the alleles at those loci; number of alleles, magnitude of effects, patterns of pleiotropy, additivity, dominance, epistasis, epigenetic effects (Holland, 2007; Tiezzi and Maltecca, 2015; Yang *et al.* 2007). These methods behave different manner to overcome the difficulties of the second challenge and providing robust estimations from simple to complex genetic architecture of quantitative traits (Daetwyler *et al.* 2010; Fernández *et al.* 2016).

However, the performance of prediction methods can be affected by genetic architecture and in particular patterns of pleiotropy, additivity, dominance and epistasis (Lidan Sun; 2015).

The aims need to assess the accuracy and sensitivity of whole genome prediction statistical methods that are linked conceptually by genetic architecture. However, selection of the best genomic prediction model with considering a specific genetic architecture of a complex traits remains one of the main goals for animal breeders and evolutionary geneticists. In other hand, the genetic architecture of most economic traits remains unknown for animal breeders and evolutionary geneticists (Stranger *et al.* 2011). Then, unravelling the performance of GS methods toward the genomic architecture of a polygenic trait is fundamental to study ability of methods for predictions.

The objective of this study was to compare different genomic prediction models includes parametric and nonparametric, and assess their abilities toward a specific trait with only additive architecture. We hypothesize that with further exploration of predictive ability of GS models toward the purely genomic architectures, we will clearly find the true accuracy of genomic prediction models under a simple genetic architecture.

MATERIALS AND METHODS

In order to evaluate the impact of purely additive genetic architectures of complex trait on the accuracy and predictive ability of GS models, we simulated a quantitative trait which influenced by only narrow sense heritability ($h^2=0.3$) and assessed predictive ability of the 14 genomic prediction models.

Population structure
A population was simulated for 2000 historical generations at an effective size of 100 ($N_e=100$). After 55 generations random mating, during the whole process, all individuals were generated with one gamete from a random father and one from a random mother. In each generation, 20 males mated with 400 females, 20 half-sib families. Therefore, the data set for the estimation of the marker effects consisted of the 4800 individuals from the last five generations and used to estimate predictive ability and sensitivity of statistical method.

The genome was assumed to consist of 5 chromosomes each 100 cM long and 2000 loci/chromosome (i.e. a total of 10000 SNP plus 300 QTL) were located at random map positions (as shown in Figure 1). Both SNP and QTL were biallelic. Mutations were generated at a rate of 2.5×10^{-5} per locus per generation at the marker loci and at the QTL loci. Similar to Meuwissen and Goddard (2001), a standard gamma distribution with shape parameter $\alpha= 0.42$ and scale parameter $\beta= 2.619$ was used to drawn allele substitution effects (α_j). The sign of an allele substitution effect was drawn at random with equal chance.

True genomic estimated breeding values
The true breeding value (TBV) for each animal was calculated as the expected genotypic value of a certain QTL genotype that carried by animal i:

$$TBV_i = \sum_{j=1}^{P}[(X_{AA} = 2) \times (2q_j\alpha_j) + (X_{Aa \text{ or } Aa}=1) \times (q_j\alpha_j - p_j\alpha_j) + (X_{aa}=0) \times (-2 p_j\alpha_j)]$$

Where:

X_{ij}: covariate indicator of the genotype of the jth QTL of the ith individual that has the values 2, 1, 0 for genotypes AA, Aa or aa, respectively.

p_j and q_j: allelic frequencies (A or a) for the jth marker in the training population.

α: average effect of substitution for the jth marker calculated as:

$$\alpha_j = a_j + d_j(q_j - p_j) \text{ with } d_j = 0.$$

Statistical methods
To predict marker effects and performance of GS models, a five-folds cross validation scheme were used and repeated 20 times per run. We divided the data into training and testing sets and the training sets were used to fit the models, and the testing sets were used to determine the performance of the particular method. The evaluated methods include parametric methods; genomic best linear unbiased prediction (GBLUP), ridge regression BLUP (RR-BLUP), least absolute shrinkage and selection operator (LASSO), elastic net (EN), Bayesian ridge regression (BRR), Bayesian LASSO (BL), Bayes A, Bayes B, Bayes C and nonparametric methods, includes; reproducing kernel hilbert space (RKHS), support vector machine (SVM), relevance vector machine (RVM) and gaussian processors (GP). The statistical software R (R Core Team, 2015) was used to run the parametric and nonparametric methods.

With respect to that the evaluations were based on the 20 replicates for each cross validated scenario, the average of the results was reported.

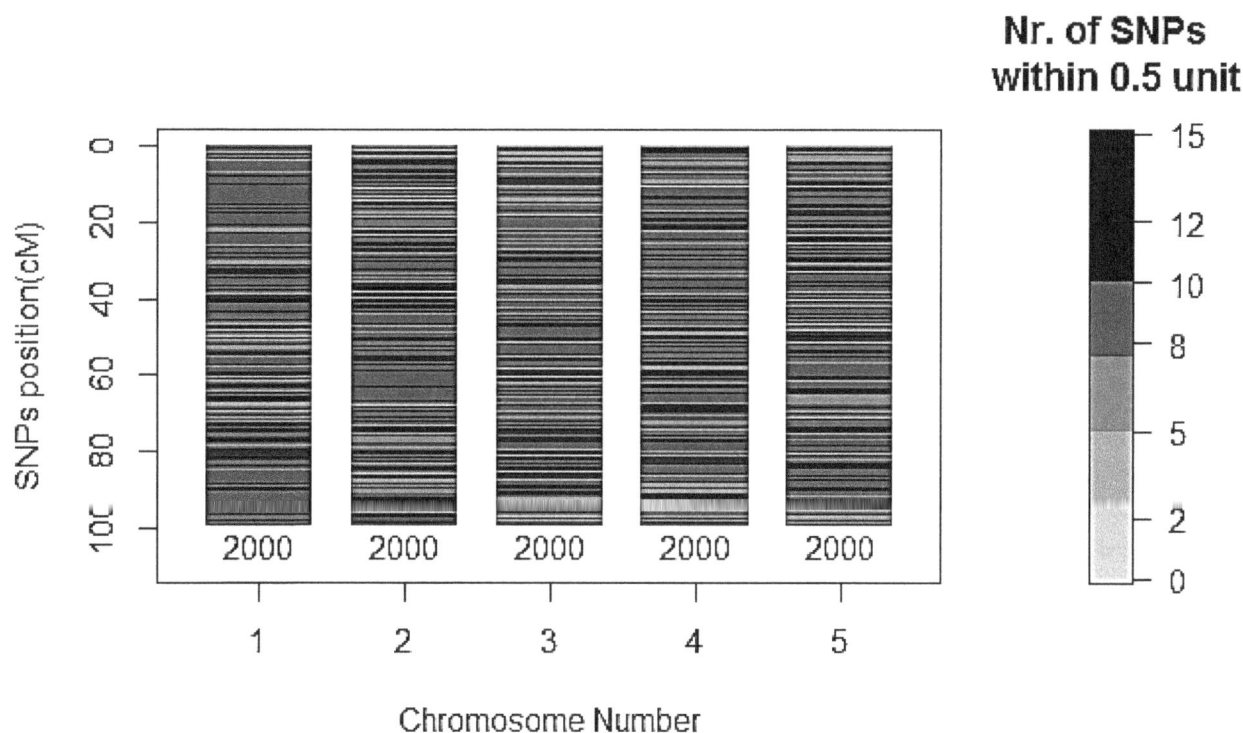

Figure 1 Distribution of randomly SNP coverage across the five simulated chromosomes

For each scenario, the sensitivity and predictive ability of the genomic prediction models was measured by four statistic criteria including: predictive correlation as the person's correlation between the true phenotypic values and the predicted estimated genomic breeding values ($\gamma_{y,GEBV}$), mean square error (MSE), empirical accuracies of genomic predictions as the correlation between GEBVs and the true breeding values ($\gamma_{GEBV,TBV}$) and the unbiasedness was assessed by regression of the simulated phenotypes on the GEBVs ($b_{y,GEBV}$). Significant differences between methods in terms of predictive ability were assessed by means of paired t-tests ($\alpha=1\%$), adjusted by bonferroni correction.

RESULTS AND DISCUSSION

The comparison of cross-validated results for different models allowed estimation of the similarities and dissimilarities of them. Averages and standard errors (SE) were computed for each statistic by considering the results of the 20 replicates available in each situation. Table 1 shows the mean and SE (sampling variabilities) of afore mentioned statistics for the nine parametric (five Bayesian and four frequentist) and five non-parametric implemented models. The predictive correlation of models ranges from 0.676 for LASSO to 0.568 for random forest regression and the range of mean square error and bias for tested models were 23.15-30.34 and 0.93-1.63, respectively.

The predictive correlation ($\gamma_{y,GEBV}$), MSE and bias of the Bayesian methods were quite similar. The predictive correlation of GBLUP, LASSO and EN models and Bayesian models was higher than RKHS and machine learning models.

Comparison of described models for MSE and bias also demonstrated high similarity between Bayesian and frequentist parametric method and a slight their differences with non-parametric and machine learning method. The bias between Bayesian and parametric method was very close to 1 for all traits except elastic net ($\gamma_{y,GEBV}$=0.93). Unbiased models are expected to have a slope coefficient of 1, whereas values greater than 1 indicate a biased overestimation in the GEBV prediction and values smaller than 1 indicate a biased underestimation of the GEBV. In our study, in terms of empirical accuracy, the five Bayesian methods and other parametric performed similarly with correlations over 0.90 and non-parametric and machine learnings had slightly downward correlations. Bayes B, Bayes C and LASSO were slightly more accurate than others, while random forest regression yielded the smallest bias (the worst), the lowest accuracy and the highest MSE. Bayesian and parametric (RR-BLUP, GBLUP, LASSO and EN) models reached a very similar predictive correlation ($\gamma_{y,GEBV}$) for the given trait. However, non-parametric methods tended to outperform than other models for MSE, bias and empirical accuracies.

Table 1 Prediction accuracies criteria means across different genomic prediction models for a purely additive trait ($h^2=0.3$) with five folds cross-validation

Method	Predictive correlation	MSE	Bias	$\gamma_{TBV,GEBV}$
Bayes A	0.666 ± 0.010^{bc}	23.15 ± 0.09^{f}	1.00 ± 0.04^{d}	0.914 ± 0.06^{c}
Bayase B	0.669 ± 0.010^{bc}	23.24 ± 0.09^{f}	0.99 ± 0.04^{d}	0.923 ± 0.06^{b}
Bayase C	0.669 ± 0.010^{bc}	23.24 ± 0.09^{f}	0.99 ± 0.04^{d}	0.923 ± 0.06^{b}
Bayase L	0.661 ± 0.010^{bd}	23.39 ± 0.09^{f}	1.00 ± 0.04^{d}	0.909 ± 0.06^{e}
Bayase R	0.661 ± 0.014^{bd}	23.41 ± 0.14^{f}	1.00 ± 0.06^{d}	0.908 ± 0.09^{e}
Least absolute shrinkage and selection operator (LASSO)	0.676 ± 0.014^{a}	22.73 ± 0.14^{h}	1.03 ± 0.06^{c}	0.926 ± 0.09^{a}
Elastic net (EN)	0.670 ± 0.014^{bc}	23.13 ± 0.14^{fg}	0.93 ± 0.06^{e}	0.915 ± 0.09^{c}
Ridge regression BLUP (RR-BLUP)	0.659 ± 0.014^{d}	26.71 ± 0.14^{b}	0.99 ± 0.06^{d}	0.907 ± 0.09^{e}
Genomic best linear unbiased prediction (GBLUP)	0.672 ± 0.014^{b}	22.78 ± 0.14^{hg}	0.99 ± 0.06^{d}	0.911 ± 0.09^{d}
Support vector machine (SVM)	0.638 ± 0.014^{e}	24.86 ± 0.14^{d}	1.15 ± 0.06^{b}	0.874 ± 0.09^{h}
Relevance vector machine (RVM)	0.647 ± 0.010^{e}	24.23 ± 0.09^{e}	0.94 ± 0.04^{e}	0.887 ± 0.06^{g}
Reproducing kernel hilbert space (RKHS)	0.608 ± 0.014^{g}	26.23 ± 0.14^{c}	1.03 ± 0.06^{c}	0.834 ± 0.09^{i}
Gaussian processors (GP)	0.650 ± 0.010^{e}	24.31 ± 0.09^{e}	1.14 ± 0.04^{b}	0.893 ± 0.06^{f}
RF	0.568 ± 0.014^{h}	30.34 ± 0.14^{a}	1.63 ± 0.06^{a}	0.778 ± 0.09^{j}

MSE: mean squared error.
The means within the same column with at least one common letter, do not have significant difference (P>0.001).

Three machine learning method (Support Vector Machine, Relevance Vector Machine and Gaussian Process) and random forest performed poorly on these datasets, even though the models parameters was optimized well, but the methods significantly different from all the other for the criteria (P<0.01).

The others pairs of non-parametric methods significantly different from each other (P<0.01), with comparing means, in terms of predictive correlation, MSE and empirical accuracy, RVM had higher values but RKHS had a best bias, very close to 1 (i.e. $\gamma_{TBV,GEBV}=1.03$).

The boxplots (Figures 1 and 2), show the distribution of the prediction accuracy values for 100 runs and the relative performance of the methods. Figure 1 contains 28 boxplots of bias (A) and predictive correlation of predicted breeding values (B) and Figure 2 shows MSE and empirical accuracy boxplots for the 14 different methods. In each figure, the first five boxplots are for the Bayesian methods, the next four box plots represent the frequentist methods and the later five plots are for non-parametric methods. These summary plots clearly show the ability and inability of using any kind of methods when only additive architecture is present.

Bias of the methods
The coefficient of regression (slope) of simulated phenotype on GEBV was calculated as a measurement of the bias of each method. Ideally a value of $\gamma_{y,GEBV}$ equal to one indicates no bias in the prediction. Figure 2(A) shows the slopes of regressed simulated phenotypes on estimated breeding values for all models.

All Bayesian method had very similar bias and very close to one (red vertical line in Figure 2(A)) and they were not significantly different than one, indicating no significant bias in the prediction. Across the frequentist method RR-BLUP and GBLUP had a bias similar to Bayesian methods and very close to one but a slight upward and downward estimation found for LASSO and elastic net, respectively. In addition, more variation and significant differences among the non parametric methods were detected. The value of $\gamma_{y,GEBV}$ derived from RKHS was slightly better across the non-parametric models. The values of SVM, GP and RF were higher than one (over estimated) and the RVM was lower (under estimated).

Predictive ability of the methods
Figure 2(B) shows the boxplot of correlation between the simulated phenotype (y) and the predicted genomic breeding value (GEBV), since each validation group had GEBV estimated from a different prediction equation and might have a different mean. This correlation represent the predictive ability ($\gamma_{y,GEBV}$) of GS to predict phenotypes (Resende et al. 2012). Overall, the ability to predict phenotype ranged from 0.56 for Random Forest to 0.67 for LASSO (Table 1). In the Figure the red line shows the total mean accuracy of all methods (0.65), although the parametric methods differ in a priori assumptions about marker effects, but their predictive ability was similar and all of them had the accuracy higher than the total mean (red line). In contrast, nonparametric methods, particularly RF, RKHS, and SVM, provided predictions that are reasonably less accurate for a traits with purely additive architecture.

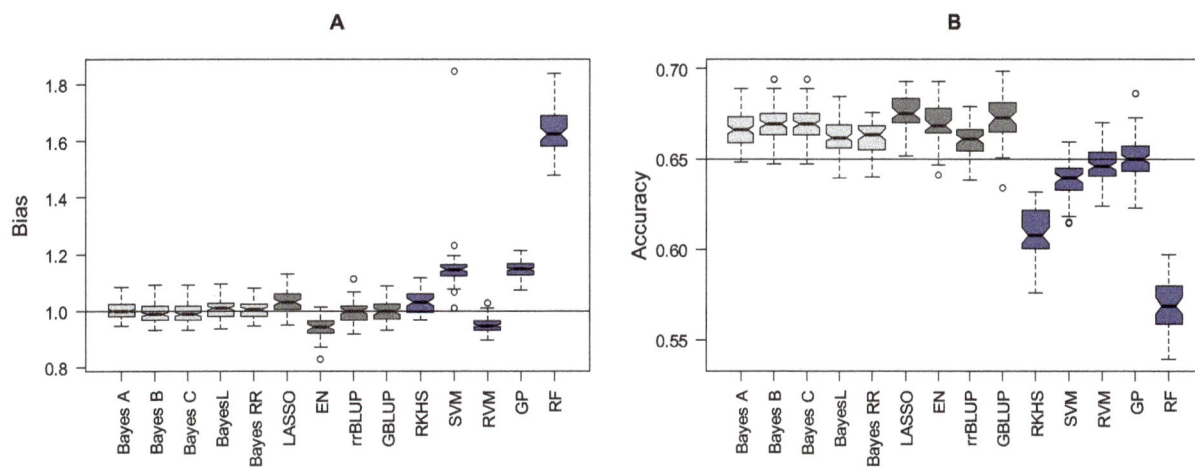

Figure 2 The boxplots of Bias and Accuracy of prediction for the trait with purely additive genetic architecture (narrow sense heritability of 0.30) The first nine boxplots (five Bayesian and four frequentist) correspond to the parametric methods, and the last five (dark blue) boxplots correspond to the non-parametric (machine learning) methods

Mean square error and empirical accuracy

Mean square error (MSE) of the estimator used for calculating as prediction error and a risk function to quantify differences between the estimator and the true value. In this study, the mean squared error of prediction between simulated phenotype and GEBV of animals in the testing set was used as the loss function to be minimized and a goodness of prediction criteria for the tested models. A lower MSE is associated with a better overall fit and also, larger estimates of $\gamma_{\text{TBV,GEBV}}$ is a criteria for more reliable predictions. The results from this study show that in terms of MSE, parametric methods performed better and MSE value downward of total MSE mean except for RR-BLUP method (Figure 3C). The five non-parametric method, showed higher MSE than the total mean. The overall fit of the models to the simulated purely additive trait, judged by the mean squared prediction error, favored parametric methods over the non-parametric regression methods. However, for empirical accuracy (i.e. $\gamma_{\text{GEBV,TBV}}$), lower estimates of the correlation were obtained for non-parametrics and all parametric performed better (Figure 3D).

Hierarchical clustering tree

Figure 4 presents the hierarchical clustering tree obtained through the averaging of the distance matrix across all methods. The clustering results showed the similarities in terms of four estimated criteria (predictive correlation, MSE, Bias and empirical accuracy) between the different models. The clustering chart clearly shows that all parametric except RR-BLUP methods placed in one cluster. The RR-BLUP and RKHS had close similarity and fallen into a separate cluster. As well as, the machine learning methods (SVM, RVM and GP) has close similarities and were placed in the same category.

The Random Forest Regression was quite different, with deep divisions in the clustering with other method. This analysis also showed strong similarity between two by two methods and conclude that Bayes B with Bayes C, LASSO with GBLUP, Bayes A with Bayes L, Bayes RR with EN, RR-BLUP with RKHS, RVN with GP have more similarities. It is also interesting to note that the elastic net clustered with Bayes RR despite being a combination of lasso and ridge regression penalty. The idea of genomic selection was initially raised more than 15 years ago, however, it was not practically used until the coming of high capacity genotyping platforms (Meuwissen *et al.* 2016). Beginning reports on advantages and disadvantages of various statistical methodologies for genomic selection have been conducted largely on simulated data sets (Daetwyler Hans *et al.* 2013). Simulation is potentially informative way to assess predict ability of genomic prediction models, especially when underlying genetic nature of complex trait and biological mechanisms are unknown. This paper describes the performance of 14 statistical approaches for the prediction of genomic breeding values using a simulated data set.

We compared nine parametric (five Bayesian and four frequentist) and five non-parametric statistical GS methods. Comparisons were based on a simulated phenotype where genotypic variability was responsible for only 30% of the phenotypic variability ($h^2=0.3$). The underlying genetic architectures responsible for the genotypic variability, consisted of 300 independently segregating biallelic QTL loci that contributed equally either in an additive manner to a quantitative trait (purely additive architecture).

Our study showed that using a parametric method in analyzing a simple additive architecture quantitative trait could provide better goodness of fit than using non-parametric method.

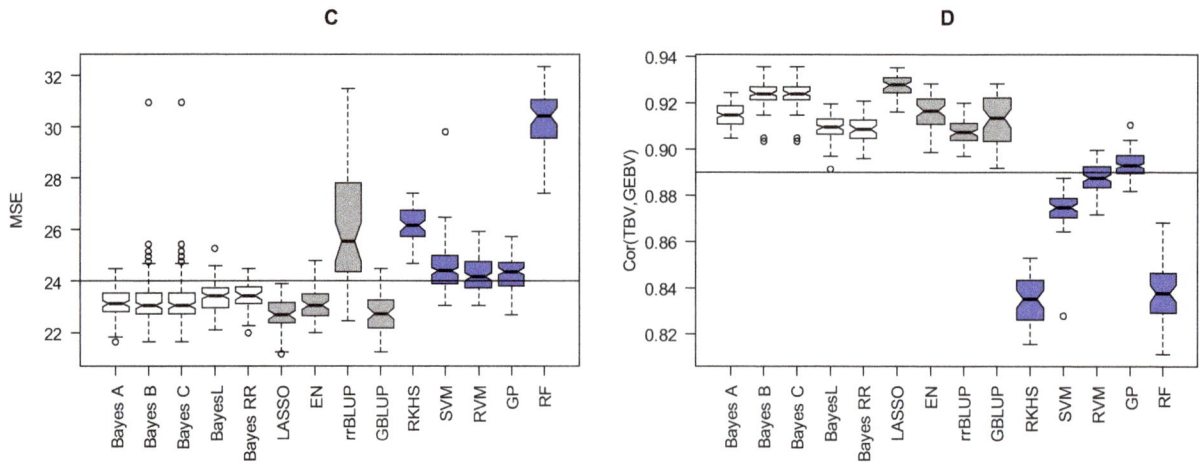

Figure 3 The boxplots of MSE and correlation of TBV and GEBV of prediction for the trait with purely additive genetic architecture (heritability of 0.30)
The first nine boxplots (five Bayesian and four frequentist) correspond to the parametric methods and the last five (dark blue) boxplots correspond to the nonparametric (Machine Learning) methods

Purely Additive Trait

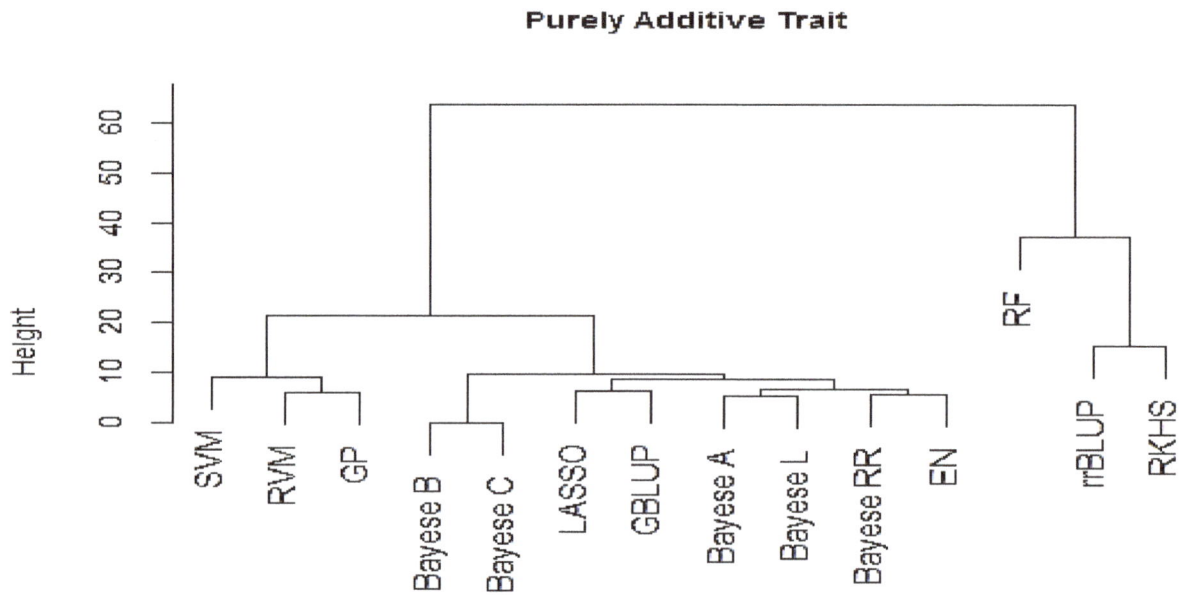

Figure 4 Hierarchical clustering of genomic selection (GS) models based on four estimated criteria, the height on the y axis refers to the value of the criterion associated with a particular agglomeration of models. Parametric (frequentist): RR-BLUP, GBLUP, LASSO, EN, parametric (Bayesian shrinkage regression): Bayes A, Bayes B, Bayes C, Bayesian LASSO, non-parametric (Machine Learnings): reproducing kernel hilbert space (RKHS), relevance vector machine (RVM), support vector machine (SVM) and gaussian processors (GP)

Our result also showed that all parametric methods, except RR-BLUP, could explain most of phenotypic variation because they showed lower MSE, higher accuracy, the regression coefficient close to one and higher $\gamma_{TBV,GEBV}$.

It seems that accuracy of statistical genomic prediction models is dependent on the genetic architecture of complex traits, the size of the training population, the number of independent chromosome segments, the heritability of complex trait and the marker density panels.

With respect to the relative performance of the prediction methods, Daetwyler *et al.* (2010) suggested that the accuracy of GBLUP is invariant to number of quantitative trait loci (QTL) affecting the trait, while the accuracy of statistical strategies taking into account the variable selection, is expected to be greater than that of GBLUP when number of independent chromosome segments are more than number of QTLs. In the present study with a highly additive quantitative trait, without involving any non-additive effects, the predictive abilities were considerably greater in the case of parametric predictions when compared to the non-parametric prediction models. Despite this, this superiority of predictive ability for parametric method, predictions varied markedly across Bayesian and frequentists methods. Another hypothesis for the differences in the superiority of parametric predictions in compare with non-parametric method, also mentioned by (Lee *et al*. 2008).

Our results are in agreement with the findings by Gianola *et al*. (2006), who state that the non-parametric methods should be able to better predict phenotypes that are based on genetic architectures consisting of epistatic interactions. Also, Howard *et al*. (2014) state that if the underlying genetic architecture is additive, then parametric GS methods are slightly better than the non-parametric methods for levels of heritability. Both the accuracy of prediction and the MSE results suggest the same about the models in terms of predictive performance. When underlying architecture of a trait is purely additive, the RR-BLUP perform the worst among the parametric methods, and the Random Forest performs the worst among the non-parametric methods (shown in Figure 2 and Figure 3). The LASSO and GBLUP had the highest prediction accuracy and the lowest MSE values among the parametric methods when only the additive is present. The other parametric methods like the EN are more power to shrink the QTL effects. Among the all methods, the Random Forest showed poor predictive ability. We can conclude that when the underlying architecture is only additive, in fact, we assume that the SNPs on genome are independent that is same as assumptions about parametric approaches. The simulation of a purely additive architecture done with the assumption that simulate additive effects for QTLS are independent. In this case, we satisfy the parametric model assumption of having independent explanatory variables, so the parametric models have a larger predictive power than the nonparametric models (Howard *et al*. 2014).

Hayes *et al*. (2010) reported that the accuracy of GEBV using parametric methods for overall type trait, which considering a normal distribution of the effects is better and conversely for fat % and proportion of back spot when predictions is based on leptokurtic distribution of the effect is better. In other study, Ober *et al*. (2012) reported that, parametric GS methods were unable to predict chill coma recovery, a quantitatively measured adaptive trait in *Drosophila*.

Two-dimensional scans of the whole genome had previously revealed that the genetic architecture of this trait is composed primarily of interactions involving many loci. Whole genome prediction is affected not only by the underlying genetic architecture but also by additional types of unpredictable genetic contributions including intra-locus dominance and genotype by environment interactions.

Here in, we have demonstrated the superior ability of the parametric method in Bayesian and frequentist context to accurately predict phenotypes for a highly additive architecture complex trait.

CONCLUSION

Fourteen regression methods proposed to calculate genomic breeding values using a complex trait which is under a highly additive genetic architecture for 10000 genome-wide SNP markers. From our evaluations, for traits controlled by large number of QTLs with only additive effects, parametric methods reached very similar predictive abilities and a clear ranking of statistical methods was observed in function of the trait analyzed. Knowledge of traits' genetic architectures can be integrated into practices of genomic prediction, which will help the approaches perform better toward different architectures.

REFERENCES

Daetwyler H., Pong-Wong R., Villanueva B. and Woolliams J. (2010). The impact of genetic architecture on genome-wide evaluation methods. *Genetics*. **185**, 1021-1031.

Daetwyler H.D., Calus M.P., Pong-Wong R., de Los Campos G. and Hickey J.M. (2013). Genomic prediction in animals and plants: simulation of data, validation, reporting and benchmarking. *Genetics*. **1932**, 347-365.

de Los Campos G., Hickey J.M., Pong-Wong R., Daetwyler H.D. and Calus M.P. (2013). Whole-genome regression and prediction methods applied to plant and animal breeding. *Genetics*. **1932**, 327-345.

Desta Z.A. and Ortiz R. (2015). Genomic selection: genome-wide prediction in plant improvement. *Trends. Plant. Sci.* **199**, 592-601.

Do D.N., Janss L.L., Jensen J. and Kadarmideen H.N. (2015). SNP annotation-based whole genomic prediction and selection: an application to feed efficiency and its component traits in pigs. *J. Anim. Sci.* **93(5)**, 2056-2063.

Fernández J., Toro M.A., Gómez-Romano F. and Villanueva B. (2016). The use of genomic information can enhance the efficiency of conservation programs. *Anim. Fronts.* **6(1)**, 59-64.

Gianola D., Fernando R.L. and Stella A. (2006). Genomic-assisted prediction of genetic value with semiparametric procedures. *Genetics*. **1733**, 1761-1776.

Gianola D. and Rosa G.J.M. (2015). One hundred years of statistical developments in animal breeding. *Ann. Rev. Anim. Biosci.* **31**, 19-56.

Hayes B. J., J. Pryce, A. J. Chamberlain, P. J. Bowman and M. E. Goddard (2010). Genetic architecture of complex traits and accuracy of genomic prediction: coat colour, milk-fat percentage and type in Holstein cattle as contrasting model traits. *PLoS Genet.* **6(9)**, 1001139.

Holland J.B. (2007). Genetic architecture of complex traits in plants. *Curr. Opin. Plant. Biol.* **10(2)**, 156-161.

Howard R., Carriquiry A.L. and Beavis W.D. (2014). Parametric and nonparametric statistical methods for genomic selection of traits with additive and epistatic genetic architectures. *G3 Bethesda.* **46**, 1027-1046.

Lee S.H., van der Werf J.H., Hayes B.J., Goddard M.E. and Visscher P.M. (2008). Predicting unobserved phenotypes for complex traits from whole-genome SNP data. *PLoS Genet.* **4**, 1000231.

Lidan Sun R.W. (2015). Mapping complex traits as a dynamic system. *Phys. Life Rev.* **13**, 194-197.

Meuwissen T., Hayes B. and Goddard M. (2016). Genomic selection: a paradigm shift in animal breeding. *Anim. Front.* **6**, 6-14.

Meuwissen T.H.E. and Goddard M.E. (2001). Prediction of total genetic value using genome-wide dense marker maps. *Genetics.* **157**, 1819-1829.

Ober U., Ayroles J.F., Stone E.A., Richards S., Zhu D., Gibbs R.A., Stricker C., Gianola D., Schlather M., Mackay T.F. and Simianer H. (2012). Using whole-genome sequence data to predict quantitative trait phenotypes in *Drosophila melanogaster. PLoS Genet.* **8(5)**, 1002685.

R Core Team. (2015). R: a language and environment for statistical computing. Vienna, Austria: the R Foundation for Statistical Computing. Available at: http://www.R-project.org/.

Resende M.D.V., Resende M.F.R., Sansaloni C.P., Petroli C.D., Missiaggia A.A., Aguiar A.M., Abad J.M., Takahashi E.K., Rosado A.M., Faria D.A., Pappas G.J., Kilian A. and Grattapaglia D. (2012). Genomic selection for growth and wood quality in Eucalyptus: capturing the missing heritability and accelerating breeding for complex traits in forest trees. *New Phytol.* **194(1)**, 116-128.

Stranger B.E., Stahl E.A. and Raj T. (2011). Progress and promise of genome-wide association studies for human complex trait genetics. *Genetics.* **187(2)**, 367-383.

Su G., Christensen O.F., Ostersen T., Henryon M. and Lund M.S. (2012). Estimating additive and non-additive genetic variances and predicting genetic merits using genome-wide dense single nucleotide polymorphism markers. *PLoS One.* **7(9)**, 45293.

Tiezzi F. and Maltecca C. (2015). Accounting for trait architecture in genomic predictions of US Holstein cattle using a weighted realized relationship matrix. *Genet. Sele. Evol.* **47(1)**, 24.

Yang J., Zhu J. and Williams R.W. (2007). Mapping the genetic architecture of complex traits in experimental populations. *Bioinformatics.* **23(12)**, 1527-1536.

5

Performance Traits of Buffalo under Extensive and Semi-Intensive Bathan System

M.M. Momin[1*], M.K.I. Khan[1] and O.F. Miazi[1]

[1] Department of Genetics and Animal Breeding, Chittagong Veterinary and Animal Science University, Khulshi, Chittagong-4225, Bangladesh

*Correspondence E-mail: mm.islam@cvasu.ac.bd

ABSTRACT

This study was conducted to investigate the scenario of buffalo production and reproduction under different farming systems at Subarno Char, in the coastal area of Bangladesh. A total of 14 farms were randomly selected and studied for various traits live weight (LW); daily milk yield (DMY); lactation length (LL); lactation production (LP); calving interval (CI); gestation length (GL); post partum heat period (PPH); age at first calving (AFC) and service per conception (SPC) of buffaloes through pre designed questionnaire, direct observation and available records. The LW (372.31±14.64 kg) and DMY (1.99±0.16 liter/day/cow) were found to be highest under semi-intensive bathan farming systems than other systems, however, the LL (275.25±2.857 days) and LP (628.80±34.49 liter) were found higher under extensive bathan farming system irrespective of breeds. On the other hand, LW (390.54±14.06 kg), DMY (2.82±0.13 liter/day/cow), LL (284.96±3.31 days) and LP (899.75±52.83 liter) were higher in River type buffaloes than other types. The GL (305.37±0.72 days), CI (640.34±51.31 days), AFC (54.72±1.57 months) and SPC (1.62±0.21) were found lowest under semi-intensive bathan farming system, but PPHP (134.04±5.30 days) was found lowest under the extensive bathan farming system. The GL (301.74±0.63 days), PPHP (123.21±7.50 days), AFC (47.00±1.35 months) and SPC (1.40±0.16) were found lowest in River type buffaloes, but CI (660.31±43.82) was lowest in crossbred buffalo cows. The birth weight was highest (28.28±0.48 kg) under semi-intensive bathan farming system. Productive and reproductive performances of buffaloes under the study area found were moderate. The profitability of buffalo rearing under extensive farming system was higher than other. The findings of this study may assist farmers and policy makers in making decisions for future buffalo farming and undertaking the genetic improvement program to increase the milk production in Bangladesh.

KEY WORDS breeding, buffaloes, farming systems, management, performance traits.

INTRODUCTION

Buffalo population in Bangladesh is about 1.47 million (FAO, 2013), which is 0.7% of total population of the world (FAO, 2010). Buffaloes are the important species in the tropical and subtropical countries of the world including Bangladesh for their uses in agricultural sector. The farmers are keeping buffaloes for milk and traction their cultivable land. Buffaloes are the second largest ruminants, reared extensively in this country and a vital part of the national economy. Buffalo not only contributes significantly to the national gross domestic product (GDP) of Bangladesh but also the provision of employment and food to the rapidly growing population of the country. Buffaloes are raised under an extensive system in the coastal and hilly areas where large-scale pasture land and enough green forages/grasses are available. In addition, buffaloes are raised under a semi-intensive system on plain land and marshy land where there is limited pasture land. Both the River and Swamp type and their crossbred buffaloes are available in Bangladesh. These buffalo are found in the Bramhaputra-Jamuna flood plain of central Bangladesh, the Ganges-Meghna flood plain of southern Bangladesh and in institutional herds (Faruque, 2000). The Subarno Char is in the coastal area of the Bay of Bengal which represents an extensive flat, coastal and delta land, situated on the tidal flood plain of the Meghna river delta, characterized by flat land and low relief.

The area is affected by diurnal tidal cycles and the tidal fluctuations vary depending on seasons, being pronounced during the monsoon season. However, there are very little information available for current buffalo production (breeding, feeding), management and economics in this area. Furthermore, the decision is required for profitable farms these are (i) the number of buffaloes to be run by a farmer; (ii) which breeds/genotype (s) are suitable; (iii) what type and level of supplementary feed is required throughout the year; (iv) the area to be cultivated for fodder; (v) the amount of feed to be conserve to meeting the shortage of feed. For running a profitable farm it is very important to identify the differences in production and efficiency to use the input (Kirk et al. 1988; Khan, 2009) this will assist the policy makers, researchers and farmers to making decision for profitable buffalo farming. Although the buffaloes are available their present scenarios are unidentified. Therefore, the current study was aim to find out solution of the above questions, evaluate the present situation of buffalo production, reproduction, economics and provide suggestion for improve the buffalo production.

MATERIALS AND METHODS

Study area and study period
The study was conducted in the coastal area (Subarno Char) from October, 2013 to November, 2014 in the Noahkhali District, which is located in the western part of Bangladesh at the bank of "the Bay of Bengal" and in the Department of Genetics and Animal Breeding at Chittagong Veterinary and Animal Sciences University (CVASU). Total number of buffaloes is shown in Table 1.

Identification of type
Different types of buffalos were seen in this area. The phenotypic and morphological features of the available buffaloes were recorded according to the criterion described by Faruque et al. (2004); Faruque and Hossain (2010). As per the phenotypic and morphological features; the available buffaloes in the study area were categorized and identified their breed/type.

Data collection
A pre designed questionnaire was used for collecting the information on buffalo production, reproduction and their management. Data was collected on various production parameters (live weight, milk yield, lactation length and lactation production) and reproduction parameters (calving interval (CI); gestation length (GL); postpartum heat period (PHP); age at first calving (AFC) and service per conception (SPC). A total of 14 households (of those farmers who have at least 10 buffalo) were surveyed directly.

Table 1 The number of various genotypes in different system

Type	Semi intensive farming system	Extensive farming system
River type	152	68
Crossbred type	181	203
Swamp type	127	411
Total	460	682

Calculation of live weight (LWT)
Body length (L) was taken from the point of shoulder to the pin bone and heart girth (G), were measured in inch using a measuring tape and the live weight of each buffalo was estimated according to the simple fairly accurately method of (Carroll and Huntington, 1988; Hossain and Akhter, 1999; Milner and Hewitt, 1969) by using the following formula:

$$LWT \ (kg) = (L \times G^2) / (300 \times 2.25)$$

Fitting the linear regression
In the linear regression equation:

$$Y = a + bx$$

Where:
Y: value of the traits.
x: lactation number.
a and b: parameters that define the shape of the curve.

The different traits (daily milk yield, lactation production, lactation length, post partum heat period and gestation length and live weight) was set as dependent and time (lactation number) was set as independent variable. The model was analyzed by microsoft excel-2007 to obtain the model parameters (a and b). Along with the fit statistic co-efficient of determination (R^2) was also obtained. Estimated values of various traits were calculated according to Van Arendok (1985) for comparing the studied values by using age adjustment factor as lactation adjustment factor.

Profitability estimation
For estimating the profitability of buffaloes under two different production systems a deterministic linear program-

ming model, context using Microsoft Excel was used according to Khan *et al.* (2010) and Khan *et al.* (2014). The profitability analyses were done based on average values of marketable products (milk and meat) and the expenses incurred in buffalo production (feed costs, health, reproduction and fixed costs). The profit was derived as the difference between income (I) and costs (C). The input parameters of buffaloes are presented in Table 5. Total metabolizable energy (ME) requirement per buffalo cow per year was the sum of ME requirement for maintenance, growth, pregnancy and production and was calculated according to AFRC (1993). The milk production per lactation was considered as milk production per calving interval and the milk yield (kg/year). The dry matter (DM) requirements were calculated by the content of ME per kg DM. It was considered that buffalo cows were consuming roughage from grazing and paddy straw and 2 to 3 kg concentrate mix (brans, oil cakes and grains) per day per buffaloes under semi-intensive and 1 to 2 kg for buffaloes under extensive system, to fulfill their energy requirements. The feed cost (roughage and concentrate mix) for buffaloes under different production systems were calculated at 0.10-0.12 US$ per kg DM. The profit was derived from the differences of the sale of milk and beef, and the cost of feed and fixed costs (operational cost).

Statistical analysis

The collected data was corrected and analyzed by using the statistical package SAS (SAS, 2008) and the following statistical model was used to obtained the least square means for each parameters.

$$Y_{ijk} = \mu + F_i + B_j + e_{ijk}$$

Where:
Y_{ijk}: traits' value.
μ: overall mean.
F_i: effects of farms.
B_j: effect of breed.
e_{ij} is the residual effect, distributed as N (0, σ^2).

The mean differences were compared using least significant difference (LSD) (Steel *et al.* 1997) at 5% level of significance.

RESULTS AND DISCUSSION

Productive performance of buffalo cows

The means with standard error values of different productive traits are shown in Table 2. All the productive parameters: LW, DMY were found significantly higher in River type than Swamp type buffalo (Table 2), however, cross-

bred showed intermediate performance. In comparison to the production system, the LW (372.31±14.64 kg) and DMY (1.99±0.16 liter/day/cow) was found to be highest under semi-intensive bathan farming system than the extensive bathan farming system irrespective to the breed. On the other hand, the LL (275.25±2.857 days) and LP (628.80±34.49 liter) of different breeds were found higher under extensive bathan farming system than the intensive bathan farming system in coastal areas (Table 2).

In comparison to breed, the LW (390.54±14.06 kg), DMY (2.82±0.13 liter/day/cow), LL (284.96±3.31 days) and LP (899.75±52.83 liter) were greatest in the case of River type buffalo cows and lowest in Swamp type, while crossbreed type shown intermediate performance irrespective to farming system. The DMY was significantly different (P<0.05) among the season within the breed, within the breed between season, between breed within a season (Table 2).

Reproductive performance of buffalo cows

Reproductive performance found better in river types than any other types. The mean ± standard error values of different reproductive traits are shown in Table 3. All reproductive traits (GL, CI, PHP, AFC and SPC) were highest in Swamp type buffalo cows and lowest in River type buffalo cows and. In comparison to breed type, GL (301.74±0.63 days), PHP (123.21±7.50 days), AFC (47.00±1.35 months) and SPC (1.40±0.16) were found to be lowest in River type buffalo cows than other type buffalo cows, but surprisingly CI (660.31±43.82) was found lowest in crossbred type cow irrespective of farming system (Table 3).

Comparing farming system, GL (305.37±0.72 days), CI (640.34±51.31 days), AFC (54.72±1.57 months) and SPC (1.62±0.21) were found to be lowest under semi-intensive bathan farming system than the extensive bathan farming system, where the values were 306.44 ± 0.58 days, 696.95 ± 35.12 days, 56.02 ± 1.90 months and 1.73 ± 0.11, respectively, whereas PHP (134.04±5.30 days) was found lowest under the extensive bathan farming system than semi-intensive bathan farming system (142.54±7.28 days) irrespective of breed.

Live weight and birth weight of male and female buffaloes

The mean with standard error values of LW and birth weight are shown in Table 4. Average LW of a male buffalo (395.39±12.45 kg) was significantly higher than for a female buffalo irrespective of the breed type (Table 4). There was no significant dissimilarity found between LW of semi-intensive bathan farming system (360.36±11.48 kg) and extensive bathan farming system (365.34±13.33 kg) irrespective of sex.

Table 2 Mean ± standard error of various productive traits in buffalo cows

Types	Semi-intensive Bathan farming system				Extensive Bathan farming system			
	LWT (kg)	DMY (lit/day)	LL (days)	LP (liter)	LWT (kg)	DMY (liter/day)	LL (days)	LP (liter)
River type	427.06[by]	3.14[by]	283.34[b]	890.56[bx]	354.01[y]	2.50[bx]	286.56[b]	908.93[by]
	±15.49	±0.23	±4.26	±47.89	±12.62	±0.12	±2.35	±57.77
Crossbred type	343.95[a]	1.60[ab]	274.58[aby]	488.31[abx]	340.43[b]	1.85[ab]	263.93[abx]	636.68[aby]
	±13.16	±0.14	±2.19	±45.78	±12.14	±0.12	±4.75	±25.47
Swamp type	345.93[ay]	1.25[a]	247.54[a]	314.70[a]	305.50[ax]	1.15[a]	243.33[a]	340.80[a]
	±15.25	±0.11	±2.62	±22.86	±8.65	±0.09	±1.48	±20.23

LWT: live weight; DMY: daily milk yield; LL: lactation length and LP: lactation production.
[a, b]: the means within the same row (within season within breed) with different letter, are significantly different (P<0.05).
[x, y]: the means within the same column (between farming system) with different letter, are significantly different (P<0.05).

Table 3 Mean ± standard error of various reproductive traits in buffalo cows

Types	Semi-intensive farming system					Extensive farming system				
	GL (days)	CI (days)	PPH (days)	AFC (months)	SPC	GL (days)	CI (day)	PPH (day)	AFC (months)	SPC
River type	301.05[ax]	580.86[x]	129.75[a]	46.00[a]	1.43	302.42[ay]	762.50[y]	116.67	48.00	1.36
	±0.57	±37.02	±9.16	±1.05	±0.20	±0.68	±50.62	±5.84	±1.65	±0.12
Crossbred	305.33[bax]	673.75	145.25[a]	58.05[b]	1.67	307.67[by]	647.36	139.50	60.00	1.88
	±0.83	±57.24	±4.39	±1.35	±0.20	±0.45	±30.40	±5.83	±1.68	±0.13
Swamp type	309.54[b]	666.41	152.63[b]	60.12[b]	1.75	309.23[ba]	681.00	145.95	60.06	1.94
	±0.90	±59.68	±8.29	±2.32	±0.22	±0.60	±24.35	±4.22	±2.45	±0.09

GL: gestation length; CI: calving interval; PPH: postpartum heat; AFC: age at first calving and SPC: service per conception.
[a, b]: the means within the same row (among season within breed within farming system) with different letter, are significantly different (P<0.05).
[x, y]: the means within the same column (between system within season within breed) with different letter, are significantly different (P<0.05).

Table 4 Live weight and birth weight of adult male and female buffaloes

Types	Male buffalo LW (kg)			Female buffalo LW (kg)		
	Semi-intensive	Extensive	Birth W (kg)	Semi-intensive	Extensive	Birth W (kg)
River type	457.80[bd]	442.82[b]		427.06[bcy]	354.01[bx]	
	±10.84	±21.72		±15.49	±12.62	
Crossbred	362.50[ax]	396.68[abdy]	28.28[y]±0.48	343.95[a]	340.43[c]	26.82[x]±0.70
	±8.86	±11.37		±13.16	±12.14	
Swamp type	359.92[ad]	352.62[a]		325.72[ac]	305.50[a]	
	±8.64	±10.46		±11.95	±8.65	
Average	**393.41[d]**	**397.37[d]**		**327.31[c]**	**333.31[c]**	
	±9.45	**±15.52**		**±13.53**	**±11.14**	

LW: live weight and W: weight.
[a, b]: the means within the same row (between breed within system and within sex) with different letter, are significantly different (P<0.05).
[c, d]: the means within the same column (between sex within system) with different letter, are significantly different (P<0.05).
[x, y]: the means within the same column (between system within breed and within sex) with different letter, are significantly different (P<0.05).

Buffalo of extensive bathan farming system attained more LW than semi-intensive bathan farming. Among the breed types, average LW of River type male buffalo (457.80±10.84) considerably differed from crossbreds and Swamp types. Among the breeds, between farming system and between sex River type males attained a maximum weight in the studied area. In case of birth weight, a significant difference (P<0.05) was observed between male and female calves.

Performance of various traits after fitting regression equation

The values of the linear regression equation of DMY, LL, LP, GL, PPH and LW of buffalo cows are shown in Table 5.

The curve shape of DMY, LL, LP, GL, PPH and LW of buffalo cows after fitting linear regression have shown in Figure 1.

Profitability of buffaloes under different production systems

Table 6 shows the costs, revenues and profit of buffalo cows under two different production system on a per cow per year basis. Table 6 indicated that the average net income of buffalo cow under semi-intensive system (US$ 100) was higher than extensive production system (US$ 40).

Total revenue was dominated by the sale of milk and beef as buffaloes were reared in these areas for milk and beef.

Table 5 Estimated model parameters (a and b) and fit statistics (R^2) of different traits of buffalo cows irrespective to breeds

| Traits | a (intercept) | b (slope) | R^2 (coefficient of determination) | Original value | Estimated value in different lactation | | | | | | | |
					1	2	3	4	5	6	7	8
DMY	0.91	0.37	0.96	1.92	1.19	1.62	2.02	2.32	2.590	2.820	3.00	3.13
LL	241.80	6.79	0.97	266.55	248.59	255.38	262.17	268.96	275.75	282.54	289.33	296.12
LP	240.00	109.60	0.97	596.67	325.13	450.01	568.80	657.66	740.72	807.84	866.19	903.96
GL	310.00	-1.18	0.80	305.87	308.82	307.64	306.46	305.28	304.10	302.92	301.74	300.56
PPH	147.00	-4.07	0.53	138.29	142.93	138.86	134.79	130.72	126.65	122.58	118.51	114.44
LW	302.30	14.66	0.48	362.85	293.88	325.24	346.28	350.11	353.06	351.23	347.44	351.54

DMY: daily milk yield; LL: lactation length; LP: lactation production; GL: gestation length; PPH: postpartum heat and LW: live weight.

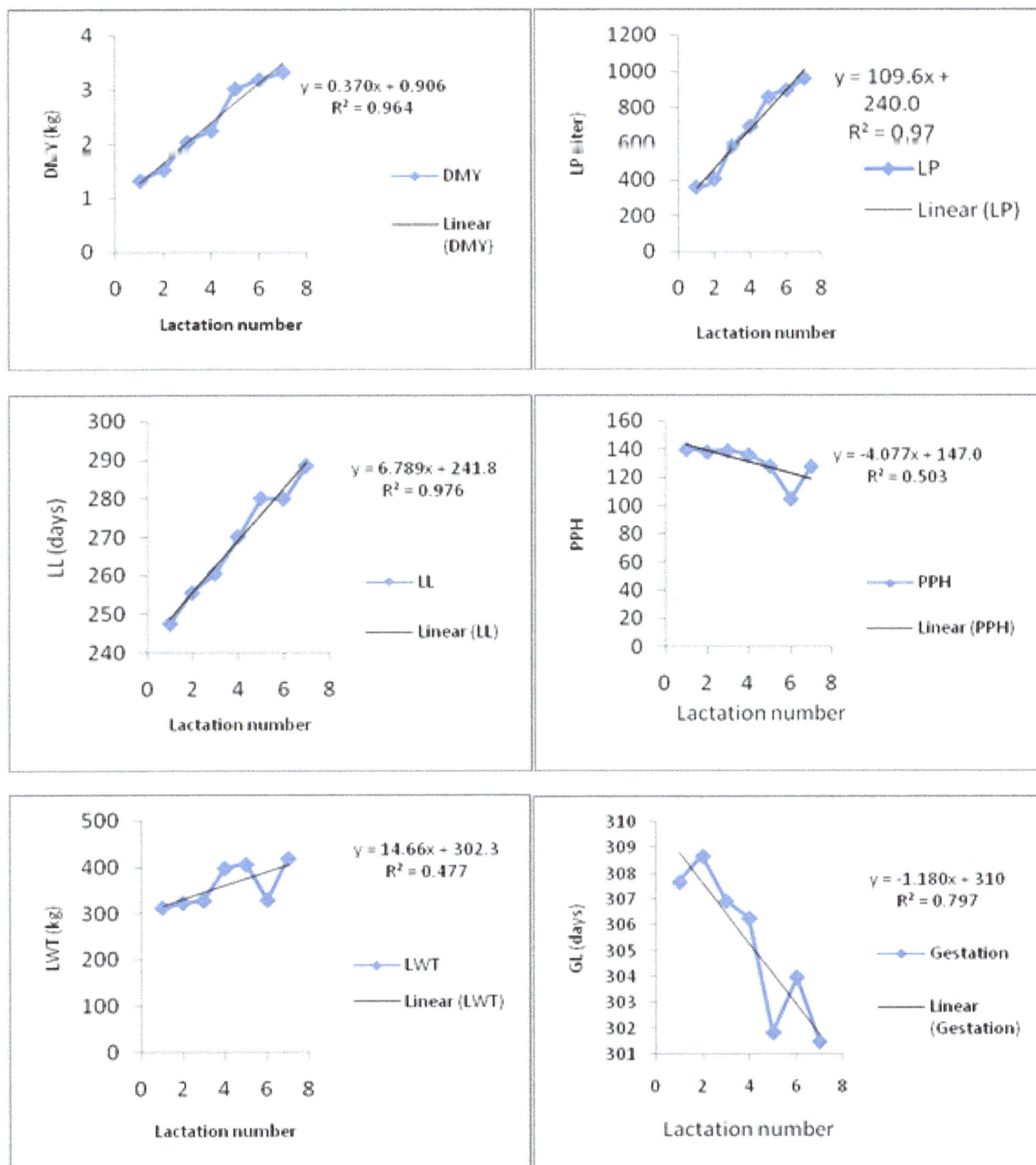

Figure 1 Curves of the different traits obtained from linear regression

Feed costs accounted about 85% of the total costs. Health costs, reproduction costs, labour costs, marketing costs and all other operational and management costs were assumed to be fixed costs. Buffaloes under both systems had similar DM requirements for maintenance, growth of replacements and lactation. However, the buffaloes had the highest milk and beef revenue and that generated highest profit in extensive system than other.

Productive performance of buffalo cows

Considering production system, semi-intensive farming system was superior farming system and river type found better performer buffalo than other type. The live weight of buffaloes irrespective of sex under semi-intensive farming systems were found to be significantly higher (P<0.05) than for extensive farming system. This higher value under semi-intensive bathan farming systems might be due to availability of heavier River type of buffalos and their crosses. In addition, this could happen due to good management factors of buffaloes. The farmer's farm buffaloes around their house and were offered the surrounding green grasses and kitchen wastes. In comparison of breed types, River type buffalo cows showed highest LW than other types. Significant variation of LW among breeds was observed due to genetics and environmental factors. Similar factors were found by another researcher (Shankar and Mandal, 2010). The findings of this current investigation were also supported by the findings of Tariq et al. (2013), who indicated average LW was 359 ± 160.9 kg in Pakistani buffalos and Ranawana (1989) reported that adult female buffaloes LW ranged from 250 to 350 kg, whereas the weight of Swamp buffalo cows ranged from 350 to 450 kg (Chantalakhana and Bunyavejchewin, 1989). The male buffaloes showed significantly more (P<0.05) LW than female buffaloes. Usually male buffaloes are heavier than females reported by Chantalakhana and Bunyavejchewin (1989). In most of the cases, LW of buffalo significantly varied between production systems, availability of green grasses and breeds. Variation of LW might be found due to genetic constituent of the buffaloes and other management factors (Shankar and Mandal, 2010).

The average DMY of buffalo cows under semi-intensive farming system found significantly higher (P<0.05) than extensive farming system. This higher average of DMY under semi-intensive bathan farming systems might be due to genetic factor (increased number of high yielding River type genotype and their crosses). A further possible cause might be good management factors (e.g. amount and quality of feed and the skill of farmer detect heat and illnesses) and factors which are beyond the farmer's control such as climatic factors: temperature and humidity, which influences milk production that leads lactation yield.

In comparison to breed types, River type of buffalo cows showed higher average DMY than other two types. Significant variation of average DMY among breeds was found as a result of genetics of cows and environmental factors. The findings of present study were not in accordance with the findings of Islam et al. (2004); Siddiquee et al. (2010); Karim et al. (2013). They reported higher value in their investigation than the present study. Lower milk yield could be attributable to improper and inadequate nutrients availability in the investigated area (Tiwari et al. 2007; Sarwar et al. 2009; Wynn et al. 2009; Pasha and Khan, 2010). Inadequate supply of quality fodder had been identified as one of the reasons for poor performances of buffalo (Sarwar et al. 2009) and the supplied fodder contains high fibrous materials resulting in poor growth, production and reproduction (Pasha, 2013).

The average LL of buffalo cows under extensive farming system was found higher than semi-intensive farming system. Comparing breed types, River types of buffalo cows showed highest average LL. Significant variation of average DMY among breeds found and this might be due to genetics and superior productive ability of River type buffalo. The findings of the present study were in line with the findings of Bingzhung et al. (2003); Khattab and Kawthar (2007); Siddiquee et al. (2010) and Karim et al. (2013), but lower than the value reported by Hussen (1990); Islam et al. (2004); Khan et al. (2007a) and Pasha and Hayat (2012). Lower LL might be due to late starting of milking after parturition and early drying off buffalo cows.

The average LP under extensive farming system was found higher than semi-intensive farming system. Considering breed types, River types of buffalo cows showed highest average LP. Significant variation of average DMY among breeds was found as a reason of genetic constituent of river type buffaloes in semi-intensive farming system. The findings of the present study were in agreement with the findings of Siddiquee et al. (2010); Karim et al. (2013) but lower than that of Hussen (1990); Islam et al. (2004); Khattab and Kawthar (2007). Lower average LP could be due to poor productive ability, shorter LL, poor nutritional management and lack of provision of housing facility.

The average birth weight differed significantly between farming system in this investigation and that might be due to availability of heavier River type of genotypes and their crosses under semi-intensive farming system. Usually, birth weight of River type buffalo found higher than those of Swamp types buffalo. The findings of this study were higher than the findings of Siddiquee et al. (2010); Karim et al. (2013) but lower than result of (Islam et al. 2004). This variation might be owing to genotypic difference of available buffalo's type and improper supply of nutrient rich feeds and fodders during pregnancy period.

Table 6 Profitability of buffaloes under two different production systems (per buffalo cow/year, US$ (1US$=BD Tk 78)

Variables	Semi-intensive Bathan farming system	Extensive Bathan farming system
Birth weight, kg	27.52	26.12
Mature live weight, kg	327.31	333.31
Gestation period, day	305	306
Lactation length (day)	269	264
Milk production/Cl (kg)	564.52	628.80
Calving interval (CI), day	640	697
Milk yield, kg/year	322	329.28
Calving rate, %	70	75
Survivability, %	80	82
Feed price per kg DM, US$	0.12	0.10
Beef price per kg live weight, US$	1.41	1.41
Price per kg milk, US$	0.58	0.58
DM requirement per cow per year, kg	1920	1920
Replacement heifer, kg	756	757
Total, kg	2676	2677
Price, US$	321.12	267.70
Non- feed costs per cow per year, milking cow, US$	36.92	36.92
Replacement heifer, US$	12.05	12.05
Total non-feed costs, US$	6.41	6.41
Total expenditure, US$	55.38	55.38
Revenue per cow per year, milk revenue, US$	195.70	200.15
Beef revenue, US$	208.38	210.74
Manure income, US$	12.82	12.82
Grand total, US$	416.90	423.71
Net income, US$	40.40	100.63

Improper and inadequate nutrients availability causes poor growth and poor reproductive performance in buffalo (Qureshi *et al.* 2002; Tiwari *et al.* 2007; Wynn *et al.* 2009; Pasha and Khan, 2010).

Reproductive performance of buffalo cows

Average GL was found to be lowest in River type buffalo cows than other types of buffalo cows. This might be due to having comparatively better reproductive efficiency of River type's buffalo. In association of farming system, GL under semi-intensive bathan farming system was lower than the extensive bathan farming system, that is due to impact of farming system and breed types of the farm. The results of this investigation were in sequence with the findings of Islam *et al.* (2004); Wangdi *et al.* (2014) but lower than the findings of Karim *et al.* (2013). The average CI was found lowest in crossbred type of buffalo cows irrespective of farming system. This might be due to benefit of vigor, production and reproductive efficiency. Comparing farming system, CI was lowest under semi-intensive bathan farming system than the extensive bathan farming system as a result of impact of farming system and types of buffalo in the farm. The average of CI in the present investigation showed higher value than the findings of Khan *et al.* (2007b); Pasha and Hayat (2012); Karim *et al.* (2013); Wangdi *et al.* (2014) in Bangladesh. This higher value might be due to poor management and poor reproductive performance (silent heat and seasonal breeder) of buffalo cows.

In comparison to type, the average postpartum heat period was lowest in River type buffalo cows compared with other type buffalo cows. Postpartum heat period was lower in the semi-intensive bathan farming system than the extensive bathan farming system irrespective to breed. The outcome of this study was similar to the outcome of Karim *et al.* (2013). Poor reproductive performance occurred due to poor nutrition quality of supplied feeds and fodders (Qureshi *et al.* 2002; Sahoo *et al.* 2004; Tiwari *et al.* 2007; Sarwar *et al.* 2009). Comparing type, the average AFC was lowest in River type buffalo cows than other types. These results might be due to genetics of the River type buffalo as they attained puberty about six months earlier than Swamp type. Between farming system, average age of first calving was found lowest under semi-intensive bathan farming system than the extensive bathan farming system, irrespective of breed type. This might be due to management factors and breed constituent of the farming system. The findings of this investigation was in line with the findings of Bhatti *et al.* (2007) and Karim *et al.* (2013). The average SPC was found lowest in River type buffalo cows than other type buffalo cows in comparison to type, whereas found lowest under semi-intensive bathan farming system than the extensive bathan farming system. The findings of the present investigation was lower than the findings of Wangdi *et al.* (2014) in Bhutan, but higher than the investigation of Islam *et al.* (2004), which is due to breeding methods. Natural breeding is the only method found in the studied area.

Performance of various traits after fitting regression equation

After fitting the regression equation with the various traits (DMY, LL, LP, GL and PPH) it was noticed that the R^2 values of DMY, LL and LP were higher than the GL, PPH and LWT. Highest values of R^2, (coefficient of determination) showed that linear regression of LP fitted more accurately than other traits. Positive values of slope (b) in case of DMY, LL, LP and LW indicates positive correlation with lactation number, where as GL and PPH shows negative values which indicate negative correlation. The higher R^2 values of the traits indicated they more closely fitted with the regression line for the increases of lactation numbers the values were also increased. If a model achieves R-squared above 90%, it indicates close agreement (Karmaker and Ray, 2011). In case of polynomial model Khan et al. (2014) considered R^2 values above 0.80 as superior.

Profitability of buffaloes under different production systems

In the study income was derived from the sale of milk, beef and manure and costs included only for feed and fixed costs. The milk payment for the farmers was based on milk volume only, which was used to calculate the profit. The net profit of buffalo cows was higher in semi-intensive production system than extensive system on per buffalo cow per year. The differences of profitability were attributed due to the differences of the prices of feed, milk, meat and the differences of breeds. Similar factors are responsible for the differences profitability was reported by Khan (2009).

The differences of feed DM consumed was found to be variable between types. The body weight of the buffalo cow is important as it affects the profitability and thereby affect on feed requirements for maintenance as well as the value of the carcass. Similar findings were observed by Lopez-Villalobos et al. (2000) and Khan (2009). In the present study, feed costs accounted for 85% of the total costs while the remaining percentage was accounted for other operational costs. Similar amount of feed costs out of the total costs for dairy farm operation were reported by Ozawa et al. (2005) in Japanese study. Under the rural condition of Bangladesh the farmers are mainly feeding their buffalo cow's straw and concentrate (brans and rice polish) and green grasses when available. The DM intakes and price per kg DM have also influenced the profitability which was also reported previously by Khan (2009) and Rahman et al. (2003).

CONCLUSION

The study revealed that most of the buffaloes under the investigated area were swamp types, irrespective to age, sex and farming system. The results showed that river type buffalo's performance and semi-intensive farming system was superior to other breeds and farming system. The live weight and daily milk yield of buffaloes was highest under semi-intensive bathan farming system and lactation length and lactation production found to be highest under extensive bathan farming system. Comparison to breeds, live weight, daily milk yield, lactation length and lactation production was found maximum for river type buffalo cows. The gestation length, calving interval, age of first calving and service per conception found lowest under semi-intensive bathan farming system except postpartum heat, which found lowest under the extensive bathan farming system. It can be concluded that this study indicates some important indication (e.g. best genotype and best farming system) which can be used by the farmers, researcher and policy makers for future improvement of buffalo in this particular area as well as in Bangladesh.

ACKNOWLEDGEMENT

I am immensely pleased to place on record my profound gratitude and heartfelt thanks to my research supervisor and teachers of the Department of Genetics and Animal Breeding, CVASU, for their encouragement and cooperation at every stage of this study from its inception to completion. I am grateful authority of CVASU for providing economical and technical support during research. Last but not the least, i would like to thank my family members and friends and farmers in the studied area for their active support.

REFERENCES

AFRC. (1993). Energy and Protein Requirements of Ruminants. CAB International, Wallingford, UK.

Bhatti S.A., Sarwar M., Khan M.S. and Hussain M.I. (2007). Reducing the age at first calving through nutritional manipulations in dairy buffaloes and cows: a review. Pakistan Vet. J. 27, 42-47.

Bingzhung Y., Zhongquan L., Xianwei L. and Caixia Z. (2003). The advance of genetic improvement and the development of the dairy industry in the Chineese water buffalo. Pp. 27-30 in Proc. 4th Asian Buffalo Cong. New Delhi, India.

Carroll C.L. and Huntington P.J. (1988). Body condition scoring and weight estimation of horses. Equine Vet. J. 20(1), 41-48.

Chantalakhana C. and Bunyavejchewin P. (1989). Buffalo production research and development in Thailand. Pp. 48-60 in Proc. Int. Symp. Buffalo Gen. Small Farms. Asia, Malaysia.

FAO. (2010). Food and Agriculture Organization of the United Nations the State of Food Insecurity in the World.

FAO. (2013). FAO Statistical Yearbook. World food and agriculture, Rome. Available at: http://www.fao.org/docrep/018/i3107e/i3107e00.htm.

Faruque M., Nomura K., Takahashi Y. and Amano T. (2004). Genetic diversity of Asian water buffalo. Pp. 61-62 in Proc. 10th NIAS Inet. Workshop. Genet. Res. Tsukuba, Japan.

Faruque M.O. (2000). Identification of the best genotype of buffalo for dairy purpose in Bangladesh and to improve their productivity. Bangladesh Agricultural Research Council. New Airport Road, Farmgate, Dhaka, Bangladesh.

Faruque M.O. and Hossain M.I. (2010). The effect of feed supplement on the yield and composition of buffalo milk. *Italian J. Anim. Sci.* **6,** 488-490.

Hossain N.M. and Akhter S. (1999). Practical Animal Science. Zaman Printers, Mymensingh, Bangladesh.

Hussen M.S. (1990). A study on the performance of indigenous buffaloes in Tangail district. MS Thesis. Bangladesh Agricultural Univ., Mymensingh, Bangladesh.

Islam M.A., Mazed M.A., Islam M.S. and Uddin M.K. (2004). Some productive performance of Nili-Ravi and crossbred (Nili-ravi×Local) buffaloes at government buffalo farm, Bagerhat, Bangladesh. *J. Anim. Vet Adv* **3,** 895-897.

Karim M.R., Hossain M.Z., Islam M.R., Parvin M.S. and Matin M.A. (2013). Reproductivity, productivity and management system of indigenous buffalo (*Bubalus bubalis*) cows in coastal areas of pirojpur and borguna district of Bangladesh. *Prog. Agric.* **24,** 117-122.

Karmaker M. and Ray R.R. (2011). Optimization of production conditions of extra cellular â-glucosidase in submerged fermentation of waterhyacinth using response surface methodology by Rhizopus oryzae MTCC 9642. *Res. J. Pharm. Biol. Chem. Sci.* **2,** 299-308.

Khan M., Ahmad N. and Khan M.A. (2007a). Genetic resources and diversity in dairy buffaloes of Pakistan. *Pakistan Vet. J.* **27,** 201-207.

Khan M.A., Lee H.J., Lee W.S., Kim H.S., Kim S.B., Ki K.S., Park S.J., Ha J.K. and Choi Y.J. (2007b). Starch source evaluation in calf starter: I. Feed consumption, body weight gain, structural growth, and blood metabolites in Holstein calves. *J. Dairy Sci.* **90,** 5259-5268.

Khan M.K.I., Miah G., Khatun M.J. and Das A. (2010). Economic values for different economic traits of Red Chittagong cow's. *Indian J. Anim. Sci.* **80,** 1138-1140.

Khan M.K.I., Blair H.T. and Lopez-Villalobs N. (2014). Economic values for traits in a breeding objective for dairy cattle in Bangladesh. *Indian J. Anim. Sci.* **84,** 682-686.

Khan M.K.I. (2009). Developments of model for the genetic improvement of dairy cattle under cooperative dairying conditions in Bangladesh. Ph D Thesis. Massey Univ., New Zealand.

Khattab A.S. and Kawthar A.M. (2007). Inbreeding and it is effects on some productive and reproductive traits in a herd of Egyptian buffaloes. *Italian J. Anim. Sci.* **6,** 275-278.

Kirk G.J., Olney G.R., Falconer D.A. and Standing W.R. (1988). The western Australian dairy farm model. *Proc. Australian Sci. Anim. Prod.* **17,** 371.

Lopez-Villalobos N., Garrick D.J., Holmes C.W., Blair H.T. and Spelman R.J. (2000). Profitability of some mating systems for dairy herds in New Zealand. *J. Dairy Sci.* **83,** 144-153.

Milner J. and Hewitt D. (1969). Weights of horses: improved estimates based on girth and length. *Canadian Vet. J.* **10(12),** 314-324.

Ozawa T., Lopez-Villalobos N. and Blair H.T. (2005). Dairy farming financial structures in Hokkaido, Japan and New Zealand. *Anim. Sci. J.* **76,** 391-400.

Pasha T.N. (2013). Prospect of nutrition and feeding for sustainable buffalo production. *Buffalo Bull.* **32,** 91-110.

Pasha T.N. and Hayat Z. (2012). Present situation and future perspective of buffalo production in Asia. *J. Anim. Plant. Sci.* **22,** 250-256.

Pasha T.N. and Khan E.U. (2010). Buffalo milk production in Pakistan. Pp. 21-25 in Proc. 9th World Buffalo Cong. Rev. Vert. Buenos Aries, Argentina.

Qureshi M.S., Habib G., Hamid H.A., Lodhu R.H. and Sayed M. (2002). Reproduction-nutrition relationship in dairy buffaloes. Effect of intake of protein, energy and blood metabolites levels. *Asian Australas J. Anim. Sci.* **15,** 330-339.

Rahman S.M.A., Alam J. and Rahman M. (2003). Economics in dairy farming in Bangladesh. *Indian J. Dairy Sci.* **56,** 245-249.

Ranawana S.S.E. (1989). Water buffalo in Sri-Lanka. Pp. 31-38 in Proc. Symp. Buffalo Gen. Small Farms. Asia, Malaysia.

Sahoo A., Elangovan A.V., Mehra U.R. and Singh UB. (2004). Catalytic supplementation of urea-molasses on nutritional performance of male buffalo (*Bubalus bubalis*) calves. *Asian-Australas. J. Anim. Sci.* **17,** 621-628.

Sarwar M., Khan M., Nisa M., Bhatti S. and Shahzad M. (2009). Nutritional management for buffalo production. *Asian-Australas J. Anim. Sci.* **22,** 1060-1068.

SAS Institute. (1996). SAS®/STAT Software, Release 6.11. SAS Institute, Inc., Cary, NC. USA.

Shankar S. and Mandal K.G. (2010). Genetic and non-genetic factors affecting body weight of buffaloes. *Vet. World.* **3,** 227-229.

Siddiquee N., Faruque M.O., Islam F., Mijan M.A. and Habib M.A. (2010). Morphometric measurements, productive and reproductive performance of buffalo in Trishal and Companiganj sub-districts of Bangladesh. *Inter. J. Biol. Res.* **1,** 15-21.

Steel R.G.D., Torrie J.H. and Dickey D.A. (1997). Principles of Statistics-a Biometrical Approach. Mac Graw-Hill Corporation. New York.

Tariq M., Younas M., Khan A.B. and Schlecht E. (2013). Body measurements and body condition scoring as basis for estimation of live weight in Nili-Ravi buffaloes. *Pakistan Vet. J.* **33,** 325-329.

Tiwari R., Sharma M.C. and Singh B.P. (2007). Buffalo calf health care in commercial dairy farms: a field study in Uttar Pradesh (India). *Livest. Res. Rural Dev.* Available at: http://www.lrrd.org.

Van Arendok J.A.M. (1985). A model to estimate the performance, revenues and costs of dairy cows under different production and price situation. *Agric. Syst.* **16,** 157-189.

Wangdi J., Bhujel P., Timsina M.P. and Sonam W. (2014). Performance of buffalo (*Bubalus bubalis*) under Bhutanese conditions. *Glob. J. Dairy Farm. Milk Prod.* **2,** 67-73.

Wynn P.C., Warriach H.M., Morgan A., McGill D.M., Hanif S., Sarwar M., Iqbal A., Sheehy P.A. and Bush R.D. (2009). Perinatal nutrition of the calf and its consequences for lifelong productivity. *Asian-Australas J. Anim. Sci.* **22,** 756-764.

6

Comparison of Different Methods of Oestrus Synchronization on Reproductive Performance of Farahani Sheep in Iran

A. Mirshamsollahi[1*]

[1] Department of Animal and Science, Markazi Agricultural and Natural Resources Research and Education Center, Arak, Iran

*Correspondence E-mail: a.mirshamsolahi@areo.ir

ABSTRACT

This experiment was carried out on 123 Farahani ewes from a herd in Delijan city in Markazi province of Iran to determine the best short-term method for oestrus synchronization. Ewes were divided into five experimental groups randomly: group 1) use of controlled intervaginal drug release devices (CIDR) for 7 days with intramuscular (IM) injection of PGF2α on zero day and IM injection of 500 IU equine chorionic gonadotropin (eCG) at the time of CIDR removal; group 2) use of CIDR for 7 days and IM injection of PGF2α on the 6th day and IM injection of 500 IU eCG at the time of CIDR removal; group 3) use of intravaginal sponge for 7 days with IM injection of PGF2α on zero day and IM injection of 500 IU eCG at the time of sponge removal; group 4) use of intravaginal sponge for 7 days with IM injection of PGF2α on the 6th day and IM injection of 500 IU eCG at the time of sponge removal and group 5) control, without any treatment. Results showed that parturition percentage on expected date was almost doubled in all treatment groups when compared to control group. Each treatment used for oestrus synchronization increased the percentage of twinning on expected date in comparison with control group. Obtained results indicated that the most ewes lambed on expected date after 7 day treatment with both the intra-vaginal sponge and CIDR, and had a favorable impact on fertility of ewes.

KEY WORDS CIDR, Farahani sheep, oestrus synchronization, PGF2α, short-term method.

INTRODUCTION

Oestrus synchronization is one of the ways for improvement of sheep reproduction management. In fact, use of oestrus synchronization resulted in lower disturbance of pastures by sheep, better planning of controlled mating and improving of breeding approaches, production of lambs of the same age and finally accessing the sheep meat in the months when its production faced to limitation (Godfrey *et al.* 1997; Niasari-Naslaji and Soukhtezari, 2005). There are two basic methods which could be applied in oestrus synchronization. The first one is based on the application of synthetic or natural progesterone to support natural corpus luteum (CL) or imitate the luteal phase of the oestrous cycle. The second method is based on the application of synthetic prostaglandin (F2α) to remove corpus luteum. Hence, the second method is depending on the presence of CL, it is used just in breeding season, whereas the first one can be used in all over the year (Wildeus, 1999; Safdarian, 2005). Usually, gonadotropins are used mostly at the intra-vaginal device withdrawal in the process of oestrus synchronization in ewes and goats. One of the most common gonadotropin is equine chorionic gonadotropin, pregnant mare serum gonadotropin (eCG, PMSG) (Barrett *et al.* 2001). It can be seen that in some breeds of sheep, application of eCG led to light induction of super-ovulation in ewes (when used in high doses of 1000 IU; Maraček *et al.* 2009) and could increase the rate of prolificacy in ewes with low prolificacy, desirably (Gordon, 1975). Regarding the decrease of fertility in ewes treated with progesterone for a long time (>12 days) due to vaginal infections, shortening this period (5-7 days) could decrease infection and vaginal disorders as well as facilitate management (Fonseca *et al.* 2005). Also, some authors indicated that short time application of progesterone in goats and ewes, out of breeding season, improved gestation rate rather than the long time application (Sadeghipanah, 2005; Fonseca *et al.* 2005).

It seems that depleting hormones of corpus luteum are essential for prediction of acceptable oestrus in short-time treatment in breeding season (Godfrey *et al.* 1997). Thus, PGF2α injection is applied at the end of short-time (5-7 days) progestin treatment. It is indicated that PGF2α injection at the end of short time treatment led to diversity in oestrus cycle process due to differences in the presence or absence of CL on the ovaries of sheep and required time for increasing the progesterone to optimum level in the blood which finally affected the fertility rate after artificial insemination. In case when depletion of CL occurred at the beginning of treatment, the level of progesterone would be similar and appropriate in all ewes. Injection of PGF2α in sponging time in modified time was not applied in artificial insemination in ewes, whereas it was successful in goats (Lida *et al.* 2004; Menchaca and Rubianes, 2004).

Sadeghipanah *et al.* (2005) studied the effect of progesterone treatment duration and levels of eCG on fertility of Mehraban ewes in breeding season. Ewes treated with progesterone for 7 days and 600 IU of eCG increased fertility rate significantly rather than 12 day treatment.

Effect of CIDR and intravaginal sponge on oestrus synchronization during the long time period (12-14 days) in ewes is well documented, but use of short times requires more studies. Regarding that progesterone treatment for the long time (>12 days) may lead to decreased fertility, shortening this time to 5-7 days reduces vaginal health risk and makes the reproduction management easier. This experiment was performed to study the effect of short time method of oestrus synchronization using CIDR and / or intravaginal sponge treatment with the injection of eCG at the time of their removal and different times of PGF2α injection on fertility, fecundity and prolificacy of Farahani ewes.

MATERIALS AND METHODS

The experiment was carried out on 123 Farahani ewes in Robat Tork village, Delijan City, Iran. Ewes were fed on natural pastures. The experiment was performed from September to October, regarding the climate changes of experimental site. Ewes were divided into 5 groups based on their age and body condition scores (BCS), randomly with

mean weight of 40.96 ± 0.29 kg. The treatments for experimental groups were as follows:

Group 1: use of CIDR (containing 40 mg of flugestone acetate) for 7 days with IM injection of PGF2α (Enzaprost, Ceva Santeanimale, England) on zero day and IM injection of 500 IU eCG (Pregnecol, Bioniche, Australia) at the time of CIDR removal; group 2: use of CIDR for 7 days and IM injection of PGF2α on the 6th day and IM injection of 500 IU eCG at the time of CIDR removal; group 3: use of intravaginal sponge for 7 days with I.M. injection of PGF2α on zero day and IM injection of 500 IU eCG at the time of sponge removal; Group 4: Use of intravaginal sponge for 7 days with IM injection of PGF2α on the 6th day and IM injection of 500 IU eCG at the time of sponge removal and group 5: control, without any treatment. In each synchronized group, dose of PGF2α was 15 mg and eCG 500 IU. The time of PGF2α injection, CIDR and sponge application were similar for all groups. After CIDR and sponge withdrawal, the rams were released in the ewe flock. After performing experimental treatment, the ewes went back to their flock and fed naturally on pastures and farm residues. After 5 months of gestation percentage of fertility, fecundity, twinning on expected time, and weight of lambs a birth time were recorded using following equations (1, 2 and 3).

The collected data were analyzed using SAS (2009). A randomized complete design was used for statistical analysis of Lamb birth weight and chi-square test f was used or the other traits.

1- % prolificacy on expected date= (number of lambs born/number of labored ewes) × 100

2- % fertility on expected date= (number of labored ewes/number of mated ewes) × 100

3- % fecundity on expected date= (number of lambs born/number of mated ewes) × 100

MATERIALS AND METHODS

The highest percentage of fertility was seen in the group 4 (intravaginal sponge for 7 days with IM injection of PGF2α on the 6th day and IM injection of 500 IU eCG at the time of sponge removal) and the lowest in the control group (Table 1). In all treated groups (1-4), the percentage of twinning enhanced by 8-12% compared to the control (P<0.05), but there was no significant difference among hormonally treated groups (Table 2).

The results indicated that the percentage of lambing (fecundity) increased in groups 1-4 in comparison with the control group on expected time (P<0.01) but there was no significant difference among hormone treatments. Application of hormones had no significant effect on the weight of lambs on birthday (Table 2).

Table 1 Percentage of parturition of Farahani ewes (fertility) treated for oestrus synchronization applying various methods and not treated at all (control)

Group	n	Parturition on expected date (%)	Parturition in 15-20 days from expected date (%)
1	23	86.95[a]	13.05[b]
2	23	86.95[a]	13.05[b]
3	27	85.18[a]	14.82[b]
4	25	92.0[a]	8.0[b]
5	25	48.0[b]	44.0[a]

n: number of ewes.
The means within the same column with at least one common letter, do not have significant difference ($P>0.01$).
Group 1: use of CIDR for 7 days with IM injection of PGF2α on zero day and IM injection of eCG on the 7th day; Group 2: use of CIDR for 7 days with IM injection of PGF2α on the 6th day and IM injection of eCG on the 7th day; Group3: use of intravaginal sponge for 7 days with IM injection of PGF2α on zero day and IM injection of eCG on the 7th day; Group 4: use of intravaginal sponge for 7 days with IM injection of PGF2α on the 6th day and IM injection of eCG on the 7th day and Group 5: without any treatment.

Table 2 Reproductive parameters of Farahani ewes treated for oestrus synchronization applying various methods and not treated at all (control)

Group	n	Twinning on the expected date (%)	Twinning in the whole group (%)	Lambing in the expected date (%)	Number of born lambs per parturition	Lamb birth weight ± SE
1	23	10[a]	8.69[a]	95.65[a]	1.1[a]	3.84±0.56[a]
2	23	10[a]	8.69[a]	95.65[a]	1.1[a]	3.66±0.36[a]
3	27	13[a]	11[a]	96.29[a]	1.13[a]	3.65±0.51[a]
4	25	13[a]	12[a]	104[a]	1.13[a]	3.78±0.58[a]
5	25	0[b]	0[b]	48[b]	1[a]	3.93±0.23[a]

n: number of ewes.
SE: standard error.
The means within the same column with at least one common letter, do not have significant difference ($P>0.01$).
Group 1: use of CIDR for 7 days with IM injection of PGF2α on zero day and IM injection of eCG on the 7th day; Group 2: use of CIDR for 7 days with IM injection of PGF2α on the 6th day and IM injection of eCG on the 7th day; Group3: use of intravaginal sponge for 7 days with IM injection of PGF2α on zero day and IM injection of eCG on the 7th day; Group 4: use of intravaginal sponge for 7 days with IM injection of PGF2α on the 6th day and IM injection of eCG on the 7th day and Group 5: without any treatment.

As shown in Table 3, application of treatments in groups 1 to 4 increased the number of lambing in trial ewes on expected date ($P<0.01$), however there was no difference among hormone treatments. Treatment for oestrus synchronization in groups 1-4 positively affected the number of twin parturition ($P<0.05$).

Table 4 shows the comparison of each experimental group with the control one. Number of parturition (fertility; $P<0.05$) and born lambs (fecundity; $P<0.01$) on expected date were higher for each of experimental groups than the control.

As shown in Table 1, application of CIDR and sponge doubled the number of parturition in a range of one week on expected times. These results indicated that the time of parturition was insignificantly different among time treatments; however it was higher in ewes treated with sponge for 7 days with the injection of PGF2α on the 6th day and eCG at the sponge removal compared with the control. All treatments increased fertility of ewes. Considering some methods of estrous synchronization in Kaboodeh ewes, Safdarian (2005) reported that percentage of parturition after intra-vaginal-sponge application was higher than application of CIDR. Also, number of parturition on expected time was recorded during 15 days and the aim of synchronization was estimated. Application of CIDR for oestrus synchronization caused that 74% of parturitions occurred within 6 days and 20%, 16 days after first parturition (Waldrona et al. 1999).

Results of Nuti et al. (1992) showed no differences in response and duration of estrous in dairy goats received cloprostenol (an analogue of PGF2α) on days 6 and 12 that is in agreement with our study.

There was no difference among lamb birth weights of different groups. It seemed that application of methods used in this study for oestrus and parturition synchronization had no effect on weight of lambs at the date of birth. These results were similar to Khaldari et al. (2005) who investigated CIDR for oestrus synchronization in Zandi ewes for 13 days followed by the application of 400 IU eCG injected to free grazing ewes. The use of above methods with application of eCG for oestrus synchronization enhanced twinning in all treated groups compared with control. Twinning in Farahani ewes was similar to the study of Niasari and Sokhteh Zari (2005).

In the present study application of CIDR or intravaginal sponge led to lambing in the range of 95-104% during one week. These results were very similar to the observations of Greyling et al. (1997) and Niasari and Sookhtehzari (2005) after using the same intravaginal devices for oestrus synchronization in ewes in the breeding season, but applied for 12 to 14 days. Fonseca et al. (2005) studied effect of duration of treatment using intravaginal sponge containing medroxyprogesterone acetate for 6 and 9 days for oestrus synchronization in non-lactating Tagn Borg goats. Their results showed that both of the 6 and 9 day treatments were the same in oestrus synchronization process (84 and 89%).

Table 3 Comparison of some reproductive traits of synchronized ewes (n=86) and control group (n=25) using χ^2 test

Item	Control group	Experimental groups
Number of parturitions on expected date	12	86
Number of parturitions after expected date	13	12
The value of χ^2 (df=1)		19.44**
Number of single parturitions	25	88
Number of twin parturitions	0	10
The value of χ^2 (df=1)		2.77*
Number of born lambs on expected date	12	95
Number of born lambs after expected date	13	13
The value of χ^2 (df=1)		20.61*

* (P<0.05) and ** (P<0.01).
The means within the same column with at least one common letter, do not have significant difference (P>0.01).
Group 1: use of CIDR for 7 days with IM injection of PGF2α on zero day and IM injection of eCG on the 7th day; Group 2: use of CIDR for 7 days with IM injection of PGF2α on the 6th day and IM injection of eCG on the 7th day; Group3: use of intravaginal sponge for 7 days with IM injection of PGF2α on zero day and IM injection of eCG on the 7th day; Group 4: use of intravaginal sponge for 7 days with IM injection of PGF2α on the 6th day and IM injection of eCG on the 7th day and Group 5: without any treatment.

Table 4 Comparison of some reproductive traits of each of different synchronization method group of ewes and control group using χ^2 test

Item	Control group	Group 1	Control group	Group 2	Control group	Group 3	Control group	Group 4
Number of parturitions on expected date	12	20	12	20	12	23	12	23
Number of parturitions after expected date	13	3	13	3	13	4	13	2
Number of ewes	25	23	25	23	25	27	25	25
The value of χ^2 (df=1)		8.181**		8.181**		8.157**		11.524**
Number of single parturitions	25	21	25	21	25	24	25	22
Number of twin parturitions	0	2	0	2	0	3	0	3
Number of ewes	25	23	25	23	25	27	25	25
The value of χ^2 (df=1)		2.268*		2.268*		2.948*		3.191*
Number of born lambs on expected date	12	21	12	22	12	26	12	26
Number of born lambs after expected date	13	4	13	3	13	4	13	2
Number of ewes	25	23	25	23	25	27	25	25
The value of χ^2 (df=1)		7.219**		9.191**		9.547**		13.097**

* (P<0.05) and ** (P<0.01).
The means within the same column with at least one common letter, do not have significant difference (P>0.01).
Group 1: use of CIDR for 7 days with IM injection of PGF2α on zero day and IM injection of eCG on the 7th day; Group 2: use of CIDR for 7 days with IM injection of PGF2α on the 6th day and IM injection of eCG on the 7th day; Group3: use of intravaginal sponge for 7 days with IM injection of PGF2α on zero day and IM injection of eCG on the 7th day; Group 4: use of intravaginal sponge for 7 days with IM injection of PGF2α on the 6th day and IM injection of eCG on the 7th day and Group 5: without any treatment.

It seems that as long as progesterone treatment used in lower duration, the fertility increased in goats. Safdarian (2005) reported the percent of lambing in expected time 144 and 141% after treatment with SIDR and Sponge for 12 days and injection of 500 UI PMSG in oestrus season, however there was no significant difference between them. Bitaraf *et al.* (2007) reported that application of CIDR or, flugestone acetate sponges with cloprostenol for oestrus synchronization in Nadooshani goats in the breeding season made no differences in the rate of gestation, fertility or fecundity. In the present study, the number of born lambs per parturition on expected date (prolificacy rate) was similar to the control and lower than it was reported by Safdarian (2005).

CONCLUSION

The overall results of this study indicated that application of hormones for oestrus synchronization in breeding season of Farahani ewes led to parturition in shorter time. The time of injection of PGF2α had no significant effect on reproductive traits of ewes. Also, application of eCG at sponge or CIDR removal increased the twinning rate and fecundity of ewes, so that most of parturitions occurred within one week. These results emphasized that the intra-vaginal sponge and CIDR in the short-term treatment had a favorable impact on both the fertility and fecundity rates of Farahani ewes.

ACKNOWLEDGEMENT

The author is thankful to "Shabestar Branch, Islamic Azad University" for their financial support of present project.

REFERENCES

Barrett D.M.W., Bartlewski P.M. and Rawlings N.C. (2001). Ultrasound and endocrine evaluation of the ovarian response to a 12-day medoxyprogesterone sponge and single injection of pregnant mare's serum gonadotropin in ewes in seasonal anestrous. *Biol. Reprod.* **64(1)**, 1-10.

Bitaraf A., Zamiri M.J., kafi M. and Izadifard J. (2007). Efficacy of CIDR, fluogestone acetate sponges and cloprostenol for estrous synchronization of Nadooshani goats during the breeding season. *Iranian J. Vet. Res.* **8(3)**, 218-224.

Fonseca J.F., Bruschi J.H., Santos I.C.C., Viana J.H.M. and Magalhaes A.C.M. (2005). Induction of estrus in non-lactating dairy goats with different estrous synchrony protocols. *Anim. Reprod. Sci.* **85**, 117-124.

Godfrey R.W., Gray M.L. and Collins J.R. (1997). A comparison of two methods of oestrus synchronization in hair sheep in the tropics. *Anim. Reprod. Sci.* **47**, 99-106.

Gordon I. (1975). Hormonal Control of Reproduction in sheep. *Proc. Soc. Anim. Prod.* **4**, 79-93.

Greyling J.P.C., Erasmus J.A. and Vander Merwe S. (1997). Synchronization of estrus in sheep using progestagen and inseminating chilled semen during the breeding season. *Small Rumin. Res.* **26**, 137-143.

Khaldari M., Tagic P., Afzalzadeh A. and Farzin N. (2005). Efficiency of of CIDR and PMSG on oestrus synchronization and twining in Zandi ewes during the breeding season. *J. Vet. Res.* **59**, 141-145.

Lida K., Kobayashi N. and Fukui Y. (2004). A comparative study of induction of estrus and ovulation by three different intrav-aginal devices in ewes during the non-breeding season. *J. Reprod. Dev.* **50(1)**, 63-69.

Maraček I., Vlčková R., Kaľatová J., Sopková D., Klapáčová K., Valocký I. and Pošivák J. (2009). Effect of assisted oestrus on the ovulation rate and reproductive performance of Tsigai sheep. *Slovak J. Anim. Sci.* **42(1)**, 51-55.

Menchaca A. and Rubianes E. (2004). New treatments associated with timed artificial insemination in small ruminants. *Reprod. Fertil. Dev.* **16(4)**, 403-413.

Niasari-Naslaji A. and Soukhtezari A. (2005). Comparison between three estrus synchronization programs using progestagens during the breeding season in the ewe. *Pajouhesh. Sazandegi.* **65**, 86-91.

Nuti L.C., Bretzlaff K.N., Elmore R.G., Meyers S.A., Rugila J.N., Brinsko S.P., Blanchard T.L. and Weston P.G. (1992). Synchronization of estrus in dairy goats treated with prostaglandin F at various stages of the estrous cycle. *Am. J. Vet. Res.* **53**, 935-937.

Sadeghipanah H., Zare-Shahneh A. and Saki A. (2005). The effect of progesterone days (SIDR) and PMSG dosage on the reproductive performance of Mehraban ewes out of breeding season. Pp. 886-889 in 1st Congr. Anim. Sci. Aquatic. Karaj, Iran.

Safdarian M. (2005). Determine of the Best Method of Estrus Synchronization. Animal Scince Research Instiute, Karaj, Iran.

SAS Institute. (2002). SAS®/STAT Software, Release 9.1. SAS Institute, Inc., Cary, NC. USA.

Waldrona D.F., Willingham T.D., Thompson P.V. and Bretzlaff K.N. (1999). Effect of concomitant injection of prostaglandin and (PMSG) on pregnancy rate and prolificacy of artificially inseminated Spanish goats synchronized with controlled internal drug release devices. *Small Rumin. Res.* **31**, 177-179.

Wildeus S. (1999). Current concepts in synchronization of estrus: Sheep and goats. *Proc. Am. Soc. Anim. Sci.* **38**, 1-14.

7

Effect of Different Progesterone Protocol and Low Doses of Equine Chorionic Gonadotropin (eCG) on Oestrus Synchronization in Arabian Ewes

F. Khalilavi[1*], M. Mamouei[1], S. Tabatabaei[1] and M. Chaji[1]

[1] Department of Animal Science, Faculty of Animal and Food Science, Ramin Agriculture and Natural Resources University Mollasani, Ahvaz, Iran

*Correspondence E-mail: fatemehkhalilavi@yahoo.com

ABSTRACT

This experiment was conducted to determine the reproductive performance of Arabian ewes treated with short and long-term progesterone devices in addition to low doses of equine chorionic gonadotropin (eCG) during the anoestrus season. A total of 36 ewes were divided into three groups: in group I vaginal sponges (60 mg medroxy progesterone acetate; (MAP)) were applied and removed after 6 days; in group II, vaginal MAP sponges were removed 12 days following insertion, while group III served as control group. The first two groups were intramuscularly injected with of 300 IU eCG, following sponge removal. Parameters such as oestrus response rate, time to onset of oestrus, duration of oestrus, pregnancy, lambing and fecundity rates were evaluated. Blood samples were collected one day before sponge insertion and two days after sponge insertion and on day of oestrus. There were significant differences between the group I and II with control group regarding the plasma estradiol and progesterone levels. There were no significant differences in oestrus response, time to onset of oestrus, pregnancy, lambing and fecundity rates between groups I and II (P>0.05). However, differences were significant when these two treatment groups were compared with the control group. In group I, duration of oestrus was significantly higher than group II (P<0.05). In addition, other factors in group I was numerically greater than group II (P>0.05). It was concluded that short-term sponge treatment (6 days) had better performance when compared with the long-term sponge treatment (12 days) in Arabian ewes.

KEY WORDS anoestrus season, Arabic ewes, MAP sponge, reproductive performance.

INTRODUCTION

Iranian Arabi sheep have a poor reproductive performance. Therefore, increasing Arabi sheep productivity by increasing lambing frequency and fecundity is considered important in the development of Arabi sheep production in Iran. On the other hand, the induction of oestrus outside the breeding season is important in supplying the market for poulterer. Oestrus synchronization is a valuable management tool that has been successfully employed to enhance reproductive efficiency (Hashemi *et al.* 2006). The ability to synchronize the time of breeding and lambing and the achievement of high fertility at first service is highly beneficial for poulterer. Several methods have been used for oestrus synchronization that including natural progesterone, synthetic progesterones, melatonin, prostaglandin F2α and isolated ram introduction (Bastan, 1995; Godfrey *et al.* 1999; Wildeus, 1999; Iida *et al.* 2004). Intravaginal devices impregnated with progesterone have been used to induce oestrus and ovulation in ewes (Godfrey *et al.* 1999; Ungerfeld and Rubianes, 2002) during breeding and non-breeding seasons. These devices are usually inserted over a period of 9-14 days that used in conjunction with eCG, particularly for the non-breeding season in ewes, injected at time of sponge removal or 48 h prior to sponge removal (Wildeus, 1999; Ungerfeld and Rubianes, 2002). However, recent studies showed that progesterone priming as short as 5-6 days is as effective as the long-term priming to induce oestrus with acceptable pregnancy rates during the non breeding season in sheep (Knights *et al.* 2001). The aim of this study was to compare effectiveness of long-term and short-term progesterone treatments combined with low dose of eCG application for oestrus induction out of the breeding season.

MATERIALS AND METHODS

Experimental design
This study was conducted during the non-breeding season (April-August) at the research farm of Ramin Agriculture and Natural Resources University, Ahwaz, Southern Iran. A total of 36 multiparous, 2-5 years old Arabic ewes were used in the present study. The ewes were randomly assigned to 3 groups; (I) intravaginal sponge containing 60 mg medroxy progesterone acetate (MAP) for 6 days (n=12), (II) intravaginal sponge containing 60 mg medroxy progesterone acetate (MAP) for 12 days (n=12) and (III) control group (without hormonal treatment) (n=12). Ewes were placed in a single open front barn and fitted with medroxy progesterone acetate (MAP, 60 mg, ESPON-JAVET®, Hipra, Spain) for 6 and 12 days and subsequently devices removed. Immediately after sponge removal, 300 IU of eCG (GONASER®, Hipra, Spain) was intramuscularly injected to the treated ewes. Oestrus was detected with six fertile Arabian rams (one ram:six ewes). The ewes were checked for signs of oestrus after sponge removal. The reproductive variables measured in experimental groups were: oestrus response= number of ewes showing oestrus / total ewes treated in each group × 100 (Akoz *et al.* 2006), time to onset of oestrus (h) (when ewes allowed a ram to mount and this was registered as the onset time of oestrus), duration of oestrus, pregnancy rate= number of pregnant ewes / number of ewes showing oestrus and mated in each group × 100 (Zeleke *et al.* 2005), lambing rate= number of ewes lambing / number of pregnant ewes in each group × 100 (Bacha *et al.* 2014) and fecundity rate= number of lambs born / number of mated ewes in each group (Bacha *et al.* 2014).

Blood sampling and hormone assay
Blood samples were collected on the one day before sponge insertion, two days after sponge insertion and day of oestrus. All blood samples were collected via jugular venipuncture into heparinized tubes and centrifuged at 3000 rpm for 15 min, thereafter plasma was stored at _ 20 ˚C until assayed.

Progesterone concentrations were measured using an ELISA kit (Monobind®; USA) with 0.1 ng/mL sensitivity. The plasma estradiol concentration was measured by ELISA kit (DRG International, GmbH, USA) with 0.625 pg/mL sensitivity.

Statistical analysis
Data were analyzed as a completely randomized design with the GLM procedure of SAS (1998). Data for oestrus responses, pregnancy rate, lambing rate and fecundity rates were analyzed using chi-square test. Time to onset of oestrus, duration of oestrus and plasma estradiol and progesterone concentrations were analyzed by using PROC GENMOD.

All results are given as mean ± SEM. Mean values were compared by the Duncan test. The values of less than 0.05 (P<0.05) were declared significant.

RESULTS AND DISCUSSION

Oestrus responses
The effect of short and long-term application of MAP sponge on oestrus response, oestrus onset and duration of oestrus in Arabian ewes are gave in Table 1. No significant differences in percentage of oestrus response and time to onset of oestrus were found between the groups I and II. However, differences in oestrus response rate and time to onset of oestrus were found to be significantly different when these groups were compared with the control group (P<0.05).

This study shows that short-term sponge treatment results in higher duration of oestrus when compared to the other two groups (Table 1; P<0.05). To our knowledge, this is the first report, which compares the effects of 300 IU eCG in Arabian ewes synchronized by long and short-term progesterone sponge.

The use of intravaginal sponge devices with eCG were found to be efficient methods for oestrus induction and synchronization in ewes during the non-breeding season. In the present study behavioral oestrus was detected in 11/12 (91.67%) and 10/12 (83.33%) ewes in response to short and long-term progesterone treatment, respectively. Both the long- and short-term treatment with progesterone devices result in a high percentage of females shown sign of oestrus.

These results indicate that using short term MAP-impregnated intravaginal sponges as a source of exogenous progesterone administered in non breeding season, had excellent effects on induction of oestrus in Arabian ewes. Knights *et al.* (2001) reported that short-term treatment of ewes with progesterone before ram introduction was adequate to induce fertile oestrus.

Table 1 Effects of short and long-term medroxy progesterone acetate (MAP) sponge treatment on oestrus performance, pregnancy, lambing and fecundity rates in Arabian ewes during anoestrus season

Variable/treatments (Mean±SE)	I (short term MAP) n= 12	II (long term MAP) n= 12	III (control) n= 12
Response of oestrus (%)	91.67[a] (11/12)	83.33[a] (10/12)	16.66[b] (2/12)
Onset of oestrus(h)	44.73±4.49[a]	45.62±3.76[a]	78±4.57[b]
Oestrus duration (h)	27.27±1.4[a]	17.38±1.8[b]	-
Pregnancy rate (%)	54.44±0.45[a]	50±0.63[a]	0[b]
Lambing rate (%)	66.66±0.16[a]	80±0.17[a]	-
Fecundity rate	0.36±0.12[a]	0.4±0.13[a]	-

The means within the same row with at least one common letter, do not have significant difference (P>0.05).
SE: standard error.

Also, Ungerfeld and Rubianes (2002) reported that short-term progesterone treatment was adequate to induce fertile oestrus and no difference in oestrus response was observed when anoestrus ewes were primed for 6 or 14 days, with intravaginal sponge treatments. No significant difference was observed in oestrus response between treatment with long and short-term progesterone sponge in this study, which concurs with Ustuner et al. (2007), Ungerfeld and Rubianes (2002).

However, Vinoles et al. (1999) reported that the oestrus response following sponge withdrawal was significantly higher in long-term than in short-term-treated ewes. This disparity may be due to differences of variables, including breed of ewes, month of years, latitude and management. Incidence of oestrus in ewes treated with either synthetic progesterone or natural progesterone in conjunction with PMSG varied from 47% (Robinson et al. 1967) to 85% or greater (Akoz et al. 2006; Timurkan and Yildiz, 2005; Zeleke et al. 2005; Hashemi et al. 2006). The oestrus response in the two group was comparable to that previously reported by Beck et al. (1993); Ozturkler et al. (2003) and Ataman et al. (2006) in short-term and Ozyurtlu et al. (2010); Ozyurtlu et al. (2011); Vinoles et al. (1999); Moradi Kor and Ziaei (2012); Koyuncu and Ozis Altıcekic (2010) in long term treatment with progesterone device. All ewes exhibited behavioral oestrus from 28 to 78 h following sponge withdrawal. In previous studies reported the onset of oestrus to occur within 6-120 h following progesterone device removal (Freitas et al. 1996; Romano, 1998; Greyling and Van der Nest, 2000).

Koyuncu and Ozis Altcekic (2010) reported that 96% of ewes exhibited oestrus 48-72 h following removals of progesterone-containing implants. Intervals to onset of oestrus resembled those reported in the previous study in which ewes were treated with progesterone-300 IU eCG (Zonturlu et al. 2008).

In this study, the time to the oestrus onset following the withdrawal of sponge in II groups was found to be 44.73 h, lower than the 46 h reported by Ungerfeld and Rubianes (2002) during non-breeding season.

The time to the oestrus onset in the I group was found to be 45.62 h which was higher than 40 h found by Romano (1996), Ungerfeld and Rubianes (2002) and Emsen and Yaprak (2006). Vinoles et al. (2001) reported that ewes treated with MAP for 6 days exhibited oestrus at 84 h and those treated for 12 days exhibited oestrus at 44 h after sponge removal. Also, Zeleke et al. (2005) reported that the ewes treated with FGA for 14 day showed oestrus at 41 h, which is lower than the values obtained in the present study (I=44.73 h and II=45.62 h). These differences may be explained by the differences in breed, season, location, nutrition, climate and presence of male after intravaginal devices removal.

The mean overall durations of the induced oestrus period were 27.27 h and 17.38 h for I and II groups respectively. Oestrus duration in the I group was higher than the II group (P<0.05). This result was lower than the duration of oestrus as reported by Nasser et al. (2012). Nasser et al. (2012) reported the duration of oestrus was 34.4 and 36.7 hours following a short (6 days) and a long (12 days) term controlled internal drug release (CIDR) treatment, respectively, and this is consistent with the results obtained by Ustuner et al. (2007). Short duration of oestrus in this study may be explained by differences in breed, age, geographical location and lower estrogen level in the blood. Stimulation of follicular growth in the ovary by eCG together with high levels and longer duration of serum estrogen concentrations could be responsible for a prolonged duration of the oestrus period (Nasser et al. 2012).

Pregnancy, lambing and the number of lambs born
In this study, ewes that did not display oestrus behavior during 35 days from mating (after second oestrus period) were considered as pregnant (Moghaddam et al. 2012).

Results for the pregnancy, lambing and fecundity rates are presented in Table 1. No significant differences in the results for pregnancy, lambing and fecundity rates were noted between the short and long-term groups. However, the pregnancy and lambing rates were found to be significantly different in the short and long-term groups as compared with the control group (P<0.05). Pregnancy rate, as determined by ewes rebred, was 45.83% (11/24). Eleven of those 24 ewes lambed within 155 days of mating. The remaining ewes probably experienced early embryonic loss. The fertility of anoestrus ewes treated with progesterone deviese ranged from 22 to 70% (Evans et al. 2001). In the present study, an injection of 300 IU eCG was used in order to induce follicular growth and increase pregnancy rate. The injection of eCG at the time of oestrus, causes a higher pregnancy rates during the non-breeding season. Vinoles et al. (1999) obtained higher pregnancy rate after short-term treatment compared to the Long-term (12 d) treatment with PMSG at the time of vaginal sponge withdrawal.

Ustuner et al. (2007) and Ozyurtlu et al. (2011) reported that there were no significant differences in terms of pregnancy, lambing and fecundity rates between the short and long-term treatment groups. No significant difference was observed in this study which agrees with (Ustuner et al. 2007; Ataman et al. 2006; Saribay et al. 2011). Lambing rate in this study was 66.66 following short-term treatment and 80% following long-term treatment. It has been shown that the lambing rate in ewes treated with short-term synthetic progesterone in combination with eCG varied from 60 to 83.3% (Ozyurtlu et al. 2011; Ataman et al. 2006; Amer and Hazza, 2009).

Estradiol concentrations

Mean plasma estradiol concentrations on the day before sponge insertion were similar among ewes of the three treatment groups (Table 2). Estradiol trend is rising and falling due to follicle activity, outside of breeding season (Menegatos et al. 2003).

Negative feedback of estradiol on tonic center in the hypothalamus affect as much as possible to prevent the occurrence of ovarian cycle outside the breeding season (Sanchez, 2005).

It can be concluded that due to high levels of estradiol and lack of optimal progesterone concentration, there is no ovarian cycle leads to ovulation in all treatments. In the most animal species oestrus behavior can be triggered by estradiol alone, but in ewes and rats, oestrus behavior is not displayed unless progesterone is present (Fajt, 2011).

Result showed that two days after initiation of the synchronization program, serum estradiol concentrations was affected by the treatments, so that use of progesterone device in non breeding season decreased estradiol concentrations in compared with the first sampling. No difference was found between the I and II groups in estradiol concentration (Table 2). Progesterone is associated with suppression of follicular growth and ovulation through exerting an inhibitory effect on the release of LH from the anterior pituitary in sheep. It inhibits the effect of estradiol secretion by granulosa cells in the ovary (Noel et al. 1994).

The mean concentration of estradiol in day of oestrus was similar among I and II groups and higher than control group (Table 2). One day before oestrus one or more follicles in the ovary grow rapidly, thereafter the concentration of estradiol in the venous blood increase generally about 10 to 20 pg/mL (Błaszczyk et al. 2004). The estradiol concentrations observed after progesterone device removal in the present study were expected to reflect estradiol secretion by the ovarian follicles. The fall in progesterone concentration stimulates estradiol secretion by the growing follicles.

Progesterone concentrations

Mean plasma progesterone concentrations on the day before sponge insertion were similar among ewes of the three treatment groups (Table 3). In non breeding season ovulation not occurs because the secretion of LH is low that not promote to development of ovarian follicles and the corpora lutea, thereafter plasma concentrations of progesterone remain very low (Chanvallon et al. 2008).

Table 2 Mean estradiol-17β concentrations (pg/mL) per examination at different days of progesterone sponge treatment

Groups	I (short term MAP)	II (long term MAP)	III (control)
Number of ewes	7	7	7
One day before sponge insertion	9.22±1.31[a]	10.20±1.14[a]	8.46±1.20[a]
Two day after sponge insertion	8.35±1.46[a]	8.89±1.63[a]	9.1±1.22[a]
Day of oestrus	11.78±2.1[a]	11.4±1.96[a]	8.88±1.36[b]

The means within the same row with at least one common letter, do not have significant difference (P>0.05).

Table 3 Mean progesterone concentration (ng/mL) per examination at different days of progesterone sponge treatment

Groups	I (short term MAP)	II (long term MAP)	III (control)
Number of ewes	7	7	7
One day before sponge insertion	0.70±0.22[a]	0.80±0.18[a]	0.96±0.27[a]
Two day after sponge insertion	6.60±0.97[a]	6.33±0.86[a]	1.11±0.14[b]
Day of oestrus	1.5±0.33[a]	1.68±0.40[a]	0.98±0.31[b]

The means within the same row with at least one common letter, do not have significant difference (P>0.05).

Obtained results from this experiment showed that the mean progesterone concentrations at the two days after sponge insertion were higher than first sampling and plasma progesterone levels increased with application progesterone device (Table 3). Following insertion, plasma progesterone levels increased rapidly and reached maximum values 2 days post insertion. Maximum values were 6.60 ± 0.97 and 6.33 ± 0.86 ng/mL for the I and II groups, respectively. Differences in maximum values were not significant (P>0.05; Table 3).

Our results are in agreement with previous studies which reported that blood plasma progesterone levels increased rapidly and reached maximum values 2 days post insertion (Husein et al. 1999; Husein and Kridli, 2002; Husein and Haddad, 2000).

Progesterone assays following 3rd blood sampling demonstrated that its concentrations in blood plasma declined following intravaginal progesterone devices withdrawal (Table3). Blood progesterone concentrations indicate the reproductive physiology of animals. The initial release of progesterone from intravaginal sponges increase during the first 48 h of treatment and reached maximum values, but decreases with the time. It has been described that, serum medroxy-progesterone acetate concentrations decrease with the time in about a 63% between the 2nd and 13th days after sponge-MAP insertion (Greyling et al. 1994), demonstrating that the progesterone supply from intravaginal sponges decreases with the time. The data presented above confirm that, in the day of oestrous, plasma progesterone values fall to very low levels (Husein and Haddad, 2006; Husein and Kridli, 2002; Kridli and Al-Khetib, 2006).

CONCLUSION

In conclusion, the use of intra-vaginal MAP sponges in a 12 day + eCG regime could adequately improve the reproductive performance of ewes during the anestrous season, with the possibility of replacing it with a 6 days MAP + eCG regime with higher efficiency. Also, the results presented in this study give evidence that a short-term progesterone protocol prolongs oestrus duration when compared to a long-term treatment.

ACKNOWLEDGEMENT

The authors gratefully acknowledge the Ramin Agriculture and Natural Resources University for research founds.

REFERENCES

Akoz M., Bulbul M., Ataman B. and Dere S. (2006). Induction of multiple births in Akkaraman cross-breed synchronized with short duration and different doses of progesterone treatment combined with PMSG outside the breeding season. *Bull. Vet. Inst. Pulawy.* **50,** 97-100.

Amer H.A. and Hazzaa A.M. (2009). The effect of different progesterone protocols on the reproductive efficiency of ewes during the non-breeding season. *Vet. Arhiv.* **79,** 19-30.

Ataman M.B., Akoz M. and Akman O. (2006). Induction of synchronized oestrus in Akkaraman cross-bred ewes during breeding and anestrus seasons: the use of short and long-term progesterone treatments. *Rev. Med. Vet.* **50,** 257-260.

Bacha S., Khiati B., Hammoudi S.M., Kaidi R. and Ahmed M. (2014). The effects of dose of pregnant mare serum gonadotropin (PMSG) on reproductive performance of algerian Rembi ewes during seasonal anoestrus. *J. Vet. Sci. Technol.* **5,** 190-198.

Bastan A. (1996). Effect of melatonin and progesterone application on reproductive performance in akkaraman ewes (Akkaraman ırkı koyunlarda melatonin ve progesterone uygulamalarının reprodu"ktif performans u"zerine etkileri). Ph D. Thesis. Ankara Univ., Ankara, Turkey.

Beck N.F.G., Davies B. and Williams S.P. (1993). Oestrous synchronization in ewes-the effect of combining a prostaglandin analogue with a 5-day progesterone treatment. *Anim. Prod.* **56,** 207-210.

Błaszczyk B., Udała J. And Gaczarzewicz D. (2004). Changes in estradiol, progesterone, melatonin, prolactin and thyroxine concentrations in blood plasma of goats following induced estrus in and outside the natural breeding season. *Small Rumin. Res.* **51,** 209-219.

Chanvallon A., Blache D., Chadwick A., Esmaili T., Hawken P.A.R., Martin G.B,. Vinoles C. and Fabre-Nys C. (2008). Sexual experience and temperament affect the response of Merino ewes to the ram effect during the anoestrous season. *Anim. Reprod. Sci.* **37,** 1110-1016.

Emsen E. and Yaprak M. (2006). Effect of controlled breeding on the fertility of Awassi and Red Karaman ewes and the performance of the off spring. *Small Rumin. Res.* **66,** 230-235.

Evans A.C., Flynn J.D., Quinn K.M., Daffy P., Quinn P. and Madgwick S. (2001). Ovulation of aged follicles does not affect embryo quality or fertility after a 14-day progesterone estrus synchronization protocol in ewes. *Theriogenology.* **56,** 923-936.

Fajt V.R. (2011). Drug laws and regulations for sheep and goats. *Vet. Clin. North America. Food Anim.* **27,** 1-21.

Freitas V.J.F., Baril G. and Saumande J. (1996). Induction and synchronization of estrus in goats: the relative efficiency of one versus two fluorogestone acetate-impregnated vaginal sponges. *Theriogenology.* **46,** 1251-1256.

Godfrey R.W., Collins J.R., Hensley E.L. and Wheaton J.E. (1999). Oestrus synchronization and artificial insemination of hair sheep ewes in the tropics. *Theriogenology.* **51,** 985-997.

Greyling J.P.C., Kotze W.F., Taylor G.J., Hangendijk W.J. and Cloete F. (1994). Synchronization of oestrus in sheep: use of different doses of progesterone outside the normal breeding season. *South African J. Anim. Sci.* **24,** 33-37.

Greyling J.P.C. and Van der Nest M. (2000). Synchronization of oestrus in goats: dose effect of progesterone. *Small Rumin. Res.* **36,** 201-207.

Hashemi M., Safdarian M. and Kafi M. (2006). Estrous response to synchronization of estrus using different progesterone treatments outside the natural breeding season in ewes. *Small Rumin. Respon.* **65**, 279-283.

Husein M.Q. and Haddad S.G. (2006). A new approach to enhance reproductive performance in sheep using royal jelly in comparison with equine chorionic gonadotropin. *Anim. Reprod. Sci.* **93**, 24-33.

Husein M.Q., Kridli R.T. and Humphrey W.D. (1999). Effect of royal jelly onestrus synchronization and pregnancy rate of ewes using flourogestone acetate sponges. *J. Anim. Sci.* **77(1)**, 431-438.

Husein M.Q. and Kridli R.T. (2002). Reproductive responses following royal jelly treatment administered orally or intramuscularly into progesterone-treated Awassi ewes. *Anim. Reprod. Sci.* **74(2)**, 45-53.

Iida K., Kobayashi N., Kohno H., Miyamoto A. and Fukui Y. (2004). A comparative study of induction of estrus and ovulation by three different intravaginal devices in ewes during the non-breeding season. *J. Reprod. Develop.* **50**, 63-69.

Knights M., Maze T.D., Bridges P.J., Lewis P.E. and Inskep E.K. (2001). Short-term treatment with a controlled internal drug releasing (CIDR) device and FSH to induce fertile estrus and increase prolificacy in anestrus ewes. *Theriogenology.* **55**, 1181-1191.

Koyuncu M. and Alticekic S. (2010). Effects of progesterone and PMSG on estrous synchronization and fertility in Kivircik ewes during natural breeding season. *Asian Australas J. Anim.* **23**, 308-311.

Kridli R.T. and Al-Khetib S.S. (2006). Reproductive responses in ewes treated with eCG or increasing doses of royal jelly. *Anim. Reprod. Sci.* **92**, 75-85.

Menegatos J., Chadio S. and Kauskoura T. (2003). Endocrine event during the pre estruos period and the subsequent estrous cycle in ewes after estrous synchronization. *Theriogenology.* **59**, 1533-1543.

Moghaddam G.H., Olfati A., Daghigh Kia H. and Rafat S.A. (2012). Study of reproductive performance of crossbred ewes treated with GnRH and PMSG during breeding season. *Iranian J. Appl. Anim. Sci.* **2(4)**, 351-356.

Moradi Kor N. and Ziaei N. (2012). Effect of PGF2 administration and subsequent eCG treatments on the reproductive performance in mature Raieni goats during the breeding season. *Asian J. Anim. Vet. Adv.* **7**, 94-99.

Nasser S.O., Wahid H., Aziz A.S., Zuki A.B., Azam M.K., Jabbar A.G. and Mahfoz M.A. (2012). Effect of different oestrus synchronizations protocols on the reproductive efficiency of Dammar ewes in Yemen during winter. *African J. Biotechnol.* **11**, 9156-9162.

Noel B., Bister J.L., Pierquin B. and Paquay R. (1994). Effect of FGA and PMSG on follicular growth and LH secretion in Suffolk ewes. *Theriogenology.* **41**, 719-727.

Ozturkler Y., Colak A., Baykal A. and Guven B. (2003). Combined effect of a prostaglandin analogue and a progesterone treatment for 5 days on oestrus synchronization in Tushin ewes. *Indian Vet. J.* **80**, 917-920.

Ozyurtlu N., AY S.S., Kucukaslan I., Gungor O. and Aslan S. (2011). Effect of subsequent two short-term, short-term and long-term progesterone treatments on fertility of Awassi ewes out of the breeding season. *Ankara Univ. Vet. Fak. Derg.* **58**, 105-109.

Ozyurtlu N., Kucukaslan I. and Cetin Y. (2010). Characterization of oestrous induction response, oestrous duration, fecundity and fertility in Awassi ewes during the non-breeding season utilizing both CIDR and intravaginal sponge treatments. *Reprod. Domest. Anim.* **4**, 464-467.

Robinson T.J., Moore N.W., Holst P.J. and Smith J.F. (1967). The evaluation of several progestogens administered in intravaginal sponges for the synchronization of estrus in the entire cyclic Merino ewe. Pp. 76-91 in Control of the Ovarian Cycle in the Sheep. T.J. Robinson, Ed. White and Bull PTY Ltd., Australia.

Romano J.E. (1996). Comparison of fluorgestone and medroxyprogesterone intravaginal pessaries for estrus synchronization in dairy goats. *Small Rumin. Res.* **22**, 216-223.

Romano J.E. (1998). The effect of continuous presence of bucks on hastening the onset of estrus in synchronized does during the breeding season. *Small Rumin. Res.* **30**, 99-103.

Sanchez E.J. (2005). Melatonin-estrogen interaction in breast cancer. *J. Pineal. Res.* **38**, 217-222.

Saribay M.K., Karaca F., Dogruer G., Ergun Y., Yavas I. and Ates C.T. (2011). Oestrus synchronization by short and long-term intravaginal sponge treatment in lactating goats during the breeding season: the effects of GnRH administrations immediately after matings on fertility. *J. Anim. Vet. Adv.* **10**, 3134-3139.

SAS Institute. (1998). SAS®/STAT Software, Release 6. SAS Institute, Inc., Cary, NC. USA.

Timurkan H. and Yildiz H. (2005). Synchronization of oestrus in Hamedani ewes: the use of different PMSG dose. *Bull. Vet. Pulawy.* **49**, 311-314.

Ungerfeld R. and Rubianes E. (2002). Short term primings with different progestogen intravaginal devices (MAP, FGA and CIDR) for eCG oestrous induction in anestrus ewes. *Small Rumin. Res.* **46**, 63-66.

Ustuner B., Gunay U., Nur U. and Ustuner U. (2007). Effects of long and short-term progestogens treatment combined with PMSG on oestrus synchronization and fertility in Awassi ewes during the breeding season. *J. Acta Vet. Brno.* **76**, 391-397.

Vinoles C., Forsberg M., Banchero G. and Rubianes E. (2001). Effect of long-termand short termprogesterone treatment on follicular development and preganancy rate in cyclic ewes. *Theriogenology.* **55**, 993-1001.

Vinoles C., Meikle A., Forsberg M. and Rubianes E. (1999). The effect of subluteal levels of exogenous progesterone on follicular dynamics and endocrine patterns during the early luteal phase of the ewe. *Theriogenology.* **51**, 1351-1361.

Wildeus S. (1999). Current concepts in synchronization of oestrus: sheep and goats. *Proc. Am. Soc. Anim. Sci.* **39**, 1-14.

Zeleke M., Greyling L.M.J. and Schwalbach T. (2005). Effect of progesterone and PMSG on oestrous synchronization and fertility in dorper ewes during the transition period. *Small Rumin. Res.* **56(1)**, 47-53.

8

Investigation of *GDF9* and *BMP15* Polymorphisms in Mehraban Sheep to Find the Missenses as Impact on Protein

A. Ahmadi[1*], F. Afraz[2], R. Talebi[1], A. Farahavar[1] and S.M.F. Vahidi[2]

[1] Department of Animal Science, Faculty of Agriculture, Bu Ali Sina University, Hamedan, Iran
[2] Department of Genomics, Animal Agricultural Biotechnology Research Institute (ABRI), North branch, Rasht, Iran

*Correspondence E-mail: ahmadi@basu.ac.ir

ABSTRACT

Utilization of fecundity genes such as *GDF9* and *BMP15* can help improve reproductive traits in sheep breeding programme. To evaluate effects of missense mutations on protein function, the polymorphisms of *GDF9* and *BMP15* genes were screened in twelve mehraban sheep using DNA sequencing, followed by protein structure modeling. Six single nucleotide polymorphism (SNPs) known as *FecG* mutations (G1-G6), were detected in exons 1 and 2 of *GDF9* gene. Mutations of G1 (*GDF9* exon 1 g.2118 G>A), G4 (*GDF9* exon 2 g.3451 T>C) and G6 (*GDF9* exon 2 g.3974 G>A) have shown amino acid substitution. None polymorphism was detected in exon 1 and exon 2 of *BMP15* gene. Based on identified polymorphisms, individuals were classified into three haplotypes of wild haplotype (without mutation), haplotype A (simultaneous mutations of G1, G2, G3 and G4) and haplotype B (simultaneous mutations of G5 and G6). The 3D-structure of GDF9 protein in A and B haplotypes was rotated 90° and 45° than wild haplotype, respectively. The missenses G1/p.Arg87His, G4/p.Glu241Lys and G6/p.Val332Ile variants were benign. However both the missenses of G7/p.Val371Met and G8/p.Ser315Phe were probably damaging. Phylogenetic tree of *GDF9* gene revealed that individuals with A and B haplotypes were distinct from wild haplotypes with bootstrapping values of 63 and 76, respectively. In conclusion, GDF9 protein in A and B haplotypes showed a higher performance than wild haplotype due to synergism effects of simultaneous mutations. These types of mutations with effect on turn and helix of *GDF9* conservative regions showed physical and functional interaction with TGFβ proteins.

KEY WORDS *BMP15*, functional interaction, *GDF9*, missense, protein structure, sheep.

INTRODUCTION

Reproduction is a complex process and fecundity traits such as ovulation rate (OR) and litter size (LS) are genetically affected by many minor genes and also some major genes, called fecundity (*Fec*) genes (Drouilhet *et al.* 2009). Utilization of major fecundity genes in sheep production can cause genetic improvement in breeding programs, and consequently high performance in reproductive traits (Notter, 2008). So far, some major genes relevant to OR and LS traits such as bone morphogenetic protein recep-tor-1B (*BMPR1B*), bone morphogenetic protein 15 (*BMP15*), growth differentiation factor 9 (*GDF9*) and beta-1, 4-N-acetyl-galactosaminyl transferase 2 or lacaune gene (*B4GALNT2*) have been reported and all of which have hyper prolificacy-associated mutations (Drouilhet *et al.* 2013; Bodin *et al.* 2007). *BMPR1B*, *BMP15* and *GDF9* genes are all part of the ovary-derived transforming growth factor-β (TGFβ) superfamily. The essential growth factors and receptors in ovarian follicular development are coded by these genes (Pramod *et al.* 2013). Causative mutations in sheep such as *FecB^B* mutation in *BMPR1B* in Booroola

Merino (Wilson *et al.* 2001) and Javanese (Bradford *et al.* 1986); mutations in *BMP15* in Romney (*FecX^I*, *FecX^H*) (Galloway *et al.* 2000), Cambridge and Belclare (*FecX^B*, *FecX^G*) (Hanrahan *et al.* 2004), Lacaune (*FecX^L*, *FecX^I*) (Bodin *et al.* 2007) and mutations in *GDF9* in Belclare and Cambridge (*FecG^H*) (Hanrahan *et al.* 2004), Norwegian White Sheep (*FecG^{NW}*) (Vage *et al.* 2013), Icelandic (*FecG^T*) (Nicol *et al.* 2009) and Santa Inê's sheep (*FecG^E*) (Silva *et al.* 2011), have been clearly identified and also their significant effect on fecundity traits have been showed in different sheep breeds.

Several methods based utilization of the physico-chemical properties of amino acids, as well as information about the role of amino acid side chains in protein structure, have been developed for assessing the effects of mutation on protein function (Reva *et al.* 2011). Mutations can influence protein folding, protein stability, protein function, protein-protein interactions, protein expression and subcellular localization (Reva *et al.* 2011). This work examined the polymorphic sites of *GDF9* and *BMP15* genes in Mehraban sheep to identify missense mutations. In addition, protein folding and physical interactions were assessed through protein modeling to interpret missense effects on protein conformation and functionality.

MATERIALS AND METHODS

Sampling

To perform the experiment, twelve Mehraban ewes, located in Hamedan province, western Iran, were selected. Blood samples (5 mL per animal) were collected from jugular vein using venoject containing EDTA. Genomic DNA was extracted from whole blood using DNP™ Kit (SinaClon BioScience Co.) and kept at -20 ˚C. Quantity and quality of gDNA were determined using NanoDrop® ND-1000 based on absorbance at A260/A280 ratio.

PCR amplification and purification of products

Specific primers were designed by Primer-BLAST tool (www.ncbi.nlm.nih.gov/tools/primer-blast/). The primers were used to amplify both exons 1 and 2 of *BMP15* (GenBank No. NC_019484.1) and *GDF9* (GenBank No. NC_019462.1) genes (Table 1). Polymerase chain reaction (PCR) was carried out on 50 ng of genomic DNA in a final 20 µL reaction containing 4 µL of 10 × PCR buffer, 2 µL of 200 µM dNTPs, 1 µL of each primer at a concentration of 0.05 µM and 0.5 unit of Taq polymerase (Promega, Wisconsin, USA).

The PCR protocol used and initial of 5 min denaturation step at 94 ˚C followed by 35 cycles at 94 ˚C for 30 s, 55 ˚C for 30s 72 ˚C for 30 s and a final 7 min extension step at 72 ˚C.

Purification of PCR products was carried out in a final 15 µL volume consisting of mixture of *T-SAP* (1 U/µL) and *Exonuclease 1* (20 U/µL) with Programs PCR as follows: digestion 37 ˚C for 45 minutes, inactivation of enzymes 80 ˚C for 30 minutes in 35 cycles.

DNA sequencing

All twelve samples were selected for DNA sequencing. The primers used for sequencing were the same primers used for the PCR. The sequencing reaction was carried out in a final 20 µL volume containing 2 µL Tampon ABI (5X), 0.5 µL Terminators Dichlororhodamine V3.1 (Life technologies Co. California, United States), 1 µL primer (10 µM), 1.5 µL H$_2$O. The program of PCR sequencing as follows: 95 ˚C for 5 min followed by 1 cycle, and then 25 cycles followed by 95 ˚C for 30 s, 55 ˚C for 30 s, 60 ˚C for 4 min. Sequencing was carried out at the National Institute of Agronomic Research (INRA), France and GenPhySE Centre on the Applied Biosystems 3730 DNA Analyzer platform.

Bioinformatics analysis of polymorphism effects on DNA and protein structures

CLC Genomics Workbench Version 7.6.4 (www.clcbio.com) was used to map all obtained read counts of *GDF9* and *BMP15* genes to the reference genomes of NC_019462.1 and NC_019484.1, respectively. Option of annotation was used with reference genome in order to find each conflict on the genome level. According to annotation analysis we carefully manifested every known polymorphisms based on other breeds and new polymorphisms based on identified conflicts. Analysis of protein characterizations was accomplished using Protean (DNASTAR Inc., Madison, WI. USA). The mutation scrutinizes in the secondary structure of proteins were envisaged using Protean (DNASTAR Inc., Madison, WI. USA). The Expasy website (http://us.expasy.org/) was used to constitute 3D structure of proteins based on homology modeling. Furthermore, the predicted structures were compared with 3D models of Protein Data Bank (www.pdb.org) TGF-ß 4YCI (Mi *et al.* 2015), then modeling of protein was carried out using modeler tools of CLC Genomics Workbench Version 7.6.4 based on protein structures of PDB and predicted structures. The sequences of amino acids for different mono-ovulatory and poly-ovulatory species were obtained from biological database such as Uniprot (http://www.uniprot.org/uniprot/). Polymorphism Phenotyping v2 (PolyPhen-2) was used to predict the possible effects of the observed missense mutations on the structure and function of proteins based on an iterative greedy algorithm (Adzhubei *et al.* 2010). To consider *GDF9* and *BMP15* in different species, at first, we decided to focus on the sequences of amino acids to achieve a general understanding about substitutions rate of amino acids. A maximum-likelihood method based on the JTT matrix-based model was applied to infer evolutionary history by CLC Genomics Workbench Version 7.6.4 (www.clcbio.com) (Jones *et al.* 1992).

Table 1 Primers used for exon 1 and 2 of *GDF9* (NC_019462.1[*]) and *BMP15* (NC_019484.1[*]) genes amplification

Gene	Coding region	Primers identity
GDF9	Exon 1 (462 bp)[a]	G9-1734 Forward: GAAGACTGGTATGGGGAAATG
		G9-2175 Backward: CCAATCTGCTCCTACACACCT
	Exon 2 (1120 bp)[a]	G9-3270 Forward: TGGCATTACTGTTGGATTGTTTT
		G9-4376 Backward: GCTCCTCCTTACACAACACACAG
BMP15	Exon 1 (762 bp)[a]	B15-112 Forward: TTCCTTGCCCTATCCTTTGTG
		B15-597 Backward: ACTTTTCTTCCCCATTTTCTCCC
		dnWT/Bar Backward: GAGGCCTTGCTACACTAGCC
		SNP50977717 Backward: CACAAAGGATAGGGCAAGGA
	Exon 2 (868 bp)[a]	B15-359 Forward: CGCTTTGCTCTTGTTCCCTC
		B15-1205 Backward: GGCAATCATACCCTCATACTCC

[*] GenBank accession numbers at NCBI.
[a] amplified fragment length.

The bootstrap consensus tree inferred from 1000 replicates was taken to represent the evolutionary history of the analyzed taxa.

RESULTS AND DISCUSSION

According to Figure 1, the results obtained by DNA sequencing showed six point mutations in exons 1 and 2 of *GDF9* gene in comparison to the reference sequence (NC_019462.1). These mutations known as G1 (g.2118 G>A, NC_019462.1), G2 (g.3451 T>C, NC_019462.1), G3 (g.3457 A>G, NC_019462.1), G4 (g.3701 A>G, NC_019462.1), G5 (g.3958 A>G, NC_019462.1) and G6 (g.3974 G>A, NC_019462.1) were previously reported by Hanrahan *et al.* (2004) (Table 2). Furthermore, sequencing results of the *BMP15* gene showed no mutations in comparison to the reference sequence (NC_019484.1).

The individuals with simultaneous mutations are surely in complete linkage disequilibrium that it was called as a haplotype situation. According to notable mutations identified in *GDF9* of Mehraban sheep, the individuals were classified in three haplotypes as, Wild haplotype (without mutation), haplotype A (simultaneous mutations of G1, G2, G3 and G4) and haplotype B (G5 and G6). It is clear that ch-

anging one nucleotide may change the amino acid that the nucleotides are coding for. As it is indicated in Table 2, only G1, G4 and G6 mutations resulted in amino acid substitution at p.Arg87His, p.Glu241Lys and p.Val332Ile positions, respectively. With changing amino acids after mutation occurrence, it impresses on the second structure of TGF-ß proteins.

The antigenic index was applied to predict the topological features of *GDF9*. Only a significant antigenic index was observed in G1 (Arg:0.45>His:-0.3). It was not seen any changes in antigenic index of G4 (Glu:0.3>Lys:0.3) and G6 (Val:-0.3>Ile:-0.3).

The degree of hydrophobicity or hydrophilicity of amino acids of GDF9 protein was quantified using hydrophilicity plot. The observed changes were belonged to G1 (Arg:1.63>His:1.49), G4 (Glu:0.422>Lys:0.467) and G6 (Val:-0.244>Ile:-0.278). It is assumed that with occurrence of missense mutations in *GDF9*, some chemical features such as aliphatic index, molecular weight, 1 microgram, isoelectric point, molar extinction coefficient, absorbance at 280[nm] and charged at pH= 7 were affected in GDF9 protein. As illustrated in Figure 2, only changed amino acid position of G1/p.Arg87His was occurred on alpha helix region of GDF9. However, none of change position of G4/p.Glu241Lys (Figure 2A) and G6/p.Val332Ile (Figure 2B) did not occurred on regions of alpha helix, Beta sheets and so on.

Because neither the BMP15 nor the GDF9 3-D structures were known (Monestier *et al.* 2014), the three types of haplotype (wild, A and B) were aligned to predict a model for GDF9 protein the three haplotypes (Figure 3). As illustrated in Figure 3, the predicted GFP9 protein structures was aligned (Figure 3A) with mouse BMP9 (PDB 4YCI by Mi *et al.* (2015)) for four accession numbers (GI:764091320, GI:764091321, GI:764091322 and GI:764091323) to manifest the turning and folding at protein structure after mutation occurrence.

The results of alignment illustrated that the structure of GDF9 protein in haplotype A (G1/p.Arg87His and G4/p.Glu241Lys, simultaneously) and hplotype B (heterozygote for G6/p.Val332Ile) has been rotated 90° and 45° than wild haplotype, respectively. According to the structural alignment of Mehraban GDF9 with 4YCI chain-A and B (Figure 3B), only chain B was entirely aligned with predicted structures of GDF9 protein and a partial part of 4YCI-D was also aligned with predicted structures of GDF9 protein (Figure 3C).

The perfect alignment of predicted structures of GDF9 protein of Mehraban ewes with mouse BMP9 (PDB 4YCI) (Figure 3D) can be inferred to responsible for physical and functional interaction between GDF9 and BMP9 proteins or other TGF-β members.

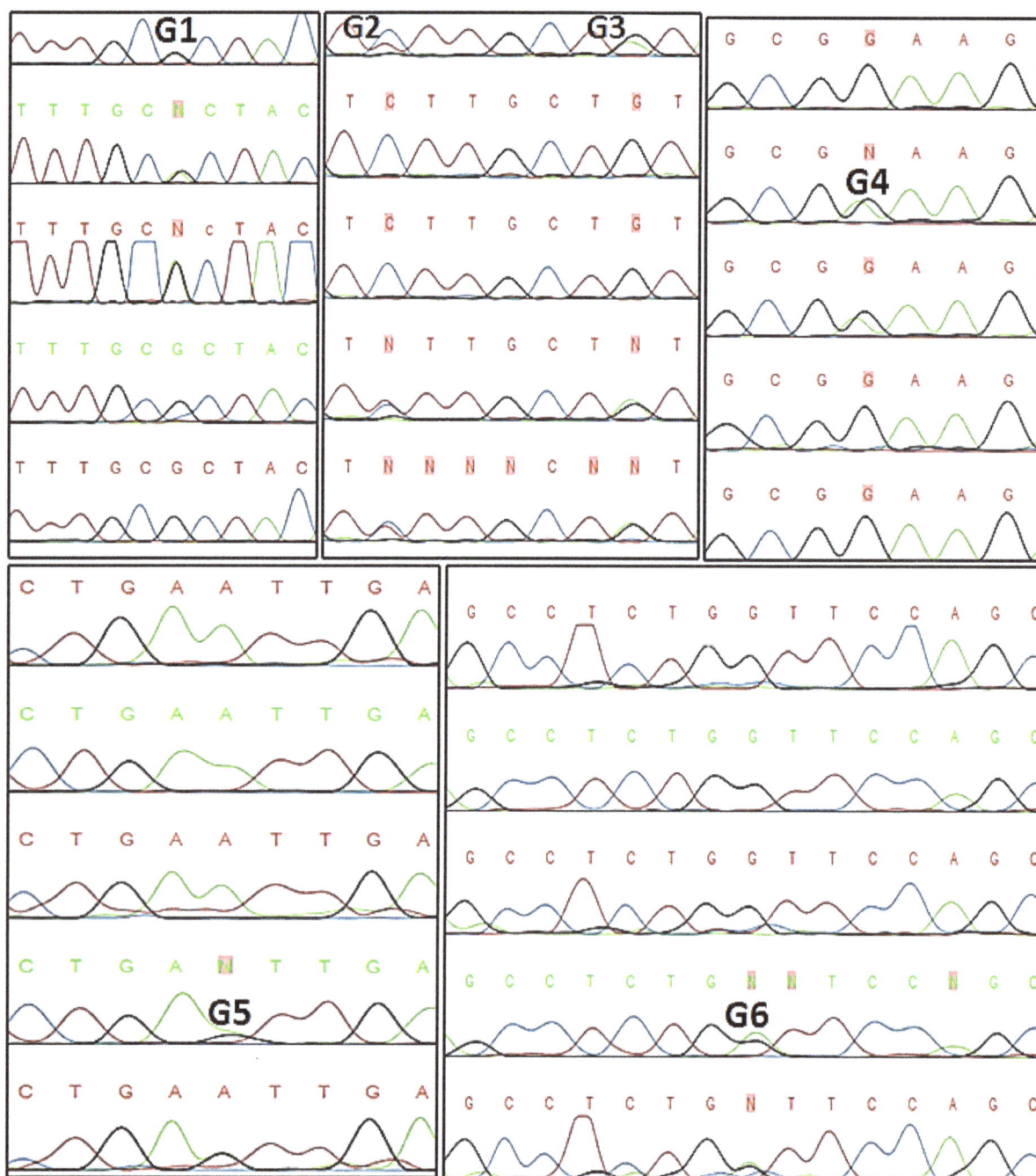

Figure 1 Identified polymorphisms of *GDF9* exon 1 and exon 2 in twelve Mehraban sheep

In comparison of reference genome NC_019462.1, four ewes revealed haplotype A (simultaneous mutations of G1, G2, G3 and G4), one ewe has haplotype B (simultaneous mutations of G5 and G6) and seven animals showed wild haplotype

Table 2 Identified mutations in exon 1 and exon 2 of *GDF9* gene

Exon's no	Variation	Position on DNA	Amino acid substitution	Position on protein
Exon1	G1	2118	R > H	87
Exon2	G2	3451	unchanged (V)	157
	G3	3457	unchanged (L)	159
	G4	3701	E > K	241
	G5	3958	unchanged (E)	326
	G6	3974	V > I	332

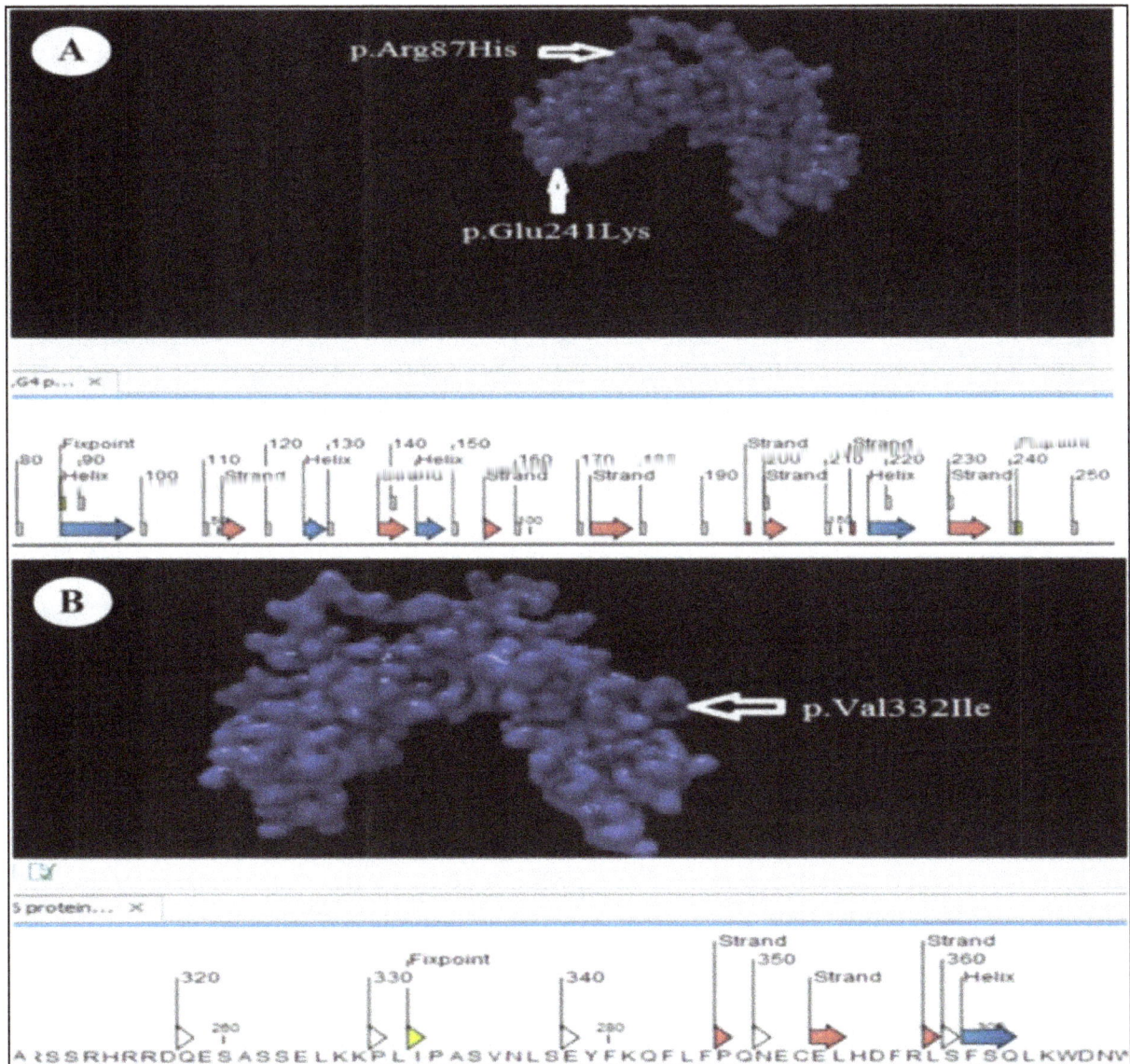

Figure 2 The predicted structures of secondary and three-dimensional of GDF9 protein in Mehraban breed

A) Three dimensional (up) and secondary structure (down) in haplotype A (simultaneous mutations of G1, G2, G3 and G4

B) Three dimensional (up) and secondary structure (down) in haplotype B (simultaneous mutations of G5 and G6)

Of course in reason of the low sequence identities between BMP15 and GDF9 on the one hand and TGFB1 on the other hand, these results need to be interpreted with caution.

Moreover, the potential impact of the identified amino acid substitution, in this study, on the straucture and functional of GDF9 was assessed using PolyPhen2. As shown in Figure 4 the amino acide substitutions of p.Arg87His, p.Glu241Lys and p.Val332Ile were benign with scores of 0.00, 0.00 and 0.031, respectively (Figure 4A and 4B).

With attention to possible impact of the missense on protein function, it can be imagined that the six point mutations of GDF9 containing G1/p.Arg87His, G2/p.Val157Val, G3/p.Lue159Lue, G4/p.Glu241Lys, G5/p.Glu326Glu, G6/p.Val332Ile are benign. However, both amino acid substitutions of G7/p.Val371Met and G8/p.Ser315Phe were probably damaging with the scores of 0.99 and 1.00 respectively (Figure 4C and 4D). They were belonged to locus of $FecG^{NW}$ and $FecG^{H}$, respectively (Vage *et al.* 2013; Hanrahan *et al.* 2004).

Phylogenetic tree of *GDF9* demonstrated that wild hapl type of Mehraban and Texel breed (O77681), human (O60383) and chimpanzee (H2QRG9), rat (Q9QYW4) and mouse (Q07105), dog (D0F1P5) and giant panda (C5HV43) were grouped together (Figure 5A).

However, haplotypes of A and B distanced from Texel breed (O77681) with bootstrapping values of 63 and 76, respectively.

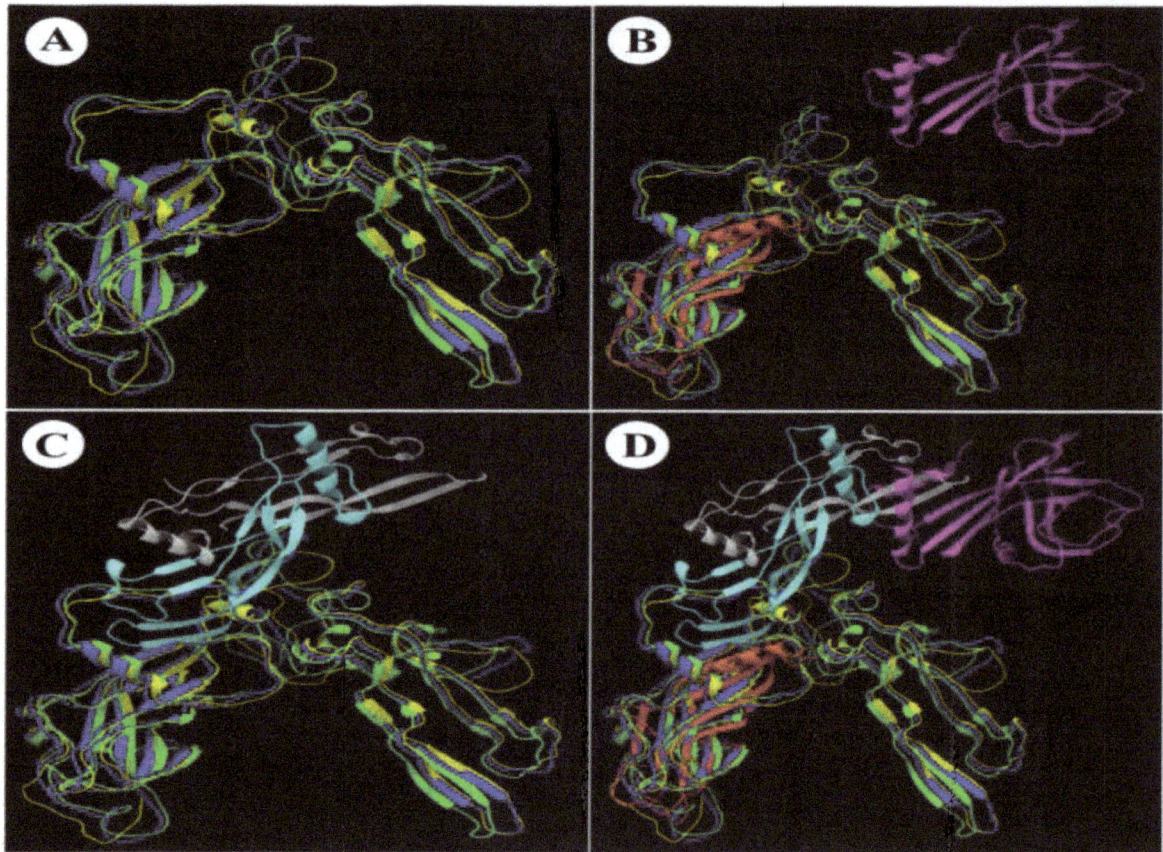

Figure 3 Alignment of predicted 3D structures of GDF9 protein via homology modeling
A) Alignment of predicted models of GDF9 protein for three type haplotype wild, A and B
B) Alignment of predicted models of GDF9 protein and mouse BMP9 (PDB 4YCI chain-A,B)
C) Alignment of predicted models of GDF9 protein and mouse BMP9 (PDB 4YCI chain-C,D)
D) Perfect structural alignment between predicted models of Mehraban GDF9 protein and mouse BMP9 (PDB 4YCI)

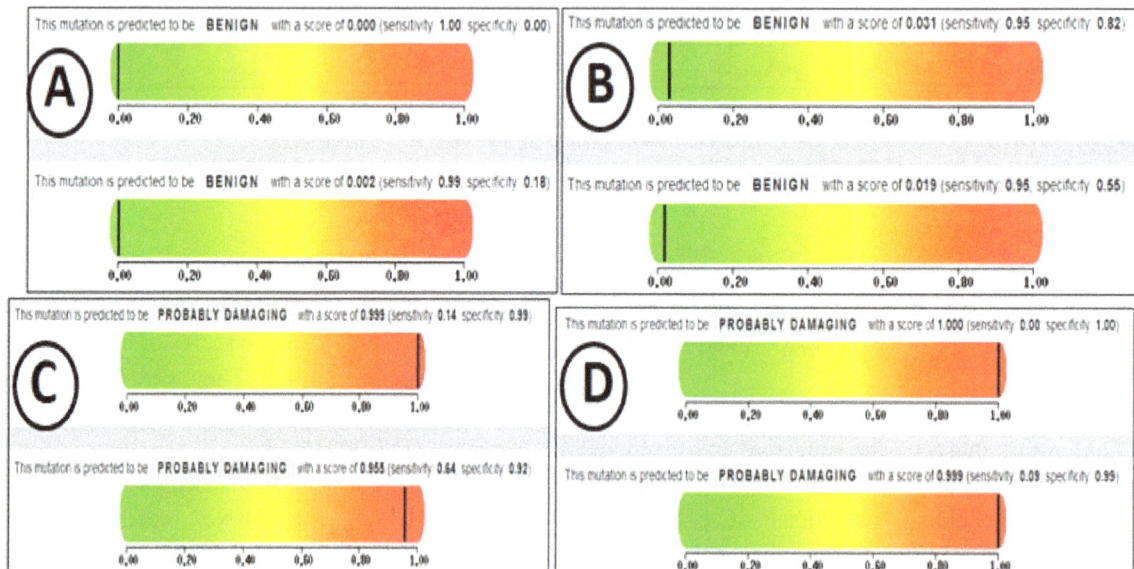

Figure 4 Impact of amino acid substitution on the structure and function of proteins
A) Impact of amino acid substitution (G1/p.Arg87His and G4/p.Glu241Lys) on GDF9
B) Impact of amino acid substitution (G6/p.Val332Ile) on GDF9
C) Impact of amino acid substitution (G7/p.Val371Met) on GDF9
D) Impact of amino acid substitution (G8/p.Ser315Phe) on GDF9

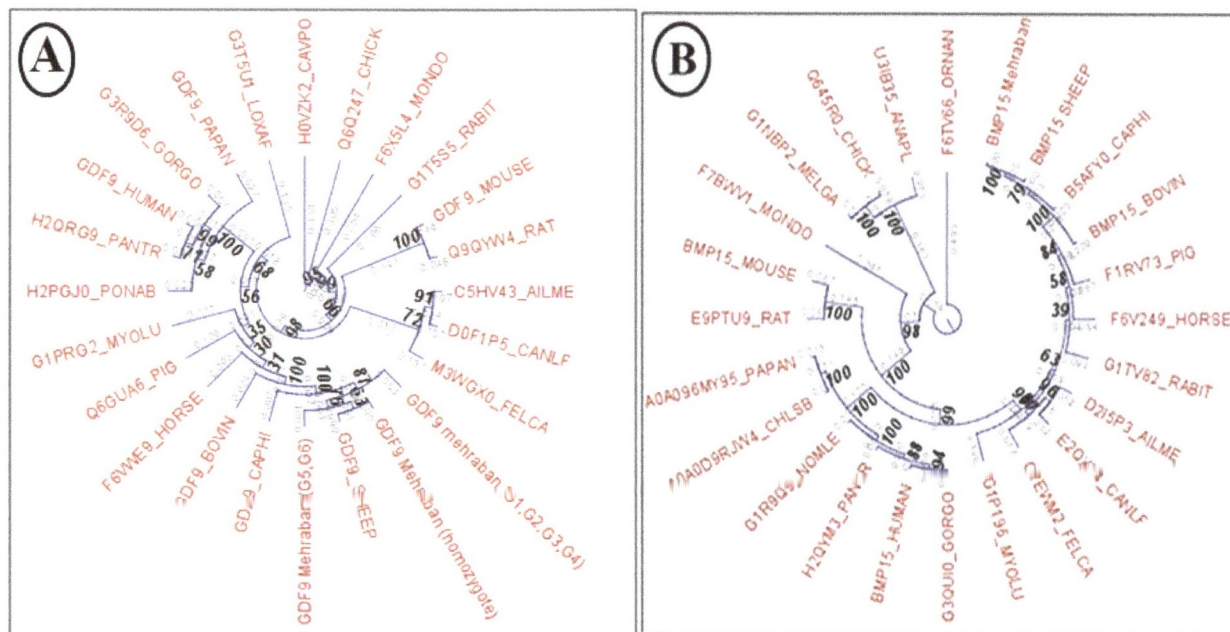

Figure 5 Molecular phylogenetic by maximum likelihood method
The tree is drawn to scale, with branch lengths (gray notes) and bootstrap values (bold and italic notes) measured in the number of substitutions per site
A) Circular cladogram of GDF9 protein in mono and poly ovulation species
B) Circular cladogram of BMP15 protein in mono and poly ovulation species

As well as, phylogenetic tree of *BMP15* in different species showed that Mehraban breed and Texel breed (Q9MZE2), human (O95972) and western lowland gorilla (G3QUI0), rat (E9PTU9) and mouse (Q9Z0L4), dog (E2QX74) and giant panda (D215P3), olive baboon (A0A096MY95) and green monkey (A0A0D9RJW4), chicken (Q645R0) and common turkey (G1NBP2) were grouped together (Figure 5B). Therefore, evolutionary origin of GDF9 and BMP15 proteins represents mono- or poly ovulatory species are not affected by protein sequence.

Functional effect of protein mutations is specified by the diversity of their impact on molecular function (Reva *et al.* 2011). GDF9 and BMP15 have the main role in the process of follicular development and oocyte maturation (Våge *et al.* 2013). It is reported eight point mutations (G1, G2, G3, G4, G5, G6, G7 and G8) occurring in exons 1 and 2 of *GDF9* gene in Cambridge/Belclare sheep, while only five mutations including G1/p.Arg87His, G4/p.Glu241Lys, G6/p.Val332Ile, G7/p.Val371Met and G8/p.Ser315Phe, led to amino acid substitution (Hanrahan *et al.* 2004). Following to study of Hanrahan *et al.* (2004), G1 was never associated to prolificacy, neither in the Cambridge/Belclare breeds nor in other breeds (Hanrahan *et al.* 2004). Anyway the role of G1 polymorphism may be retained unknown G1 polymorphism was roughly illustrated in some of the Iranian native breeds such as Mehraban (present study and Abdoli *et al.* (2013)), Baluchi (Moradband *et al.* 2011), Moghani and Farahani (Potki *et al.* 2015), Lori (Zamani *et*

al. 2015), Afshari and Shal (Eghbalsaied *et al.* 2014), Moghani and Ghezel (Barzegari *et al.* 2010). But there are conflicting results on significant association of G1 with prolificacy (Barzegari *et al.* 2010; Javanmard *et al.* 2011; Moradband *et al.* 2011; Eghbalsaied *et al.* 2014; Zamani *et al.* 2015; Abdoli *et al.* 2013), Eghbalsaied *et al.* (2014) indicated the existence of simultaneous/dual mutation (G1 and G4) in Afshari and Shal sheep breeds. Similar observations have been reported in Davisdale flock (Juengel *et al.* 2004). Among all the polymorphisms identified in *GDF9*, only five mutations that are FecG[H] (also known as G8 in Hanrahan *et al.* 2004), FecG[T] in Icelandic Thoka sheep (Nicol *et al.* 2009), FecG[E] in Brazilian Santa Ines (Silva *et al.* 2011), FecG[NW] (also known as G7) in White Norwegian sheep (Vage *et al.* 2013) and FecG[V] in Brazilian Île-de-France (Souza *et al.* 2014) are proven to be associated with an altered activity of GDF9 that affect the ovarian function in European sheep breeds. Moreover, the *BMP15* gene regulates proliferation and granulosa cells differentiation and promoting ligand expression, which plays an important role in female fertility in mammals. Also as an important role of *BMP15* in early follicle growth is relating to mono- and poly-ovulatory animals (Moore and Shimasaki, 2005).

Although any point mutations were not identified in *BMP15* gene for Mehraban breed in the present study, but several missenses in *BMP15* gene with positive effect on reproduction traits, have been identified in different sheep breeds (Galloway *et al.* 2000; Bodin *et al.* 2002; Hanrahan *et al.* 2004; Monteagudo *et al.* 2009). Phenotypically changes in ovarian function due to alterations in *BMP15* and *GDF9* genes emerge to differ among species and may be relevant to their mono- or poly ovulating status (Monestier *et al.* 2014). Screening of G8 mutation (*FecG^H*) in GDF9 and Booroola SNP in *BMPR*-1B in various Iranian sheep breeds including Shal, Ghezel, Baluchi and Sangsari, did not indicated the presence of these SNPs (Akbarpour *et al.* 2008; Ghaffari *et al.* 2009; Moradband *et al.* 2011; Kasiriyan *et al.* 2011). The ovulation rate observed in double heterozygotes (*FecX^G/FecX$^+$* and *FecG^H/FecG$^+$*) reflects an essentially additive effect of these mutations on prolificacy. However, homozygous carriers of *FecG^H* or of either of the mutations in *BMP15* and ewes with the genotype *FecX^G/FecX^B*, are sterile due to arrested follicle development (Mullen *et al.* 2013).

Without any functional testing, we cannot conclude that the observed mutations of GDF9 in Mehraban sheep are associated with ovarian function. The direct prediction of a mutation's impact on molecular function based on first principles is currently impossible for a number of reasons: e.g. lack of data (3D structures and complexes) and lack of accurate and efficient approaches for de novo modeling of protein structure and function on the molecular level. However, evolutionary analysis does provide a powerful tool, as natural selection of a particular sequence variant by definition reflects the aggregate effect of molecular changes on cell, tissue and organ physiology (Reva *et al.* 2011). In spite impact of mutation on performance is proved the based on association analysis with traits, but it must be proven through biological models as well. Demars *et al.* (2013), have been done a genome-wide association studies (GWAS) to identify genetic variants responsible for the highly prolific phenotype in two sheep flocks of Grivette and Olkuska, also they have done functional analyses based on altered the protein signaling activity *in vitro*. They identified two novel non-conservative *BMP15* mutations, that both mutations altered the BMP15 signaling activity, suggesting a novel kind of *BMP15* variant responsible for an atypical high prolificacy, in contrast to all other *BMP15* variants described so far (Demars *et al.* 2013). Therefore, to acknowledge effect of mutations on protein functionality, it should be proven as statistical model with followed up by genotyping a large number of ewes with known reproductive performance, bioinformatics model (mathematical algorithms) and biological model (*in vitro* functional assay).

Despite the identified mutations in current article were benign at alone. But the based on protein modeling, simultaneous mutations in genetic situations of Mehraban sheep will be able to create a particular conformation for protein structure and synergistic interactions with TGFβ superfamily such as *BMP* families.

CONCLUSION

Missense mutation is a kind of point mutation that changes a codon to indicate a different amino acid. This may or may not affect protein function, depending on whether the change is conservative or non-conservative, and what the amino acid actually does. According to sequencing results of Mehraban flock, point mutations of *GDF9* are surely in complete linkage disequilibrium. Although neither function wasn't found for identified point mutations of *GDF9* in Mehraban sheep, but we illustrated those four individuals are haplotype A (simultaneous mutations of G1, G2, G3 and G4), one animal was haplotype B (simultaneous mutation of G5 and G6) and seven of them were at wild haplotype (without mutation). Thus we deciphered a perfect correlation between G1/p.Arg87His and G4/p.Glu241Lys in Mehraban sheep. It was verified that G7/p.Val371Met and G8/p.Ser315Phe are probably damaging. However, we haven't seen any variants of G7 and G8 in Mehraban sheep. Protein modeling of GDF9 at haplotype A and haplotype B individuals had revealed a synergism interaction with TGFβ superfamily. Identified missenses with impressed on turns and helixes of conservative regions of GDF9, caused a synergism to create effects on functions at molecular level.

ACKNOWLEDGEMENT

We would greatly appreciate from Dr Stéphane Fabre, Ph D –HDR, I.N.R.A Centre de recherche de Toulouse, France due to technical assistance, provided reagents or equipment and sequencing of DNA samples without any charges reception. As well as we thank Julien Sarry and Florent Woloszyn for their assistances in the lab.

REFERENCES

Abdoli R., Zamani P., Deljou A. and Rezvan H. (2013). Association of *BMPR1B* and *GDF9* genes polymorphisms and secondary protein structure changes with reproduction traits in Mehraban ewes. *Genetic.* **524,** 296-303.

Adzhubei I.A., Schmidt S., Peshkin L., Ramensky V.E., Gerasimova A., Bork P., Kondrashov A.S. and Sunyaev S.R. (2010). A method and server for predicting damaging missense mutations. *Nat. Methods.* **7,** 248-249.

Akbarpour M., Houshmand M., Ghorashi A. and Hayatgheybi H. (2008). Screening for *FecG^H* mutation of growth differentiation factor 9 gene in Iranian Ghezel sheep population. *Int. J. Fertil. Steril.* **2,** 139-144.

Barzegari A., Atashpaz S., Ghabili K., Nemati Z., Rustaei M. and Azarbaijani R. (2010). Polymorphisms in *GDF9* and *BMP15* associated with fertility and ovulation rate in Moghani and Ghezel sheep in Iran. *Reprod. Domest. Anim.* **45,** 666-669.

Bodin L., Di Pasquale E., Fabre S., Bontoux M., Monget P., Persani L. and Mulsant P. (2007). A novel mutation in the bone morphogenetic protein 15 gene causing defective protein secretion is associated with both increased ovulation rate and sterility in Lacaune sheep. *Endocrinology.* **148,** 393-400.

Bodin L., SanCristobal M., Lecerf F., Mulsant P., Bibe B., Lajous D., Belloc J.P., Eychenne F., Amigues Y. and Elsen J.M. (2002). Segregation of a major gene influencing ovulation in progeny of Lacaune meat sheep. *Genet. Sel. Evol.* **34,** 447-464.

Bradford G.E., Quirke J.F., Sitorus P., Inounu I., Tiesnamurti B., Bell F.L., Fletcher I.C. and Torell D.T. (1986). Reproduction in Javanese sheep: evidence for a gene with large effect on ovulation rate and litter size. *J. Anim. Sci.* **63,** 418-431.

Demars J., Fabre S., Sarry J., Rossetti R., Gilbert H., Persani L., Tosser-Klopp G., Mulsant P., Nowak Z., Drobik W., Martyniuk E. and Bodin L. (2013). Genome-wide association studies identify two novel BMP15 mutations responsible for an atypical hyperprolificacy phenotype in sheep. *PLoS Genet.* **9,** 1003482.

Drouilhet L., Lecerf F., Bodin L., Fabre S. and Mulsant P. (2009). Fine mapping of the *FecL* locus influencing prolificacy in Lacaune sheep. *Anim. Genet.* **40,** 804-812.

Drouilhet L., Mansanet C., Sarry J., Tabet K., Bardou P., Woloszyn F., Lluch J., Harichaux G., Viguié C., Monniaux D., Bodin L., Mulsant P. and Fabre S. (2013). The highly prolific phenotype of Lacaune sheep is associated with an ectopic expression of the *B4GALNT2* gene within the ovary. *PLoS Genet.* **9,** 1003809.

Eghbalsaied S., Amini H., Shahmoradi S. and Farahi M. (2014). Simultaneous presence of G1 and G4 mutations in growth differentiation factor 9 gene of Iranian sheep. *Iranian J. Appl. Anim. Sci.* **4,** 781-785.

Galloway S.M., McNatty K.P., Cambridge L.M., Laitinen M.P., Juengel J.L., Jokiranta T.S., McLaren R.J., Luiro K., Dodds K.G., Montgomery G.W., Beattie A.E., Davis G.H. and Ritvos O. (2000). Mutations in an oocyte-derived growth factor gene (*BMP15*) cause increased ovulation rate and infertility in a dosage-sensitive manner. *Nat. Genet.* **25,** 279-283.

Ghaffari M., Nejati-Javaremi A. and Rahimi G. (2009). Detection of polymorphism in *BMPR-IB* gene associated with twining in Shal sheep using PCR-RFLP method. *Int. J. Agric. Biol.* **11,** 97-99.

Hanrahan J.P., Gregan S.M., Mulsant P., Mullen M., Davis G.H., Powell R. and Galloway S.M. (2004). Mutations in the genes for oocyte-derived growth factors *GDF9* and *BMP15* are associated with both increased ovulation rate and sterility in Cambridge and Belclare sheep (*Ovis aries*). *Biol. Reprod.* **70,** 900-909.

Javanmard A., Azadzadeh N. and Esmailizadeh A.K. (2011). Mutations in bone morphogenetic protein 15 and growth differentiation factor 9 genes are associated with increased litter size in fat-tailed sheep breeds. *Vet. Res. Commun.* **35,** 157-167.

Jones D.T., Taylor W.R. and Thornton J.M. (1992). The rapid generation of mutation data matrices from protein sequences. *Comput. Appl. Biosci.* **8,** 275-282.

Juengel J.L., Bodensteiner K.J., Heath D.A., Hudson N.L., Moeller C.L., Smith P., Galloway S.M., Davis G.H., Sawyer H.R. and McNatty K.P. (2004). Physiology of GDF9 and BMP15

signalling molecules. *Anim. Reprod. Sci.* **83,** 447-460.

Kasiriyan M.M., Hafezian S.H. and Hassani N. (2011). Genetic polymorphism *BMP15* and *GDF9* genes in Sangsari sheep of Iran. *Int. J. Gen. Mol. Biol.* **3,** 31-34.

Mi L.Z., Brown C.T., Gao Y., Tian Y., Le V.Q., Walz T. and Springer T.A. (2015). Structure of bone morphogenetic protein 9 procomplex. *Proc. Natl. Acad. Sci.* **112,** 3710-3721.

Monestier O., Servin B., Auclair S., Bourquard T., Poupon A., Pascal G. and Fabre S. (2014). Evolutionary origin of bone morphogenetic protein 15 and growth and differentiation factor 9 and differential selective pressure between mono and polyovulating species. *Biol. Reprod.* **91,** 83-91.

Monteagudo L.V., Ponz R., Tejedor M.T., Lavina A. and Sierra I. (2009). A 17 bp deletion in the bone morphogenetic protein 15 (*BMP15*) gene is associated to increased prolificacy in the Rasa Aragonesa sheep breed. *Anim. Reprod. Sci.* **110,** 139-146.

Moore R.K. and Shimasaki S. (2005). Molecular biology and physiological role of the oocyte factor, *BMP15*. *Mol. Cell. Endocrinol.* **234,** 67-73.

Moradband F., Rahimi G. and Gholizadeh M. (2011). Association of polymorphism in fecundity genes of *GDF9*, *BMP15* and *BMPR1B* with litter size in Iranian Baluchi sheep. *Asian Australas. J. Anim. Sci.* **24,** 1179-1183.

Mullen M.P., Hanrahan J.P., Howard D.J. and Powell R. (2013). Investigation of prolific sheep from UK and Ireland for evidence on origin of the mutations in *BMP15* (*FecX^G, FecX^B*) and *GDF9* (*FecG^H*) in Belclare and Cambridge sheep. *PLoS One.* **8,** 53172.

Nicol L., Bishop S.C., Pong-Wong R., Bendixen C., Holm L.E., Rhind S.M. and McNeilly S.A. (2009). Homozygosity for a single base-pair mutation in the oocyte specific *GDF9* gene results in sterility in Thoka sheep. *Reproduction.* **138,** 921-933.

Notter D.R. (2008). Genetic aspects of reproduction in sheep. *Reprod. Domest. Anim.* **43,** 122-128.

Potki P., Mirhoseini S.Z., Afraz F. and Vahidi S.M.F. (2015). Study of polymorphism in *GDF9* gene in Moghani and Farahani sheep breeds using PCR-RFLP technique. Pp. 24-26 in Proc. 1st Int. and 9th Natl. Biotechnol. Congr. Islamic Repub. Isfahan, Iran.

Pramod K.R., Sharma S.K., Kumar R. and Rajan A. (2013). Genetics of ovulation rate in farm animals. *Vet. World.* **6,** 833-838.

Reva B., Antipin Y. and Sander C. (2011). Predicting the functional impact of protein mutations: application to cancer genomics. *Nucleic. Acids. Res.* **39,** 118.

Silva B.D., Castro E.A., Souza C.J., Paiva S.R., Sartori R., Franco M.M., Azevedo H.C., Silva T.A.S.N., Vieira A.M.C., Neves J.P. and Melo E.O. (2011). A new polymorphism in the growth and differentiation factor 9 (*GDF9*) gene is associated with increased ovulation rate and prolificacy in homozygous sheep. *Anim. Genet.* **42,** 89-92.

Souza C.J.H., McNeilly A.S., Benavides M.V., Melo E.O. and Moraes J.C.F. (2014). Mutation in the protease cleavage site of *GDF9* increases ovulation rate and litter size in heterozygous ewes and causes infertility in homozygous ewes. *Anim.*

Genet. **45,** 732-739.

Vage D.I., Husdal M., Matthew P.K., Klemetsdal G. and Boman I.A. (2013). A missense mutation in growth differentiation factor 9 (*GDF9*) is strongly associated with litter size in sheep. *BMC Genet.* **14,** 1-9.

Wilson T., Wu X.Y., Juengel J.L., Ross I.K., Lumsden J.M., Lord E.A., Dodds K.G., Walling G.A., McEwan J.C. and O'Connell A.R. (2001). Highly prolific Booroola sheep have a mutation in the intracellular kinase domain of bone morphogenetic protein IB receptor (ALK-6) that is expressed in both oocytes and granulosa cells. *Biol. Reprod.* **64,** 1225-1235.

Zamani P., Abdoli R., Deljou A. and Rezvan H. (2015). Polymorphism and bioinformatics analysis of growth differentiation factor 9 gene in Lori sheep. *Ann. Anim. Sci.* **15,** 337-348.

9

The Expression of Myogenin and Myostatin Genes in Baluchi Sheep

K. Forutan[1], M. Amin Afshar[1*], K. Zargari[2], M. Chamani[1] and N. Emam Jome Kashan[1]

[1] Department of Animal Science, Science and Research Branch, Islamic Azad University, Tehran, Iran
[2] Department of Agronomy, Varamin-Pishva Branch, Islamic Azad University, Varamin, Iran

*Correspondence. E mail. aminafshar@gmail.com

ABSTRACT

Myogenin gene (MYoG) affects the synthesis of muscle myofibrillar growth and increase of meat production. The myostatin (MSTN) gene is identified as a specific negative regulator of skeletal muscle growth. Reduction of the expression level of MSTN through mutation in the sequence of this gene leads to an increase of myogenesis and regeneration of muscle cells during the postnatal growing period of sheep. The Baluchi sheep are among the most popular breeds of sheep for breeding in Iran and have an important portion in meat production industry of the country. In present work the relative expression level of the two candidate genes have been studied at two age intervals (9 and 12 months) in male and female Baluchi sheep. In order to analyze the relative expression level of MYoG and MSTN gene in Baluchi sheep's longissimus muscles, quantitative real-time polymerase chain reaction (qRT-PCR) reaction has been applied. Results of RT-PCR for sex effect showed MSTN and MYoG expression were not highly expressed in ram' longissimus dorsi compared to ewe at the same age stages (P>0.05) and there were no significant differences between male 12 months comparing to female 12 months (P>0.05) and male 9 months with female 9 months (P>0.05). For the effect of age, relative expression of MYoG and MSTN genes on Baluchi sheep longissimus muscles did not show significant differences between males or females (P>0.05).

KEY WORDS Baluchi, gene expression, myogenin (MYoG), myostatin (MSTN), RT-PCR, sheep.

INTRODUCTION

The Baluchi sheep is one of the popular breeds in eastern Iran and southwest Pakistan and has an important portion in meat production industry of Iran with a body size varies between 35-38 kg in adult ewes and 45-48 kg in rams (Yazdi *et al.* 1997). The mayogenin gene (MYoG) is a positive regulator and a member of myogenic differentiation (MYoD) genes (MYoG, MYF-5, MYF-6 and MYoD1). The MYoG is known as myogenic factor 4 and has direct effect on myogenesis in skeletal muscle myofibril. MYoG is effective on synthesis and myogenesis of skeletal muscles's myofibrils (Zhihong *et al.* 2009; Femanda *et al.* 2013). MYoG is closely associated with the number of muscle fibers at birth time, which is most important in determination of maximal lean meat growth capacity in pigs (Handel and Stickland, 1988). The myogenesis synthesizes myofibrillar proteins in the skeletal muscles and regulates the number of their myofibers. It has positive effect on meat percentage and number of myofibrils (Sun *et al.* 2012). The myostatin (MSTN) gene has negative direct effect on development and regeneration of the skeletal muscles. The gene is known as a negative regulator. The MSTN protein is one of the transforming growth factor-β (TGF-β) superfamily that includes a group of development, differentiation and growth factors. The gene is an important inhibitor on the embryonic development and meat percentage (Lin *et al.* 2002).

MSTN involved in early birth, maturation age for improvement and development of skeletal muscles. MSTN gene also known as growth / differentiation factor 8. Myostatin in early birth, growth and maturation age for improvement many of muscles (Florent *et al.* 2007). The gene production (i.e. MSTN protein) inhibits growth of the animal and mediates the formation of muscle fiber. It worth mentioning that feeding and situation of rearing have direct effect on the gene expression at growing stages (Yingying *et al.* 2015). During the life of the animal the gene can be active or deactive (Kobolok *et al.* 2002). Therefore, this study was aimed to investigate the relative expression level of MSTN and MYoG genes in male and female Baluchi sheep's longissimus muscles at two age intervals (9 and 12 months) .

MATERIALS AND METHODS

Animals and tissue collection

A number of 18 (9 male and 9 female) nine months old Baluchi lambs with initial body in rams (29±0.21 kg) and female (21±0.35 kg) reared for 3 months with the same conditions. They were fed adequately and their diet were assigned with vital nutrients and important supplements for growing period (Sun *et al.* 2012). Samples were taken from the dorsi muscles tissues of the lambs at the start and end of experimental period. The animals were locally anesthetized by the injection of 0.06 mg/kg of Xylazine and then 2 cm incision has been made at the point of injection to separate skin from meat tissues. All tissue samples were immediately frozen in liquid nitrogen tank and stored until RNA isolation (Sun *et al.* 2012).

RNA isolation and quantitative real-time PCR

Total RNA was extracted from muscle tissue using Trizol reagent according to Trizol Regent Kit instructions (Qiagen, TRI Reagent, RNeasy Plus Mini Kit). The concentration of total RNA isolated was determined by absorbance at 260 nm and purity (A260/A280) of > 1.8. The first strand complementary DNA (cDNA) was synthetized from 5 µg of total RNA using Cinnagen reverse transcript kit (Cinnagen, Tehran, Iran). After reverse transcription analysis of gene, expression of myostatin and myogenin were performed through qRT-PCR reaction by making use of Eppendrof apparatus (Applied Biosystem Inc). Total volume of PCR reaction system was 20 µL including 10 PCR-master mix (2X) of Cinnagen, 1 µL cDNA, 1 µL primer (10 pmol) and 8 µL dH$_2$O. PCR condition for myogenin and myostatin genes were as following: The real-time PCR reaction was carried out for 1 cycle 15 min at 95 °C, 40 cycle 25 sec at 95 °C and 40 cycle 60 sec at 62 °C. A housekeeping gene RPL19 was used as normalizing control. The primers were design by primer express software (Applied

Biosystems) (Table 1). After amplification, 6 µL of each PCR product was analyzed on agarose gel electrophoresis 2% and vitalized with red gel (Sigma) under UV light (Figure 1).

Statistical analysis

Difference in relative gene expression levels were analyzed using the Ct value and standard error between samples (Joshua *et al.* 2006). Myogenin and myostatin genes were analyzed based on ∆Ct as Ct (MYoG/MSTN gene)-Ct (RPL19 gene). Gene expression data were expressed as means ± SE.

Two-factor analysis of variance ANOVA (P<0.05) was used to measure the interaction between gender (∆Ct 12 month female–∆Ct 9 month female), (∆Ct 12 month male–∆Ct 9 month male) (Figure 2) and various developmental stages after 6 months (∆Ct male–∆Ct female), (∆Ct 12 mo male–Ct 12 mo female), (∆Ct 9 mo male–∆ Ct 9 mo female) and no significant interactions were observed in MYoG and MSTN (Figure 3).

Myostatin expression

The expression of the MSTN gene in Baluchi sheep showed no significant difference between males and females at total number of animals (P>0.05), 9 months of age and 12 months of age (Figure 3) and showed no significant difference between male animals and between female ones at 12 months age (P>0.05), (Table 2, Figure 2).

This shows that MSTN gene expression in male ram's longissimus muscles was not higher than the females at two ages (Figure 3). In addition, 12 month old males showed no significant differences when compared with 12 months old females (P>0.05).

Myogenin expression

The expression of the MYOG gene in Baluchi sheep showed no significant difference between males and females at the total animals (P>0.05), 9 mo of age and 12 mo of age (Figure 3) and no difference between male and female at 12 mo of age (P>0.05), (Tables 3 and 4, Figure 2).

RESULTS AND DISCUSSION

The myostatin gene is a regulator factor in the muscle that causes to achieve highest amount of muscle mass. The gene is expressed in many tissues. If the MSTN gene undergoes mutation, the negative regulating function of the gene does not work (Kobolok *et al.* 2002). Myogenin (MYoG) is a transcription factor that has direct effect on skeletal muscle which means that with expressed MYoG gene the muscle mass increases. The variations of MYoG gene may be relative to myogenesis process and cause variations in muscle quality (meat percentage).

Table 1 Sequence of real-time PCR primers used in this study

Gene	Gen bank accession No.	Forward primer	Reverse primer	Product size (bp)
Myogenin	443185	GGAGAAGCGCAGACTCAAGAAG	CTATGGGAGCTGCATTCACTGG	231
Myostatin	443449	ATCCGATCTCTGAAACTTGACAT	AGTCCTTCTTCTCCTGGTTCTG	182
RPL19	100270789	AGCCTGTGACTGTCCATTCC	ACGTTACCTTCTCGGGCATT	126

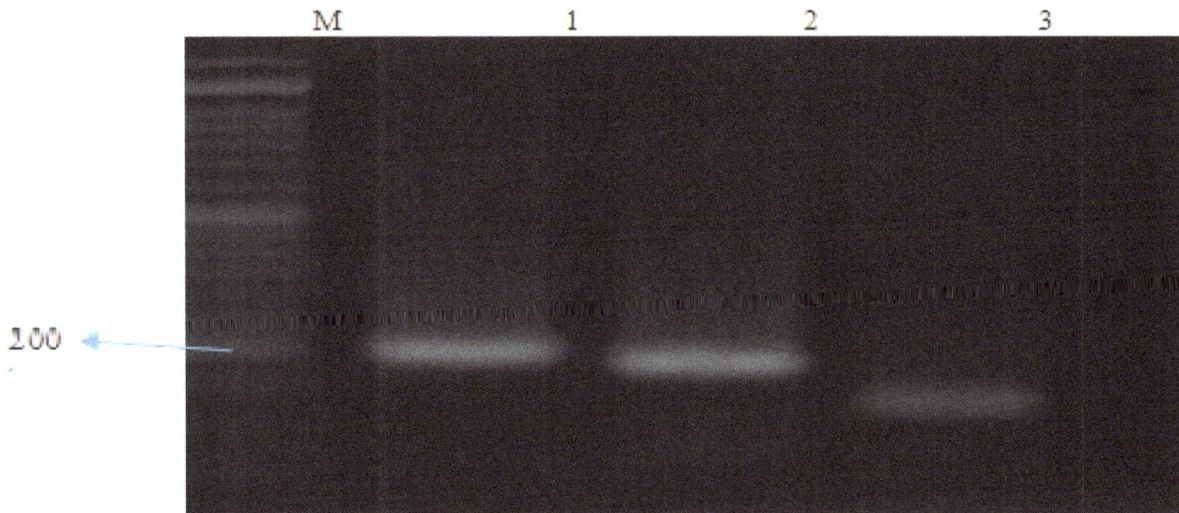

Figure 1 Agarose gel electrophoresis of 1) MSTN, 2) MYoG and 3) RPL19 gene
M: molecular ladder

Table 2 Variance analysis of MSTN expression among different developmental stages in the same sex ($2^{-\Delta\Delta Ct}$)

Gene	Gender	Total	9 mounts old	12 months old
MSTN	Female	1	1	1
	Male	0.41±0.17	0.42±0.08	0.63±0.41

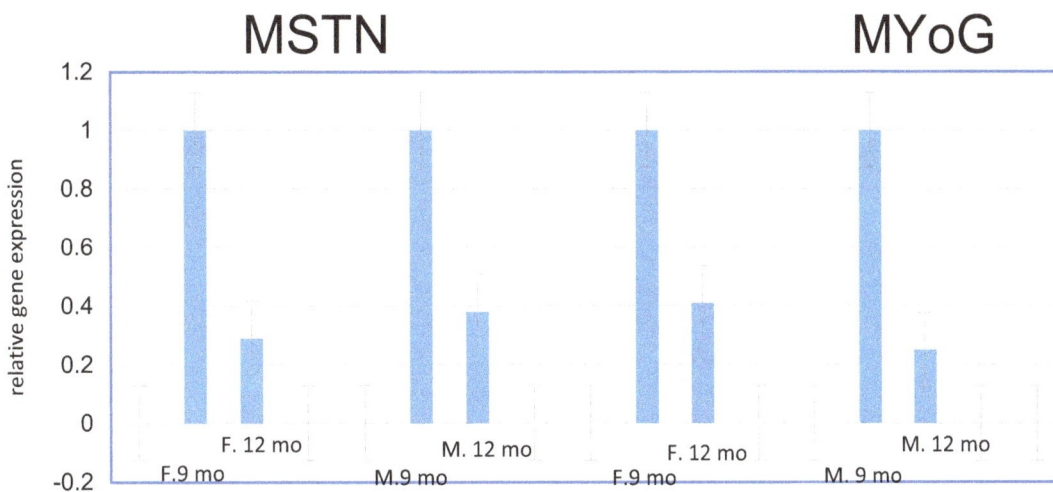

Figure 2 MYoG and MSTN genes expression in 12 month of age of Baluchi sheep in male and female sex ($2^{-\Delta\Delta Ct}$) method, 9 months for the control group)

There was no significant difference (P>0.05) between males and females at two growth stages (9 and 12 months old) in expression of the MYoG gene.

Yang *et al.* (2006) determined the developmental changes of MSTN and MYoG genes expression in longissimus dorsi muscle of Erhualian and Large white pigs.

Table 3 Variance analysis of MYoG expression among different developmental stages in the same sex ($2^{-\Delta\Delta Ct}$ method)

Gene	Gender	Total	9 mounts old	12 months old
MYoG	Female	1^m	1^m	1^m
	Male	0.86 ± 0.16	0.51 ± 0.587	0.11 ± 0.07

Table 4 Variance analysis of MSTN and MyoG expression among different developmental stages in the same sex ($2^{-\Delta\Delta Ct}$ method)

Gene	Gender	9 months old	12 months old	Gene	Gender	9 months old	12 months old
MSTN	Male	1	0.38 ± 0.09	MYoG	Male	1	0.25 ± 0.01
	Female	1	0.29 ± 0.04		Female	1	0.41 ± 0.02

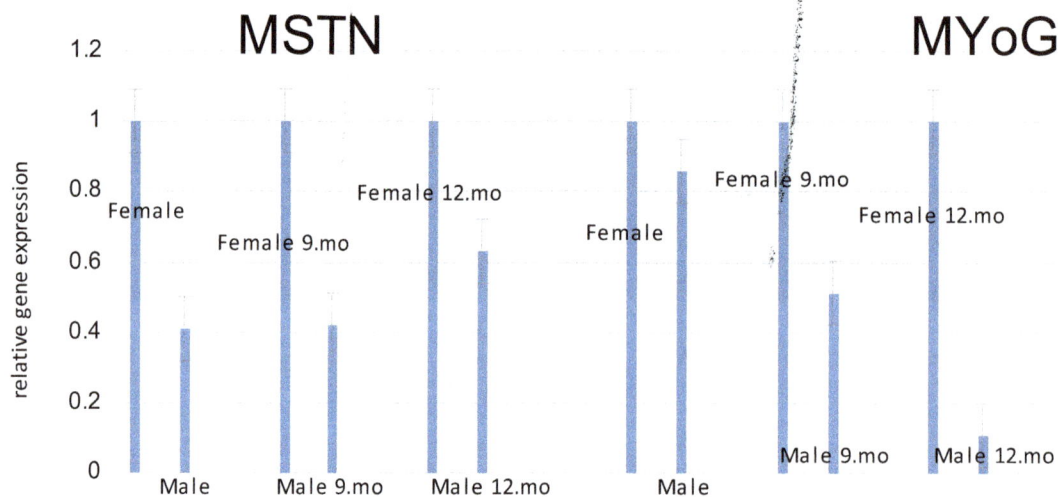

Figure 3 MYoG and MSTN genes expression in males of Baluchi sheep in the same growth stage (12 and 9mo) and total population ($2^{-\Delta\Delta Ct}$ method, ewe for the control group)

They indicated that risen level of myogenin and myostatin genes expression might have important regulation's effect on maturation of myofibrils during postnatal stage and no observed meat percentage with changing age. In the present study, the results for MSTN and MYoG genes expression is similar to Yang *et al.* (2006). Hasty *et al.* (1993) reported that the expression of the MYoG gene has negative relation with lean meat percentage and age in Landrace pigs and breed had no effect on the result and the result of (Sun *et al.* 2012) showed that the expression of MYoG in male sheep' longissimus dorsi muscle was non-significantly (P>0.05) higher than female after birth. Both of these results are similar to our results about comparison of MYoG expression for between 12 and 9 mo of age at the same sex (P>0.05). The variation in MSTN and MYoG mRNA levels were related neither to age nor to age of the animals within the ranges covered in the study. In the present study, with the comparison of the male and female, myostatin and myogenin expression were not highly expressed in ram' longissimus dorsi compared to ewe at different growth stages after 3 months, it seems sex and age has no direct effect on MSTN and MYoG expression in this breed.

Shibata *et al.* (2006) showed that myostatin gene could increase adipogenesis in muscle marbling, and that the myostatin and MRF genes might have an effect at an early stage of skeletal muscle regeneration. Extremely significant (P<0.01) or significant (0.01<P< 0.05) differences were observed between males and females at the same growth stage in Erhualian pigs and Hu sheep (Yang *et al.* 2006; Sun *et al.* 2012). (Shan *et al.* 2009) reported the MYoG expression in longissimus dorsi muscle (LDM) of Jinhua pig and Landrace pig (35, 80, 125 days of age) and determined that with an increase in the age of pigs the expression of the myogenin gene increased and caused to increase lean meat percentage, the expression of myogenin and meat percentage had positive correlation, in contrast the expression of MYoG gene with an increase in age decreased and lean meat percentage in Landrace pigs and observed negative correlation in the expression of MYoG and muscle percentage in Landrace breed. Results of MYoG expression in longissimus dorsi muscle of Baluchi sheep breed were different from MYoG mRNA expression LDM Jinhua pig. (Su *et al.* 2014) determined association between IGF-I and DLK1 gene expression and meat quality in Hu sheep.

Results showed that just growth stage have significantly affected IGF-I and DLK1 expression (P<0.01) while sex had no effect on meat quality. (Su *et al*. 2014) suggested that in Hu sheep different growth ages had significant effect on DLK1 and IGF-I relative gene expression in Hu sheep's muscle.

In another research (Sun *et al*. 2014) analyzed developmental changes in myogenin and IGF-I gene expression and their association with meat traits in Hu sheep. Compression expression of the myostatin gene in Hu sheep were significantly different between rams and ewes at the 2 day (P<0.05), 1 month (P<0.05) and 3 month (P<0.01) ages.

Myostatin gene expression in rams longissimus muscles was higher than that of ewes at all ages (P<0.05), except for the 3 month age (P 0.05) and there were no significant difference (P>0.05) between rams and ewes at any age in expression of the myogenin gene. The result of (Su *et al*. 2014) and (Sun *et al*. 2014) are different from our results about MSTN gene expression in Baluchi sheep and about myogenin gene expression, the result of (Sun *et al*. 2014) were similar .

CONCLUSION

The expression of the myogenin gene was generally equal in male Baluchi sheep at 12 and 9 months of age. Similar expression of the myostatin and myogenin in male and female Baluchi sheep at 12 mo of age, showed a new expression pattern of these genes between different sexes which may indicate the regulatory role of contribution between MSTN and MYoG gene expression on skeletal muscle formation. While in other researches MSTN and MYoG genes have positive correlation, and the expression of these two genes in myofibrils have direct effect on meat percentage, in our results about direct effect of expression of these genes on growth stages and sex of sheep, the differences might be related to the experimental situation and selection of ages and measurement methods may be the reason of these difference. We guess that the change in algorithms of body in higher age in adult sheep (9-12 mo) inclines to save fat in adipocyte instead of increasing myofibrils (Yingying *et al*. 2015). Many important genes have been discovered to be involved in controlling meat traits in livestock like MSTN and MYoG genes which can subsequently affect other gene expression and change their expression in comparison. These results about MSTN and MYoG gene expression of Baluchi sheep can provide a theoretical basis for further research and provide genetic data for Baluchi sheep. The results of present study can also be helpful to provide information for other genes that affect the breeding strategies for meat percentage and carcass traits.

ACKNOWLEDGEMENT

This study was funded by Science and Research Branch, Islamic Azad University, Tehran, Iran.

REFERENCES

Femanda A., Costa A., Cassiane M., Barbosa C.M., Aguiar A. and Maeli S. (2013). Morphometry and expression of MyoD and myogenin in white and red skeletal muscles of juvenile fish Colossoma macropomum (Cuvier 1818). *Acta Zool*. **95(4)**, 430-437.

Florent C., Carine M.P., Jean B. and Jean-Francois O. (2007). Target genes of myostatin loss-of-function in muscles of late bovine fetuses. *BMC Genomics*. **8**, 63-68.

Handel S.E. and Stickland N.C. (1988). Catch-up growth in pigs: A relationship with muscle cellularity. *Anim. Prod*. **44**, 311-317.

Hasty P., Bradley A., Morris J.H. and Edmondson D.G. (1993). Muscle deficiency and neonatal death in mice with a targeted mutation in the myogenin gene. *Nature*. **364**, 501-506.

Joshua S., Yuan A., Reed C., Neal C. and Stewart J. (2006). Statistical analysis of real-time PCR data. *BMC Bioinformatic*. **7**, 85-91.

Kobolok J. and Elen G. (2002). The role of the myostatin protein in meat quality: a review. *Arch. Tierz. Dummerstorf*. **45(2)**, 159-170.

Lin C.S., Wu Y.C., Sun Y.L. and Huang M.C. (2002). Postnatal expression of growth/differentiation factor-8 (GDF-8) gene in European and Asian pigs. Asian-Australas J. Anim. Sci. **9**, 1244-1249.

Shan L., Zhan G., Miao Z.M., Wan Z.G. and Xu R. (2009). Myogenin mRNA abundance in longissimus dorsi muscle of Jinhua pig and Landrace pig. *Chinese J. Vet. Sci*. **29(3)**, 374-377.

Shibata M., Matsumoto K., Aikawa K., Muramoto T., Fujimura S. and Kadowaki M. (2006). Gene expression of myostatin during development and regeneration of skeletal muscle in Japanese Black cattle. *J. Anim. Sci*. **84**, 2983-2989.

Su R., Sun W., Li D., Wang Q., Lv X., Musa H., Chen L., Zhang Y. and Wu W. (2014). Association between DLK1 and IGF-I gene expression and meat quality in sheep. *Genet. Mol. Res*. **13(4)**, 10308-10319.

Sun W., Li D., Wang P., Musa H., Ding J., Li B. and Ma Y. (2012). Postnatal expression of myostain (MSTN) and myogenin (MYoG) genes in Hu sheep of China. *African J. Biotechnol*. **11(58)**, 12246-12251.

Sun W., Su R., Li D., Musa H., Kong Y., Ding T., Ma Y., Chen L., Zhang Y. and Wu W. (2014). Developmental changes in IGF-I and MyoG gene expression and their association with meat traits in sheep. *Genet. Mol. Res*. **13(2)**, 2772-2783.

Yang X., Chen J., Xu Q. and Wei X. (2006). Development changes of myostatin and myogenin genes expression in longissimus dorsi muscle of Erhualian and Large White pigs. *J. Nanjing Agric. Univ*. **29**, 64-68.

Yazdi M., Engstrom G., Nasholm A., Johansson K., Jorjani H. and Liljedah L. (1997). Genetic parameters for lamb weight at different ages and wool production in Baluchi sheep. *J. Anim. Sci.* **65(2),** 247-255.

Yingying L., Fengna L., Lingyun H., Bie T., Jinping D., Xiangfeng K., Yinghui L., Meimei G., Yulong Y. and Guoyao W. (2015). Dietary protein intake affects expression of genes for lipid metabolism in porcine skeletal muscle in a genotype-dependent manner. *Br. J. Nutr.* **113(7),** 1069-1077.

Zhihong Y., Kyle L., Quarrie M. and Erwin A. (2009). From an arrested myoblast phase to a differentiated state MyoD and E-protein heterodimers switch rhabdomyosarcoma cells. *Gene. Dev.* **23,** 694-707.

10

Use of Microsatellite Polymorphisms in *Ovar-DRB1* Gene for Identifying Genetic Resistance in Fat-Tailed Ghezel Sheep to Gastrointestinal Nematodes

R. Hajializadeh Valilou[1]*, S.A. Rafat[1], M. Firouzamandi[2] and M. Ebrahimi[1]

[1] Department of Animal Science, Faculty of Agriculture, University of Tabriz, Tabriz, Iran
[2] Department of Pathobiology, Faculty of Veterinary Medicine, University of Tabriz, Tabriz, Iran

*Correspondence E-mail: hajializadeh20@gmail.com

ABSTRACT

This study was designed to identify genetically resistant animals to gastrointestinal nematode (GIN) infections using microsatellite polymorphisms of *Ovar-DRB1* gene in Iranian Ghezel sheep breed lambs. In the present study 120 male Ghezel lambs were at 4 to 6 months of age randomly selected from six different sheep flocks in East Azerbaijan province (n=20 per flock). These lambs were naturally infected with GINs, and individual fecal samples were collected twice with a week interval to evaluate fecal egg counts (FEC). Blood samples were also collected for DNA isolation and PCR was performed to amplify the second exon and microsatellites within the second intron of the *Ovar-DRB1* gene. The data were analyzed using a mixed model of SAS software. The present study identified 24 genotypes and 20 alleles on *Ovar-DRB1* gene. Results indicated that the presence of 510 bp (base pair) allele (called allele F) in both homozygote and heterozygote animals had a strong association (P<0.01) with lower FEC; while, presence of 506 bp allele (called allele E) in homozygote animals was significantly associated (P<0.01) with higher FEC. Thus, this study showed a strong association between microsatellite polymorphism of *Ovar-DRB1* gene and resistance to GIN infections in Ghezel sheep lambs.

KEY WORDS gastrointestinal nematodes, Ghezel sheep, microsatellite polymorphisms, *Ovar-DRB1* gene.

INTRODUCTION

Ghezel, an Iranian native fat-tailed sheep breed, is mainly inhabited in North West regions of Iran (east and west side of Azerbaijan provinces), (Baneh *et al.* 2010). Sheep rearing is based on pasture grazing; therefore, animals are very susceptible to the widespread gastrointestinal nematode (GIN) infections (Baneh *et al.* 2010). Thus, GIN infections pose a high economic loss in terms of medical expenses for the treatment, high mortality rate and low productivity in young lambs (Amarante *et al.* 2009). Although anthelmintic drugs are common in use for the treatment of internal parasite infections, development of anthelmintic resistance by

GINs worldwide and also in Iran, has driven the investigation toward finding a new strategy to overcome GIN infections in order to improve the economy of farmers in the area (Le Jambre, 1976; Hazelby *et al.* 1994; Gholamian *et al.* 2006). GIN infections vary among animals based on their genetic resistance. The genetically resistant animals to the infection have higher viability and productivity than the animals with lower degree of genetic resistance (Amarante *et al.* 2009). Accordingly, Matika *et al.* (2011) identified QTL regions for resistance to gastrointestinal nematode on ovine chromosomes (OAR) 3 and 14 in Suffolk and Texel sheep breeds using the multiplex microsatellite panels. Also, Salle *et al.* (2012) found four QTL regions on ovine chromosomes (OAR) including 5, 12, 13 and 21 which were responsible for resistance to *Haemonchus contortus* nematode in Romane × Martinik Black Belly backcross lambs. Other studies using the ovine 50 k SNP Chip identified association between QTLs on chromosomes 1, 6, 7 and 14 and resistance to gastrointestinal nematodes in Scottish Blackface, Romney, Perendale, Red Maasai × Dorper and Churra sheep breeds (Riggio *et al.* 2013; McRae *et al.* 2014; Benavides *et al.* 2015). The major histocompatibility complex (MHC) is a cell surface molecules which are encoded by a large gene family placed on chromosome 20 in sheep. There are two main groups of MHC gene family. MHC class I and class II. Ovar-DR gene is one of MHC class II genes and codes heterodimeric proteins (DRα and DRβ, each one of these proteins including α1, α2 and β1, β2 domains) in the membrane of B cells and antigen presenting cells such as macrophages, dendritic cells and Langerhans cells (Tizard, 2013). Charon *et al.* (2002) showed a relationship between microsatellite polymorphisms in the second intron of *MHC-DRB1* gene and resistance to *Teladorsagia circumcincta* in Polish Heath sheep. Dominik (2005) in a study indicated a list of QTLs and candidate genes for resistance to gastrointestinal nematodes in sheep. Furthermore, he found a strong relationship between MHC gene loci and genetic resistance to GINs. Among all the candidate genes (interferon γ, interleukin-4, MHC-I and MHC-II) responsible for GIN resistance, *Ovar-DRB1* gene (MHC class II) plays an important role in immunity to GINs and other internal parasites (Dominik, 2005). Several studies have proved the role of polymorphisms in *Ovar-DRB1* gene on resistance to numerous internal parasites and bacterial diseases in several sheep breeds (Hajializadeh *et al.* 2015; Schwaiger *et al.* 1995; Castillo *et al.* 2011; Larruskain *et al.* 2012; Shen *et al.* 2014). In our previous study, a significant association was found between gastrointestinal fecal egg counts (FEC) and the second exon of *Ovar-DRB1* gene using polymerase chain reaction-restriction fragment length polymorphism (PCR-RFLP), (Hajializadeh *et al.* 2015). Since PCR-RFLP technique has some limitations for identifying all variations within an amplified gene fragment (Ferreira *et al.* 2014), new simpler and more accurate technique is required. Simple tandem repeats (STRs) or microsatellites are repeated short sequence motifs (Zane *et al.* 2002) which are a very powerful technique for studying the genetic diversities within and between populations and pedigree analysis in farm animals and livestock breeding (Abdul Muneer, 2014).

Moreover, high rate of polymorphism and relatively simple scoring and data analysis, are important features that makes microsatellite markers of large interest for many genetic studies (Abdul Muneer, 2014). As selecting resistant animals is usually based on low FEC (as a phenotypic trait), thus can be used as a standard method for assessing the level of resistance to GIN (Eady *et al.* 2003). Based on the above information and as mechanisms responsible for genetic resistance to GIN infections are not totally understood and also because of environmental resemblance between natural challenges with gastrointestinal nematodes and actual rearing conditions of sheep in pastures, the present study was designed to detect resistant animals to GIN infections (based on natural challenges) using microsatellite polymorphisms in the *Ovar-DRB1* gene in Iranian Ghezel sheep lambs.

MATERIALS AND METHODS

Animals and sampling

This study was conducted during the period from May to June, 2014 when animals were potentially at the highest risk of becoming infested with GINs. For this reason, 120 male weaned Ghezel lambs were at 4 to 6 months of age randomly selected from lambs in six flocks (n=20 per flock) in east Azerbaijan province, Iran. All procedures used in this experiment were warranted based on University of Tabriz Animal Care and Ethics Committee. Before the start of experiment on selected lambs deworming procedure was performed to ensure elimination of all the parasites. After 28 days, lambs feces were examined to confirm their parasite-free conditions. Afterward, all selected lambs from six flocks were allowed 28 days grazing on contaminated pastures during which they did not receive any deworming drugs (in order to naturally parasitize animals). On day 31 post infection, individual blood and fecal samples were collected twice with one week interval. Blood samples were obtained from jugular vein using sterile vacuum tubes coated with anticoagulant (EDTA) and were frozen at -20 °C for future DNA isolation. Fecal samples were collected individually from the rectum of each lamb for determining fecal egg counts (FEC) using modified McMaster (Clayton Lane) method (Anonymous, 1977; Zajac and Conboy, 2012). During FEC evaluation, nematode eggs were classified into four major species as following: 1- Strongyles (including: *Haemonchus contortus*, *Teladorsagia circumcincta*, *Ostertagia occidentalis* and *Trichostrongylus axei*, *colubriformis*, *vitrinus* and *rugatus*), 2- *Nematodirus* spp., 3- *Trichuris ovis*. and 4- *Marshallagia marshalli*. The total summations of four parasite classes counted in each lamb's feces were reported as FEC.

Genotype and molecular analysis

Blood DNA was isolated using Samadi Shams *et al.* (2011) protocol. Polymerase chain reaction (PCR) was performed using PCR master mix kit (Ampliqon Company) in a T-personal thermo-cycler (Biometera Personal Cycler Version

3.26 co., Germany). The 25 µL PCR mixture contained: 50-100 ng of DNA, 2.5 µL of 10X PCR buffer (200 mM (NH$_4$) 2SO$_4$), 0.1 mM Tween 20%, 750 mM Tris-HCl (pH=8.8), 2.5 mM MgCl$_2$, 200 µM dNTPs and 3 µL mix of oligo nucleotides (10 p mol from each primer), 1U Taq DNA polymerase (Dream Taq polymerase, Ampliqon company) and 11 µL ddH$_2$O. Forward and reverse sequences of primers (Bioneer, Korea) used for amplification of the second exon and microsatellite sequences in the second intron of *Ovar-DRB1* gene included F: 5'-TCTCTGCAGCACATTTCCTGG-3' (Ammer *et al.* 1992) and R: 5'-CGTACCCAGAGTGAGTGAAGTATC-3' (Schwaiger *et al.* 1993). PCR program included 36 cycles of amplification which were as following: 1 cycle of initial denaturation at 94 °C for 5m, 35 cycles included denaturation at 94 °C for 30 s, annealing at 55 °C for 60 s, extension at 72 °C for 60 s; followed by termination at 72 °C for 5 m. Then, PCR products were electrophoresed at 1200 V with 25 mA for 35 min in 4% acryl amide gel using Gel-Scan™ 3000 automated DNA Sequencer on a Real-Time Gel system (Corbett Robotic co., Australia). Size of the PCR products was determined based on a 25 bp DNA standard ladder (Thermo Scientific). The UVIdoc software (version 99.02 for windows) was used to identify microsatellite alleles in the second intron of *DRB1* gene.

Statistical analysis

For data analysis, total nematode egg count was considered as FEC (fecal egg count). Then, FEC values were considered as a residual deviation of flock × time interactions. Also, the distribution of residuals was tested for skewness (ω) and kurtosis (κ). Normality test was performed for FEC; then, the data which were not normally distributed, were transformed using Box–Cox transformation [(FECλ-1)/λ], (Box and Cox, 1994). Optimum values of λ between the range of -2 and 2 were determined using a maximum-likelihood criterion (Draper and Smith, 1981) in trans regression procedure of SAS (2002). Blood samples of 120 lambs in six flocks were used for evaluating microsatellite polymorphisms of Ovar-DRB1. During analysis of association between each allele and FEC, genotypes with frequencies lower than 2% were deleted. The effects of each genotype and also each allele on transformed FEC were analyzed by the mixed model (repeated measures analysis of variance) of SAS software (Hajializadeh *et al.* 2015; Castillo *et al.* 2011). The statistical model is described as:

$$Y_{ijklm} = \mu + Flock_i + Genotype_j + Time_k + Lamb_l (Flock_i \times Genotype_j) + e_{ijklm}$$

Where:

Y_{ijklm}: dependent variable.

μ: overall mean.

$Flock_i$: fixed effect of flocks (1, 2,...,6).

$Genotype_j$: fixed effect of genotypes (genotypes coded as 0, 1 and 2, based on number of copies in the determined alleles).

$Time_k$: fixed effect of times (sampling time 1 and 2).

$Lamb_l$: random effect of lamb (1, 2, 3,...,120).

$Flock_i \times Genotype_j$: interaction effect between Flock and Genotype.

e_{ijklm}: experimental error.

In the primary model, we considered the interaction effect between Flock and Time, but since this interaction was not significant, Flock × Time was deleted from the final statistical model presented in the above.

RESULTS AND DISCUSSION

The descriptive statistics regarding fecal egg count (FEC) of various classes of gastrointestinal nematodes from five flocks of Ghezel lambs are presented in Table 1. Based on these statistics, the prevalence of *Trichuris ovis* nematode in the investigated lambs was the lowest (6%); therefore, it was excluded from the data at the time of statistical analysis. On the other hand, *Nematodirus* nematode showed the highest prevalence in Ghezel lambs of six flocks (44.64%). Therefore, these results showed that pastures in east Azerbaijan province (Iran) had the highest level of contamination with *Nematodirus* nematode.

Analysis of microsatellite polymorphisms in Ovar-DRB1 showed 20 alleles and 24 genotypes. The frequency of each observed alleles are presented in Table 2. Based on these results, 10 genotypes had frequencies higher than 2% including: AA, BB, DD, EE, FF, HH, JJ, MM, OO and RR. Among these genotypes, F (20%), E (15%) and O (12.5%) alleles had the highest frequencies of microsatellite loci in *Ovar-DRB1* gene.

Results also showed an association between genotypes (with 0, 1 and 2 allele copy numbers) of the most frequent alleles and FEC (Table 3). Based on the results presented in Table 3, F and E allele frequencies had significant effects on FEC (P<0.01). Accordingly, the presence of E allele (506 bp) in homozygote (EE) animals, as a GIN susceptible allele, was accompanied with a significant increase in the means of FEC (P<0.01). Whereas the presence of F allele (510 bp), as a GIN resistant allele, in homozygote (FF) and heterozygote (FA) animals was accompanied with a significant decrease in the means of FEC in resistant animals (P<0.01). Meanwhile, other observed alleles with their various genotypes showed no significant association with FEC.

Moreover, the comparison of association related to FEC means in different flocks are presented in Table 4.

Table 1 The descriptive statistics of fecal egg count (FEC) for various classes of gastrointestinal nematodes

Nematode classes	Mean	SD	Minimum value	Maximum value
Strongyles FEC	83.33	121.37	0	612.50
Nematodirus FEC	108.53	147.15	0	805.00
Trichuris ovis FEC	1.47	7.57	0	70.00
Marshallagia marshali FEC	51.28	75.75	0	315.00
Total nematode FEC	243.14	251.33	0	1505.00

SD: standard deviation.

Table 2 Allelic and genotypic frequencies

Ovar-DRB1				
Allele	Allelic frequencies	Genotypes	N	Genotypic frequencies (%)
A	0.062	AA	7	5.83
B	0.041	BB	5	4.16
C	0.012	CC	1	0.83
D	0.054	DD	3	2.5
E	0.15	EE	18	15
F	0.204	FF	24	20
G	0.012	GG	1	0.83
H	0.083	HH	10	8.33
I	0.016	II	2	1.66
J	0.05	JJ	6	5
K	0.016	KK	2	1.66
L	0.016	LL	2	1.66
M	0.037	MM	4	3.33
N	0.016	NN	2	1.66
O	0.125	OO	15	12.5
P	0.020	PP	2	1.66
Q	0.016	QQ	2	1.66
R	0.054	RR	5	4.166
S	0.016	SS	2	1.66
T	0.008	FA	1	0.833
-	-	MC	1	0.833
-	-	PD	1	0.833
-	-	RG	1	0.833
-	-	TR	2	1.66

Analysis results showed a significant effect of flock (as a fixed effect) on means of FEC (P<0.01). While, no significant effect of time on means of FEC was observed.

In the present study microsatellite polymorphisms of the second intron in *Ovar-DRB1* gene of Iranian Ghezel sheep breed was evaluated to identify intra-breed variation in genetic resistance to GIN infections. In a previous study, using PCR-RFLP technique in Ghezel sheep showed polymorphisms in the second exon of DRB1 gene and its association with FEC for the first time in Iran (Hajializadeh *et al*. 2015). Since using PCR-RFLP technique has some limitations, in the present study microsatellites were used as a simpler, more applicable and more accurate technique for identifying genetic variations.

Simple sequence repeats (STRs) play important roles in natural evolution (Awadalla and Ritland, 1997) and genome evolution (Moxon and Wills, 1999). Also microsatellites were mainly used in genomic selections of farm animals and livestock breeding as well as determining the location

of mutations in genetic disorders with the help of disease markers (Eady *et al*. 2003; Abdul Muneer *et al*. 2009; Abdul Muneer, 2014; Teneva *et al*. 2013). Researchers also showed significant functional role of STRs in regulation of gene expression, binding to nuclear proteins and its function as transcriptional activating elements (Li *et al*. 2002).

In the present study, a total of 24 genotypes and 20 microsatellite alleles were identified in the second intron of Ovar *MHC-DRB1* gene and these results are in line with previous studies including Schwaiger *et al*. (1993), Charon *et al*. (2002) and Castillo *et al*. (2011).

It was also showed that the presence of allele called F (510 bp) in homozygote (FF) and heterozygote (FA) animals was strongly (P<0.01) associated with lower means of FEC in resistant animals.

On the other hand, the presence of E allele (506 bp) in homozygote (EE) animals was positively (P<0.01) associated with higher means of FEC in susceptible animals to GINs.

Table 3 Association between genotypes (allele copy numbers) and means of fecal egg count (FEC)

Allele	Genotype	Allele copy number	Least squares means of FEC ± SE	P-value
A	-	0	5.00±0.21	NS
	FA	1	5.12±0.21	NS
	AA	2	5.00±0.21	NS
F	-	0	5.20±0.05	
	FA	1	5.05±0.49	
	FF	2	4.48±0.09	(P<0.01)[1]
E	-	0	4.92±0.05	
	EE	2	5.56±0.13	(P<0.01)[2]
B	-	0	5.02±0.05	NS
	BB	2	5.17±0.25	NS
H	-	0	5.00±0.05	NS
	HH	2	5.25±0.18	NS
J	-	0	5.01±0.05	NS
	JJ	2	5.34±0.22	NS
M	-	0	5.02±0.05	NS
	MC	1	5.60±0.57	NS
	MM	2	4.98±0.28	NS
R	-	0	5.03±0.05	NS
	RT	1	4.63±0.58	NS
	RR	2	5.02±0.27	NS
D	-	0	5.01±0.05	NS
	PD	1	5.65±0.57	NS
	DD	2	5.14±0.33	NS
O	-	0	4.91±0.09	NS
	OO	2	5.04±0.09	NS

NS: non significant.
SE: standard error.
[1] Comparison with considering 0 and 1 allele copy numbers.
[2] Comparison with considering 0 allele copy number.

Table 4 Association between the fixed effect of flock and means of fecal egg counts (FEC)

Means of FEC ± SE (Flock)		Means of FEC (Flock)	P-value
4.35 0.17 (1)	vs.	4.04 (2)	NS
4.35±0.17 (1)	vs.	5.03 (3)	**
4.35±0.17 (1)	vs.	4.62 (4)	NS
4.35±0.17 (1)	vs.	5.70 (5)	**
4.35±0.17 (1)	vs.	6.12 (6)	**
4.04±0.14 (2)	vs.	5.03 (3)	**
4.04±0.14 (2)	vs.	4.62 (4)	**
4.04±0.14 (2)	vs.	5.70 (5)	**
4.04±0.14 (2)	vs.	6.12 (6)	**
5.03±0.14 (3)	vs.	4.62 (4)	*
5.03±0.14 (3)	vs.	5.70 (5)	**
5.03±0.14 (3)	vs.	6.12 (6)	**
4.62±0.14 (4)	vs.	5.70 (5)	**
4.62±0.14 (4)	vs.	6.12 (6)	**
5.70±0.13 (5)	vs.	6.12 (6)	*
6.12±0.14 (6)	-	-	-

* (P<0.05) and ** (P<0.01).
NS: non significant.
SE: standard error.

Thus, it seems that F allele is an effective allele for reducing FEC in Ghezel breed lambs. In a similar study, Schwaiger *et al.* (1995) discovered an association between microsatellite polymorphisms of the second intron in Ovar-DRB1 and resistance to *Ostertagia circumcincta* in naturally infected Scottish Blackface sheep. They also identified nineteen alleles in which the presence of two alleles (G2 and I) was accompanied with lower FEC in sheep. In another similar study, Outteridge *et al.* (1996) using MHC-DRB1 microsatellite polymorphism in Merino sheep indicated eight alleles with high frequency in genotype of animals in which two allele were associated with low FEC. Furthermore, Paterson *et al.* (1998) reported an association between microsatellite polymorphisms in the second intron of *MHC-DRB1* gene and resistance to Teladorsagia circumcincta. They also indicated that the presence of 257 bp allele was associated with low parasite resistance in lambs, while the presence of 263 bp allele was associated with high parasite resistance in yearling Soay sheep (Paterson *et al.* 1998). Charon *et al.* (2002) also showed that the presence of 468 bp, 482 bp and 530 bp microsatellite alleles in the second intron of *MHC-DRB1* gene was associated with lower FEC, while the presence of 568 bp was associated with higher FEC of *Teladorsagia circumcincta* in Polish Heath sheep. In another study in Mexico, Castillo *et al.* (2011) showed association between MHC (Ovar-MHC1, Ovar-DRB1 and Ovar-DRB2) microsatellites and FEC, blood packed cell volume, and blood eosinophilia in Pelibuey sheep which were artificially infected with *Haemonchus contortus* larvae. They also identified twenty alleles and found that alleles having 482 and 500 base pairs in length were associated with lower FEC, while a simultaneous higher blood eosinophilia and antibody levels in resistant lambs (Castillo *et al.* 2011). In the studies, Schwaiger *et al.* (1995), Charon *et al.* (2002) and Castillo *et al.* (2011) indicated significant association between microsatellites polymorphisms in the second intron of *DRB1* gene and lower FEC in Scottish Blackface sheep, Polish Heath sheep and Mexican Pelibuey sheep. In accordance with these results, in the present study the association between microsatellites polymorphisms in the second intron of *DRB1* gene and lower FEC was detected in Iranian Ghezel sheep.

Contradictory to the results of the present study, Cooper *et al.* (1989), Blattman *et al.* (1993) and Hulme *et al.* (1993) found no relationship between polymorphisms in MHC locus and susceptibility or resistance to GINs.

Davies *et al.* (2006) identified QTL regions on ovine chromosomes of 2, 3, 14 and 20 which were associated with FEC of GINs in Scottish Blackface sheep. Moreover, Riggio *et al.* (2014) used a joint (Meta) analysis for identifying genome-wide significant regions on OAR 4, 12, 14, 19 and 20. As OAR20 is near MHC regions, it is considered an important candidate gene for resistance to GINs in three European sheep populations. With considering these results (Davies *et al.* 2006; Riggio *et al.* 2014) and results of the present study, it seems that MHC regions still have more potential for finding the best candidate gene against GINs in sheep population. Furthermore, some complimentary studies using genome-wide association study (GWAS) found QTL regions on 1, 3, 6, 7, 14, 15 and 26 chromosomes which were also associated with resistance to gastrointestinal nematodes in different worldwide sheep breeds (Matika *et al.* 2011; Riggio *et al.* 2013; McRae *et al.* 2014; Benavides *et al.* 2015; Pickering *et al.* 2015); then, these new QTL regions can be considered for future studies in Ghezel sheep breed to find best candidate gene for resis-

tance to gastrointestinal nematodes.

Since the presence of moderate heritability (0.2–0.6) of resistance to GINs in sheep (Baker, 1998; Stear *et al.* 2007), selective breeding can be used for developing a resistant sheep population in the area based on the presence of resistant alleles. Also, detected molecular markers responsible for resistance to GINs in sheep can be used in breeding programs in order to build a resistant population.

CONCLUSION

Results of the present study showed that microsatellite markers can be used as a powerful and accurate tool in recognizing resistant animals to GIN infections. Furthermore, our results showed that *Ovar-DRB1* gene can be used as a useful candidate gene in Ghezel sheep breeding programs for improving genetic resistance to GINs. According to the strong association detected between the presence of 510 bp allele (called allele F) and genetic resistance to GIN infections, this finding can be expanded to similar strains of this breed like Morkaraman in Turkey and Afshari in Iran and even in other related breeds.

ACKNOWLEDGEMENT

We would like to gratitude Dr Seyed Abolghasem Mohammadi, Dr Arash Javanmard and Amir Kahnamooyi for their help in this research.

REFERENCES

Abdul Muneer P.M. (2014). Application of microsatellite markers in conservation genetics and fisheries management: recent advances in population structure analysis and conservation strategies. *Genet. Res. Int.* **1,** 1-11.

Abdul Muneer P.M., Gopalakrishnan A., Musammilu K.K., Mohindra V., Lal K.K., Basheer V.S. and Lakra W.S. (2009). Genetic variation and population structure of endemic yellow catfish, *Horabagrus brachysoma* (Bagridae) among three populations of Western Ghat region using RAPD and microsatellite markers. *Mol. Biol. Rep.* **36,** 1779-1791.

Amarante A.F.T., Susin I., Rocha R.A., Silva M.B., Mendes C.Q. and Pires A.V. (2009). Resistance of santaines and crossbred ewes to naturally acquired gastrointestinal nematode infections. *Vet. Parasitol.* **165,** 273-280.

Ammer H., Schwaiger F.W., Kammer Baver C., Gomolka M., Arriens A., Lazary S. and Epplen J.T. (1992). Exonic polymorphism *vs.* intronic simple repeat hypervariability in MHC DRB genes. *Immunogenetics.* **35,** 332-340.

Anonymous. (1977). Manual of veterinary parasitological laboratory techniques. Technical Bulletin, Ministry of Agriculture, Fisheries and Food, London.

Atlija M., Arranz J.J., Martinez Valladares M. and Gutierrez Gil B. (2016). Detection and replication of QTL underlying resistance to gastrointestinal nematodes in adult sheep using the ovine 50K SNP array. *Genet. Sel. Evol.* **48**, 1-16.

Awadalla P. and Ritland K. (1997). Microsatellite variation and evolution in the *Mimulusguttatus* species complex with contrasting mating systems. *Mol. Biol. Evol.* **14**, 1023-1034.

Baker R.L. (1998). Genetic resistance to endoparasites in sheep and goats: a review of genetic resistance to gastrointestinal nematode parasites in sheep and goats in the tropics and evidence for resistance in some sheep and goat breeds in sub humid coastal Kenya. *Anim. Genet. Res. Inform. Bull.* **24**, 13-30.

Baneh H., Hafezian S.H., Rashidi A., Gholizadeh M. and Rahimi G.H. (2010). Estimation of genetic parameters of body weight traits in Ghezel sheep. *Asian-Australas J. Anim. Sci.* **23**, 149-153.

Benavides M.V., Sonstegard T.S., Kemp S., Mugambi J.M., Gibson J.P., Baker R.L., Hanotte O., Marshall K. and Van Tassell C. (2015). Identification of novel loci associated with gastrointestinal parasite resistance in a Red Maasai x Dorper backcross population. *PLoS. One.* **10**, 1-20.

Blattman A.N., Hulme D.J., Kinghorn B.P., Woolaston R.R., Gray G.D. and Beh K.J. (1993). A search for associations between major histocompatibility complex restriction fragment length polymorphism bands and resistance to Haemonchus contortus infection in sheep. *Anim. Genet.* **24**, 277-282.

Box G.E.P. and Cox D.R. (1994). An analysis of transformations. *J. R. Stat. Soc. Series B. Stat. Methodol.* **26**, 211-252.

Castillo J.A., Medina R.D., Villalobos J.M., Gayosso-Vazquez A., Ulloa-Arvizu R., Rodriguez R.A., Ramirez H.P. and Morales R.A. (2011). Association between major histocompatibility complex microsatellites, fecal egg count, blood, packed cell volume and blood eosinophilia in Pelibuey sheep infected with Haemonchus contortus. *Vet. Parasitol.* **177**, 339-344.

Charon K.M., Moskwa B., Rutkowski R., Gruszczyñska J. and Swiderek W. (2002). Microsatellite polymorphism in *DRB1* gene (MHC class II) and its relation to nematode fecal egg count in Polish Heath Sheep. *J. Anim. Feed. Sci.* **11**, 47-58.

Cooper D.W., Van Oorschot R.A.H., Piper L.R. and Le Jambre L.F. (1989). No association between the ovine leucocyte antigen (OLA) system in the australian merino and susceptibility to Haemonchus contortus infection. *Int. J. Parasitol.* **19**, 695-697.

Davies G., Stear M.J., Benothman M., Abuagob O., Kerr A., Mitchell S. and Bishop S.C. (2006). Quantitative trait loci associated with parasitic infection in Scottish Blackface sheep. *Heredity.* **96**, 252-258.

Dominik S. (2005). Quantitative trait loci for internal nematode resistance in sheep. *Genet. Sel. Evol.* **37**, 83-96.

Draper N.R. and Smith H. (1981). Applied Regression Analysis. John Wiley and Sons, New York.

Eady S.J., Woolaston R.R. and Barger I.A. (2003). Comparison of genetic and nongenetic strategies for control of gastrointestinal nematodes of sheep. *Livest. Prod. Sci.* **81**, 11-23.

Ferreira M., Bressane K.C.O., Moresco A.R.C., Moreira-Filho O., Almeida-Toledo L.F. and Garcia C. (2014). Comparative application of direct sequencing, PCR-RFLP and cytogenetic markers in the genetic characterization of Pimelodus (*Siluriformes, Pimelodidae*) species: possible implications for fish conservation. *Genet. Mol. Res.* **13**, 4529-4544.

Gholamian A., Eslami A., Nabavi L. and Rasekh A.R. (2006). A field survey on resistance of gastrointestinal nematodes to Levamisole in sheep in Khuzestan province of Iran. *J. Vet. Res.* **61**, 7-13.

Hajializadeh Valilou R., Rafat S.A., Notter D.R., Shojda D., Moghaddam G.A. and Nematollahi A. (2015). Fecal egg counts for gastrointestinal nematodes are associated with a polymorphism in the *MHC-DRB1* gene in the Iranian Ghezel sheep breed. *J. Front. Genet.* **6**, 1-11.

Hazelby C.A., Probert A.J. and Rowlands D.A.P.T. (1994). Anthelmintic resistance in nematodes causing parasitic gastroenteritis of sheep in the UK. *J. Vet. Pharmacol. Ther.* **17**, 245-252.

Hulme D.J., Nicholas F.W., Windon R.G., Brown S.C. and Beh K.J. (1993). The MHC class II region and resistance to an intestinal parasite in sheep. *J. Anim. Breed. Gen.* **110**, 459-472.

Larruskain A., Minguijón E., Garcia-Etxebarria K., Arostegui I., Moreno B., Juste R.A. and Jugo B.M. (2012). Amino acid signatures in the Ovar-DRB1 peptide binding pockets are associated with Ovine pulmonary adenocarcinoma susceptibility /resistance. *Biochem. Biophys. Res. Commun.* **428**, 463-468.

Le Jambre L.F. (1976). Egg hatch as an *in vitro* assay of Thiabendazole resistance in nematodes. *Vet. Parasitol.* **2**, 385-391.

Li Y.C., Korol A.B., Fahima T., Beiles A. and Nevo E. (2002). Microsatellites: genomic distribution, putative functions and mutational mechanisms. *Mol. Ecol.* **11**, 2453-2465.

Matika O., Pong Wong R., Woolliams J.A. and Bishop S.C. (2011). Confirmation of two quantitative trait loci regions for nematode resistance in commercial British terminal sire breeds. *Animal.* **5**, 1149-1156.

McRae K.M., McEwan J.C., Dodds K.G. and Gemmell N.J. (2014). Signatures of selection in sheep bred for resistance or susceptibility to gastrointestinal nematodes. *BMC Genom.* **15**, 637-650.

Moxon E.R. and Wills C. (1999). DNA microsatellites: agents of evolution?. *Sci. Am.* **280**, 94-99.

Outteridge P.M., Andersson L., Douch P.G.C., Green R.S., Gwakisa P.S., Hohenhaus M.A. and Mikko S. (1996). The PCR typing of *MHC-DRB* genes in the sheep using primers for an intronic microsatellite: application to nematode parasite resistance. *Immunol. Cell. Biol.* **74**, 330-336.

Paterson S., Wilson K. and Pemberton J.M. (1998). Major histocompatibility complex variation associated with juvenile survival and parasite resistance in a large unmanaged ungulate population. *Proc. Natl. Acad Sci. USA.* **95**, 3714-3719.

Pickering N.K., Auvray B., Dodds K.G. and McEwan J.C. (2015). Genomic prediction and genome-wide association study for dagginess and host internal parasite resistance in New Zealand sheep. *BMC Genom.* **16**, 958-969.

Riggio V., Matika O., Pong Wong R., Stear M.J. and Bishop S.C. (2013). Genome-wide association and regional heritability mapping to identify loci underlying variation in nematode resistance and body weight in Scottish Blackface lambs. *Heredity.* **110**, 420-429.

Riggio V., Pong Wong R., Salle G., Usai M.G., Casu S., Moreno C.R., Matika O. and Bishop S.C. (2014). A joint analysis to identify lociunderlying variation in nematode resistance in three European sheep populations. *J. Anim. Breed. Genet.* **131**, 426-436.

Salle G., Jacquiet P., Gruner L., Cortet J., Sauvé C., Prévot F., Grisez C., Bergeaud J.P., Schibler L., Tircazes A., François D., Pery C., Bouvier F., Thouly J.C., Brunel J.C., Legarra A., Elsen J.M., Bouix J., Rupp R. and Moreno C.R. (2012). A genome scan for QTL affecting resistance to *Haemonchus contortusin* sheep. *J. Anim. Sci.* **90**, 4690-4705.

Samadi Shams S., Zununi Vahed S., Soltanzad F., Kafil V., Barzegari A., Atashpaz S. and Barar J. (2011). Highly effective DNA extraction method from fresh, frozen, dried and clotted blood samples. *Bioimpacts.* **1**, 183-187.

SAS Institute. (2002). SAS®/STAT Software, Release 9.1. SAS Institute, Inc., Cary, NC. USA.

Schwaiger F.W., Buitcamp J., Weyers E. and Epplen J.T. (1993). Typing of artiodactyl *MHC-DRB* genes with the help of intronic simple repeated DNA sequences. *Mol. Ecol.* **2**, 55-59.

Schwaiger F.W., Gostomski D., Stear M.J., Duncan J.L., Mckellar Q.A., Epplen J.T. and Buitcamp J. (1995). An ovine major histocompatibility complex DRB1 allele is associated with low faecal egg counts following natural, predominantly *Ostertagia circumcincta* infection. *Int. J. Parasitol.* **25**, 815-822.

Shen H., Han G., Jia B., Jiang S. and Du Y. (2014). MHC-*DRB1/DQB1* gene polymorphism and its association with resistance/susceptibility to cystic Echinococcosis in Chinese Merino sheep. *J. Parasitol. Res.* **1**, 1-7.

Stear M.J., Doligalska M. and Donskow Schmelter K. (2007). Alternatives to anthelmintics for the control of nematodes in livestock. *Parasitology.* **134**, 139-151.

Teneva A., Dimitrov K., Petrović Caro V., Petrović M.P., Dimitrova I., Tyufekchiev N. and Petrov N. (2013). Molecular genetics and SSR markers as a new practice in farm animal genomic analysis for breeding and control of disease disorders. *Biotechnol. Anim. Husb.* **29**, 405-429.

Tizard I.R. (2013). Veterinary Immunology. Sanders WB, Philadelphia.

Zajac A.Z. and Conboy G.A. (2012). Veterinary Clinical Parasitology. Wiley-Blackwell, US.

Zane L., Bargelloni L. and Patarnello T. (2002). Strategies for microsatellite isolation: a review. *Mol. Ecol.* **11**, 1-16.

11

Inbreeding and Inbreeding Depression on Body Weight in Iranian Shal Sheep

Z. Patiabadi[1], S. Varkoohi[1*] and S. Savar-Sofla[2]

[1] Department of Animal Science, Faculty of Agriculture, Razi University, Kermanshah, Iran
[2] Animal Science Research Institute of Iran, Agricultural Research, Education and ExtensionOrganization (AREEO), Karaj, Iran

*Correspondence E-mail: s.varkoohi@gmail.com

ABSTRACT

The aim of this study was to estimate amount of inbreeding coefficient in Shal sheep and its impact on growth performance. Pedigree information and body weight at different ages (birth weight, 3 month weight, 6 month weight, 9 month weight and 12 month weight) were used from 6692 lambs from 90 rams and 1007 ewes. Data were collected on Ghazvin sheep breeding station during 1997-2013. Estimation of inbreeding coefficient was done by CFC program and quantifying the individual inbreeding regression of traits was run by wombat software. Number of inbred animals at pedigree was 1616 lambs, equal to 24.15% of total population. The average of inbreeding coefficient on whole population and inbred population were 1.51% and 6.28%, respectively. Regression coefficients per 1% inbreeding for birth weight, 3 month weight, 6 month weight, 9 month weight and 12 months weight were estimated as -0.001, -0.017, -0.005, -0.019 and -0.019 kg, respectively. The highest inbreeding coefficient was 31.25% and most of inbred animals had inbreeding coefficients lower than 5%. These results confirmed the low level of inbreeding in the population. Annual trend of inbreeding coefficient on population average was 0.07 and non significant statistically. Applying a designed mating system like crossbreeding could be a suitable method to avoid inbreeding depression.

KEY WORDS growth traits, inbred population, inbreeding depression, regression coefficients.

INTRODUCTION

Any genetic improvement programs applied for livestock are based on two main approaches: selection and crossbreeding. By contrast to crossbreeding, intensive selection within a single population reduces genetic diversity and increases the inbreeding rate (Barczak *et al.* 2009). A definition for inbreeding is given by mating of individuals whose relatedness between them is greater than the average degree of relationship existing in population and capable changing genotypic frequencies on a population without modifying the gene frequencies (Lush, 1945). The rate of inbreeding needs to be limited to maintain diversity at an acceptable level, so that genetic variation will ensure that future animals can respond to changes in the environment and to selection. Without genetic variation, animals cannot adapt to these changes (Van Wyk *et al.* 2009). Heterozygosity and allelic diversities can be lost from small, closed, selected populations at a rapid rate. The loss of diversity and resulting increase in homozygosity might result in decreased productions and / or fitness of inbred animals (Lamberson and Thomas, 1994; Ercanbrack and Knight, 1991; Analla *et al.* 1998; Dario and Bufano, 2003). Furthermore, inbreeding depression in domestic animals can lead to a decrease in selection response and in potential genetic gains in economic traits. Measuring the effect of inbreeding on productive traits is important in order to estimate the magnitude of change associated with increases in inbreeding (Negussie *et al.* 2002; Barczak *et al.* 2009).

The initial consequence of inbreeding is inbreeding depression, which reduces the performance of growth, production, health, fertility and survival traits (Fernandez and Toro, 1999). Furthermore, inbreeding depression in domestic animals can lead to a decrease in selection response and genetic gains potential on economic traits. The emergence of disorders due to recessive gene action might occur, as well. It is apparent that different breeds and populations, as well as different traits vary in their response to inbreeding. Some populations might show a very pronounced effect of increased inbreeding for a trait, whereas others might not demonstrate much of an effect (Negussie *et al.* 2002; Barczak *et al.* 2009).

Many studies have reported the rate of inbreeding and inbreeding depression in sheep. Pedrosa *et al.* (2010) reported that average of inbreeding was 2.33% in Santa Inês sheep in Brazil. Van Wyk *et al.* (2009) and Selvaggi *et al.* (2010) reported that inbreeding rates was 16% in Elsenburg Dormer sheep and 8.1% in Leccese sheep, respectively. Akhtar *et al.* (2000) showed that inbreeding depression in Hissardale sheep in Pakistan was -0.093, -0.130 and -0.190 kg for BW_6, BW_9 and BW_{12}, respectively. Dorostkar *et al.* (2012) found that inbreeding depression for body weight traits in Iranian Moghani sheep at birth, 3, 6, 9 and 12 months of age was 0.7, -0.291, -0.260, -0.180 and -0.410 kg, respectively, (per 1% increase in individual coefficient).

Therefore, inbreeding is an important parameter to monitor and control in breeding programs. The aim of this study was to evaluate the effects of inbreeding on body weight at the ages of birth (BW_0), 3 (BW_3), 6 (BW_6), 9 (BW_9) and 12 months (BW_{12}) in Iranian Shal sheep.

MATERIALS AND METHODS

Data description

Pedigree of 6692 animals from 90 sires and 1007 dams that were collected on Shal Breed Station in Ghazvin, during 1997 to 2013 years, were used to estimate inbreeding coefficients. In the pedigree, 16.40% of animals had unknown sire, 12.19% of animals had unknown dam and 12.13% both parents were unknown. Inbreeding coefficients for animals with unknown parents considered as zero. The modified algorithm of Colleau was used to estimate individuals inbreeding coefficient CFC software, (Sargolzaei *et al.* 2006). Description of pedigree is presented in Table 1.

Details of used data for estimation of inbreeding depression are given in Table 2, where the number of records is shown after editing i.e. animals with weight > weight at the same month ± 2 SD are deleted.

There were fewer records at 9 months than 6 months of age, possibly because the heavier animals at 8 or 9 months of age were sent to the market.

Table 1 Data description of the studied Shal sheep flock

Items	n	% of total	The mean of inbreeding (%)
Total number of animals	6692	100	1.51
Non inbreed	5076	75.85	0
Inbreed	1616	24.15	6.28
Number of animals with unknown sires	1098	16.40	-
Number of animals with unknown dam	816	12.19	-
Number of animals with both parents unknown (foundation animals)	812	12.13	-

Table 2 Number of observations, mean and standard deviation of traits

Items	BW_0	BW_3	BW_6	BW_9	BW_{12}
No. of records	6690	6654	6662	6599	6528
Mean body weight (kg)	4.31	20.90	34.13	47.42	60.46
Standard deviation (SD), (kg)	0.92	3.46	3.92	4.21	4.28
Coefficient of variation (CV) (%)	21.34	16.55	11.48	8.87	7.07
Minimum (kg)	1.50	9.36	18.60	30.06	42
Maximum (kg)	7.30	33.21	50.80	64.71	78.80

BW_0: birth weight; BW_3: body weight in 3 month of age; BW_6: body weight in 6 month of age; BW_9: body weight in 9 month of age and BW_{12}: body weight in 12 month of age.

Data analysis

Data were analyzed by least squares analysis of variance using the general linear model (GLM) procedure of the SAS software package (SAS, 2004). The fixed effects were including: sex of lambs in two classes (male-female), type of birth in four classes (single, twins, triplets, quadruplet), age of the dam at lambing in seven classes (2 to 8 years old) and year of birth in 17 classes (1997 to 2013), respectively.

Therefore, these effects were excluded from the final model. Moreover, the age of lambs was placed in the model as a covariate factor. By excluding or including various random effects, six univariate linear animal models were fitted for each trait. Direct additive genetic effect was presented in all models and only random effect in Model 1. Models 2 and 3 included maternal permanent environmental effect and maternal additive genetic effect, respectively.

There was an additional effect [direct-maternal genetic covariance ($\sigma_{a,m}$)] in model 4 compared to model 3. Models 5 and 6 included both maternal effects and also with and without covariance between animal effects. Six univariate models were described as below:

$$y = Xb + Z_1a + e \quad \text{Model 1}$$
$$y = Xb + Z_1a + Z_3c + e \quad \text{Model 2}$$

$y = Xb + Z_1a + Z_2m + e$ Model 3
$Cov\ (a,m) = 0$

$y = Xb + Z_1a + Z_2m + e$ Model 4
$Cov\ (a,m) \neq 0$

$y = Xb + Z_1a + Z_2m + Z_3c + e$ Model 5
$Cov\ (a,m) = 0$

$y = Xb + Z_1a + Z_2m + Z_3c + e$ Model 6
$Cov\ (a,m) \neq 0$

Where:

y: $n \times 1$ vector of observations in each considered trait,
b: vector of fixed effects with a significant effect on related traits. Overall, fixed effects were included: lamb's sex (male and female, 2 classes), year of birth (1997 to 2013, 17 classes), birth type (single, twins, triplets, quadruplet, 4 classes) and dam age (2-8 years and older ewes, 7 classes), maternal permanent environmental effects, and residual effects, respectively.

a, m, c, and e: vectors of direct genetic effects, maternal genetic effects, maternal permanent environmental effects, and residual effects, respectively. It is assumed that these random effects are normally distributed with a mean of zero and variances $A\sigma^2_a$, $A\sigma^2_m$, $I_d\sigma_c^2$ and $I_d\sigma_e^2$, respectively. Also, σ_a^2, σ_m^2, σ_c^2 and σ_e^2 are direct additive genetic variance, maternal additive genetic variance, maternal permanent environmental variance, and residual variance, respectively. A is the additive numerator relationship matrix that is created using pedigree information. I_d and I_n are identity matrices with dimensions equal to the number of dams and observations, respectively.

X, Z_1, Z_2 and Z_3: design matrices (0 and 1) that are related to fixed effects, direct additive genetic effects, maternal additive genetic effects, and maternal permanent environmental effects to observations.

Log-likelihood ratio (Log L) tests were performed to determine significant random effects and consequently the most appropriate model for each considered traits. By inclusion of a random effect in the model, a significant increase was seen in the Log L compared to the reduced model (model without this effect). However, when the difference between the values of Log L was not greater than a critical value of χ^2, the simplest model was considered to be the best model. Statistical significance for models set at 5% probability level. The best model for BW was the full model (Model 6) and for BW_3, BW_6, BW_9 and BW_{12} was model 4.

RESULTS AND DISCUSSION

With the use of dense genomic marker data it is now possible to estimate inbreeding levels using the data alone, thus avoiding the problems of incomplete pedigree and also accounting for Mendelian segregation, but we do not have data to do this. Distribution of animals in different classes of inbreeding was shown in Table 3.

Table 3 Distribution of animals in different classes of inbreeding

Classes of F	Number of animals	% of total
F=0	5076	75.84
0 < F ≤ 5	879	13.13
5 < F ≤ 10	318	4.75
10 < F ≤ 15	285	4.27
15 < F ≤ 20	70	1.04
20 < F ≤ 25	49	0.73
F > 25	10	0.20

F: inbreeding coefficients.

Based on the distribution of inbreeding coefficients, the animals were divided into 7 classes of inbreeding (F=0, 0<F≤5, 5<F≤10, 10<F≤15, 15<F≤20, 20<F≤25 and F>25). Inbreeding coefficients for the animals in the founder population (year 0) and the animals brought into the flock during the period under study were considered zero because their parents were unknown and there was no pedigree information. The results showed that inbreeding among the groups, most inbred animals (13.13%) of the animals with inbreeding coefficients zero to 5 percent that these results are confirmed low levels of inbreeding in the herd. In the herd, the only 0.20 percent of all animals, inbreeding coefficient greater than 25 percent and 75.84 percent of the population has an inbreeding coefficient is zero. Accordingly the maximum number of animals were considered as the first class of inbreeding (F=0) and minimum number of animals as seventh class for all studied weights.

Descriptive statistics of inbreeding coefficients for whole population and inbred population are shown at Table 4. The mean of inbreeding coefficient in females and males were 1.40 and 1.58 %, respectively. Totally, 24.15% of animals were inbred with mean inbreeding coefficient of 6.28%.

This illustrated that low mating of close relatives was occurred in this population. Inbreeding coefficient in this study was lower than other published results (Rzewuska *et al.* 2005; Norberg and Sorenson, 2007). This estimates were higher than studies on Baluchi sheep (Mehmannavaz *et al.* 2002), Moghani sheep (Dorostkar *et al.* 2012) and other Iranian sheep breeds. These low estimates were due to low accuracy of data recording on the station made low pedigree completeness. The highest inbreeding coefficient was 31.25% and most of inbred animals had inbreeding coefficients lower than 5%. Some animals of the studied population had presented high levels of inbreeding, reflecting the intensive use of few sires. Increasing trend for mean of inbreeding in whole animals, females and males by 17 years were shown at Figure 1.

Table 4 Descriptive statistics for inbreeding coefficients for the studied population of Shal sheep

Items	All population			Inbred population		
	Female + male	Female	Male	Female + male	Female	Male
Number of animals	6692	2628	4064	1616	630	986
Mean inbreeding coefficients (%)	1.51	1.40	1.58	6.28	5.86	6.55
Standard deviation (SD) (%)	4.01	3.77	4.15	6.06	5.77	6.23
Minimum (%)	0	0	0	0.01	0.01	0.01
Maximum (%)	31.25	30.27	31.25	31.25	30.27	31.25

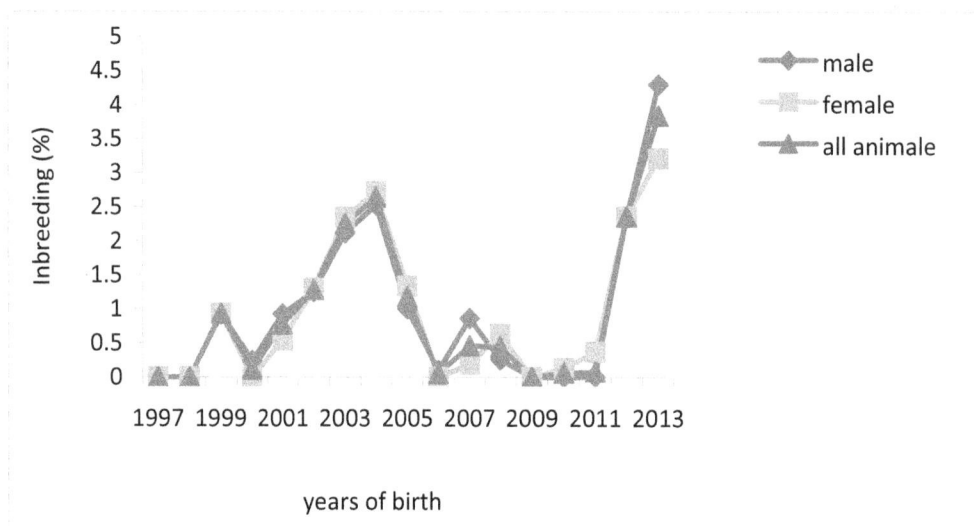

Figure 1 Annual mean of inbreeding for all, female and male animals

The proportion of inbred animals increased from zero in 1997 to 3.84 in 2013. The proportion of inbred animals in 2013 may be a cause for concern, although the average level of inbreeding was still very low. The mean of inbreeding was zero in early years of studied period. The maximum inbreeding was observed in 2013 for male animals. The increased values of inbreeding in some years may be due to poor controlling on close relative mating and excessive using of some individuals as breeding rams. Mean of inbreeding in females was zero in 2006. The mean level of inbreeding was decreased in whole animals at 2006 and 2009. The reason of this decrease was probably because of ram admittance in herd and the prevention of closed mating in sheep by the breeder by not using very few sires, and using them fairly equally. In those years, the station began to perform synchronization of ovulation and some female and male animals were imported to the station. This decrease was observed in all females and males and it could say that breeders selected non-related animals for mating. In 2010, the mean of inbreeding level for all animals was 0.06%.

This percentage was very low, but it illustrated that the mean of inbreeding had been increased compared to the base year (1997). Annual inbreeding rate for whole animals was 0.07% per year during 17 years of study.

It was observed that the average inbreeding coefficient increased due to the reason that inbred males and female individuals belonging to the same population or flock are mated together. This estimate of inbreeding rate was less than 0.40, 1.00 and 1.53% reported by Huby et al. (2003), Norberg and Sorenson, (2007) and Van Wyk et al. (2009), respectively.

Totally the inbreeding coefficient of 17 years was a non-significant and positive trend, so that in some years of decline but increase again, these fluctuations could be due to various factors such as the ram productive ewe percentage, rams herd displacement levels pedigree, the evolution of the parent changes in the number of sheep center and management methods different over the years. In this study, the average of inbreeding total sheep population Shal was born consistent throughout the year 1997 to 2013 was equal to 1.51 percent. Possibly resulting amount at causes due to lack of specific information regarding the number of parents and grandparents animals common, under-estimated. But inbreeding coefficients of inbred animals according to the number of these animals show in the population, the Sexual Intercourse of targeted largely and been controlled. Furthermore, the number of animals with high inbreeding coefficients in population indicates a lack of understanding in control mating close relatives in the population.

The result of variance analysis showed that the year of birth had significant effects on all studied traits ($P<0.01$). Sex of lamb had significant effect on all traits ($p<0.01$). The significant effect of fixed factors in these characters could be assigned partly to the differences in the endocrine system of female and male lambs. Also, age of dam had significant effect on birth weight, BW_3, BW_6, BW_9 and BW_{12} ($P<0.05$). Type of birth had a significant effect on weight changes in all traits ($P<0.01$). Single born lambs had higher body weights and pre-weaning growth rate than twins and triplets.

Due to climate conditions, feedstuff availability and ewe nutrition, especially during late pregnancy in sheep, it is expected that the birth year affects growth traits. The effect of sex and type of birth can also be caused by differences in the endocrine system, possible loci related to growth on sex chromosome and competition between twins for uterine space, milk consumption and other maternal ability compared to single-born lambs.

Single-born lambs were weighty than twins, which may be due to intense competition between twins; low milk production by ewe will not provide feed requirement of lambs and consequently they cannot express their potential capacity. It seems that increase in dam age had no effect on milk production and nursing of ewe of this breed. Nevertheless, there is a relationship between age of dam and BW because uterine environment will be better with increasing age.

Regression coefficients per 1 % increase of inbreeding for birth weight, BW_3, BW_6, BW_9 and BW_{12} were -0.001, -0.017, -0.005, -0.019 and -0.019 kg, respectively (Table 5).

Table 5 Inbreeding depression for studied traits per 1 percent increase in inbreeding coefficient

Trait	Regression coefficient of all animals (kg)
BW_0	-0.00±0.0009
BW_3	-0.017±0.0048
BW_6	-0.005±0.001
BW_9	-0.019±0.0051
BW_{12}	-0.019±0.0051

BW_0: birth weight; BW_3: body weight in 3 month of age; BW_6: body weight in 6 month of age; BW_9: body weight in 9 month of age and BW_{12}: body weight in 12 month of age.

These regression coefficients show no significant inbreeding depression. These estimates for BW were higher than -0.0005 kg estimated for Baluchi sheep (Mehmannavaz et al. 2002), but, lower than those was reported by some other researchers like as Van Wyk et al. (2009) for Dormer sheep (-0.006 kg/1% inbreeding); Dorostkar et al. (2012) for Moghani sheep (-0.007 kg/1% inbreeding); Mandal et al. (2005) for Muzaffarnagari sheep (-0.01 kg/1% inbreeding).

Also, individual regression coefficients for BW were estimated -0.0001, -0.00008 and -0.00009 kg per 1% inbreeding for Texel, Shropshire and Oxford Down, respectively

(Norberg and Sorensen, 2007). Reason of variation in inbreeding coefficients could be due to differences among breeds in alleles segregating, amount of genetic variation in the base population, location, management, and diversity of the founders in the tested flock (MacKinnon, 2003).

Dorostkar et al. (2012) reported that inbreeding coefficient was -0.007 kg/1% inbreeding; Mandal et al. (2005), Mehmannavaz et al. (2002) and Yavarifard et al. (2014) reported that inbreeding effect for Muzaffarnagari, Baluchi, and Mehraban breeds are -0.048, -0.026 and -0.014 kg per 1% inbreeding, respectively; current result of this paper for BW_3 was lower than the mentioned reports.

Inbreeding depression for BW_6 and BW_9 per 1% increase in inbreeding coefficient were -0.005, -0.019 kg, respectively. Estimation of inbreeding depression for BW_6 per 1% inbreeding was -0.260 kg that reported by Dorostkar et al. (2012) in Moghani sheep and Akhtar et al. (2000) reported that -0.093 kg in Hissardale sheep; likewise, estimation of inbreeding depression for BW_9 was -0.129 kg that reported by Mandal et al. (2005) in Mozafarnagari sheep, -0.180 kg by Dorostkar et al. (2012) in Moghani sheep and -0.130 kg by Akhtar et al. (2000) in Hissardale sheep that was lower than result of current study.

The average of regression coefficient for body weight in 10 to 12 months age per 1% inbreeding was -0.112 kg (Mandal et al. 2005), -0.410 kg (Dorostkar et al. 2012) and -0.190 kg (Akhtar et al. 2000) in Mozafarnagari sheep, Moghani sheep and Hissardale sheep, respectively, which are in disagreement with the regression coefficients that found in this study (-0.019 kg % for 12 month weight).

Inbreeding is generally associated with deterioration in growth in reproductive traits in small ruminants (Lamberson and Thomas, 1994; Wocac, 2003) and level of inbreeding may be an important factor for such effects to appear. Level of inbreeding was generally low (1.51%), mainly due to periodic introduction of unrelated ewes and rams which helped in controlling the rate of increase in the level of inbreeding. Although, some of the animals introduced may be relatives, but were assumed unrelated because of lack of pedigree recording at filed level from where such animals were purchased. This may be one of the factors that resulted in estimation of low level of inbreeding in the present flock. With the exception of few years, most of the animals were always inbred across different years but level of inbreeding was low. The inbreeding may accumulate quickly for a flock of this size due mainly to small effective population size as indicted by Ilkin (1979), where inbreeding increased to 28 percent over a period of about 15 years in a flock of British Alpine goats. Increasing number of breeding males for each breeding season would help to improve the effective population size. The level of inbreeding was comparatively low in the flock under study, due mainly to twice introduction of unrelated ewes and rams during the study period. The continuous rise in the level of inbreeding over the years however, warns that matings in the future should be more planned to avoid matings of close relatives. Increase in number of breeding males and their more frequent replacement would help the level of inbreeding to be reduced.

There are several methodological and biological factors which determine the estimated inbreeding impact on performance traits. It is well known that both negative effects and positive ones exist. Hence, in a given population, »bad« and »good« inbreeding effects are mixed (Barczak *et al.* 2009). Reasons of variation in inbreeding effects could be due to differences between the breeds in allele separation, amount of genetic variation in the base population, management, and diversity of the founders of the flocks examined (MacKinnon, 2003). The inbreeding level estimates are strongly determined by the two main factors: depth and completeness of pedigree and selection intensity. Selection intensity is often increased by the reproductive technologies being focused on a few superior animals (especially sires) and the application of advanced methods of genetic evaluation. Embryo transfer and artificial insemination technology currently allow the intensive use of the same sires, leading to increase in the relationship coefficient between animals and therefore inbreeding in the population. A high inbreeding level is observed for populations rebuilt from small number of founders, but on the other hand the accuracy is strongly improved despite the incompleteness of pedigrees (Barczak *et al.* 2009).

Animal breeding emphasis on the genetic breeding values of traits as criteria for sires and dams selection can also raise the inbreeding coefficient, since relationship between animals tend to present similar genetic values, having as a consequence the selection of the most frequent relatives (Pedrosa *et al.* 2010). Breeders should be aware that inbreeding levels can increase rapidly and become a problem in their flocks, therefore, monitoring the inbreeding situation may be of benefit. The following can be used to reduce the increase in inbreeding:

1) purchasing the sires (or ewes) that are not related or only remotely related to the flock.

2) not mating close relatives such as half sibs or sire and the daughter.

3) reduction in generation interval by replacing all sires after two or three years' use and replacing older ewes with new ones.

4) reduce variation in family size by mating each ram to a similar number of ewes and selecting one ram progeny from each sire used. Artificial insemination and other de

velopments in reproductive technology need to be used cautiously in meat sheep flocks.

This technology can lead to rapid increases in inbreeding through a dramatic reduction in effective population size within individual flocks and also within a breed as whole.

CONCLUSION

The average of inbreeding coefficient on whole population and inbred population in Iranian Shal sheep were 1.51% and 6.28%, respectively, which was comparatively low. An increasing trend for inbreeding was observed over the years. A negative effect of inbreeding was seen on all body weight ranges. Regression coefficients per 1% inbreeding for birth weight, 3 month weight, 6 month weight, 9 month weight and 12 months weight were estimated as -0.001, -0.017, -0.005, -0.019 and -0.019 kg, respectively. Although inbreeding depression was not generally a possible cause of reduction in growth performance of Iranian Shal sheep in the current situation, but caution needs to be taken in the utilization of designed mating system to maintain the level of inbreeding under control.

ACKNOWLEDGEMENT

The authors thank the Ghazvin sheep breeding station for providing the pedigree data and records of weight traits.

REFERENCES

Akhtar P., Ahmad Z., Mohiuddin G. and Abdullah M. (2000). Effect of inbreeding on different performance traits of Hissardale sheep in Pakistan. *Pakistan Vet. J.* **20(4)**, 169-172.

Analla M., Montilla J.M. and Serradilla J.M. (1998). Analyses of lamb weight and ewe litter size in various lines of Spanish Merino sheep. *Small Rumin. Res.* **29**, 255-259.

Barczak E., Wolc A., Wojtowski J. and Slosarz P. (2009). Inbreeding and inbreeding depression on body weight in sheep. *J. Anim. Feed. Sci.* **18**, 42-50.

Dario C. and Bufano G. (2003). Effect of inbreeding on milk production in Altamurana sheep breed. *J. Anim. Prod.* **55**, 270-273.

Dorostkar M., Faraji Arough H., Shodja J., Rafat S.A., Rokouei M. and Esfandyari H. (2012). Inbreeding and inbreeding depression in Iranian Moghani sheep breed. *J. Agric. Sci. Technol.* **14**, 249-556.

Ercanbrack S.K. and Knight A.D. (1991). Effects of Inbreeding on Reproduction and wool production of rambouilet, targhee and Columbia ewes. *J. Anim. Sci.* **69(12)**, 4734-4744.

Fernandez J. and Toro M.A. (1999). The use of mathematical programming to control inbreeding in selection schemes. *J. Anim. Breed. Genet.* **116**, 447-466.

Huby M., Griffon L., Moureaux S., De Rochambeau H., Danchin

Burge C. and Verrier E. (2003). Genetic variability of six french meat sheep breeds in relation to their genetic management. *Genet. Sel. Evol.* **35,** 637-655.

Ilkin T.L. (1979). Genetic analysis of the British Alpine goat flock in Australia. Pp. 125-130 in Proc. 2nd Nat. Goat Breed. Conf, Perth. Australia. Faisalabad, Pakistan.

Lamberson W.R. and Thomas D.L. (1994). Effects of Inbreeding in sheep: a review. *Anim. Breed. Abstr.* **52,** 287-297.

Lush J.L. (1945). Animal Breeding Plans. Iowa State College Press, Ames, Iowa, USA.

MacKinnon K.M. (2003). Analysis of Inbreeding in a Closed Population of Crossbred Sheep. MS Thesis. University of Blacksburg, Virginia, USA.

Mandal A., Pant K.P., Notter D.R., Rout P.K., Roy R., Sinha N.K. and Sharma N. (2005). Studies on inbreeding and its effects on growth and fleece traits of Muzaffarnagari sheep. *Asian-Australas. J. Anim. Sci.* **10,** 1363-1367.

Mehmannavaz Y., Vaez Torshizi R., Salehi A. and Shorideh A. (2002). Inbreeding and its effect on production traits in Iranian Baluchi sheep. Pp. 263- 268 in Proc. 1st Iranian Conf. Genet. Breed. Appl. Livest. Poult. Aquat. Tehran, Iran.

Negussie E., Abegaz S. and Rege J.E.O. (2002). Genetic trend and effects of inbreeding on growth performance of tropical fat-tailed sheep. Pp. 25-35 in Proc. 7th World Congr. Gen. Appl. Livest. Prod. Montpellier, France.

Norberg E. and Sorensen A.C. (2007). Inbreeding trend and inbreeding depression in the Danish populations of Texel, Shropshire and Oxford Down. *J. Anim. Sci.* **85,** 299-304.

Pedrosa V.B., Santana J.M.L., Oliveira P.S., Eler J.P. and Ferraz J.B.S. (2010). Population structure and inbreeding effects on growth traits of Santa Inês sheep in Brazil. *Small Rumin. Res.* **93,** 135-139.

Rzewuska K., Klewiec J. and Martyniuk E. (2005). Inbred effect on reproduction and body weight in a closed flock of Booroola sheep. *Anim. Sci. Pap. Rep.* **23(4),** 237-247.

Sargolzaei M., Iwaisaki H. and Colleau J.J. (2006). A tool for monitoring genetic diversity. Pp. 27-28 in Proc. 8th World Congr. Genet. Appl. Livest. Prod. Horizonte, Brazil.

SAS Institute. (2004). SAS®/STAT Software, Release 9.1. SAS Institute, Inc., Cary, NC. USA.

Selvaggi M., Dario C., Peretti V., Ciotola F., Carnicella D. and Dario M. (2010). Inbreeding depression in Leccese sheep. *Small Rumin. Res.* **89,** 42-46.

Van Wyk J.B., Fair M.D. and Cloete S.W.P. (2009). Case study: the effect of inbreeding on the production and reproduction traits in the elsenburg Dormer sheep. *Livest. Sci.* **120(3),** 218-224.

Wocac R.M. (2003). On the importance of inbreeding at Tauern-schecken goats (German language). *Arch. Tierz. Dummerstorf.* **46,** 455-469.

Yavarifard R., Ghavi Hossein-Zadeh N. and Shadparvar A.A. (2014). Population genetic structure analysis and effect on inbreeding on body weights at different ages in Iranian Mehraban sheep. *J. Anim. Sci. Technol.* **56,** 31-40.

12

Comparison of Artificial Neural Network and Multiple Regression Analysis for Prediction of Fat Tail Weight of Sheep

M.A. Norouzian[1*] and M. Vakili Alavijeh[2]

[1] Department of Animal Science, College of Abouraihan, University of Tehran, Tehran, Iran
[2] Department of Mathematics, Faculty of Mathematical Science, Shahid Beheshti University, Tehran, Iran

*Correspondence E-mail: manorouzian@ut.ac.ir

ABSTRACT

A comparative study of artificial neural network (ANN) and multiple regression is made to predict the fat tail weight of Balouchi sheep from birth, weaning and finishing weights. A multilayer feed forward network with back propagation of error learning mechanism was used to predict the sheep body weight. The data (69 records) were randomly divided into two subsets. The first subset is the training set comprising of 75 percent data (52 records) to build the neural network model and test data set comprising of 25 percent (17 records), which is not used during the training and is used to evaluate performance of different models. The mean relative error was significantly ($P<0.01$) lower for ANN than the MLR model. The coefficient of determination (R^2) values computed for the body measurements were generally higher (0.93) using ANN model than the multiple linear regression (MLR) model (0.81). The ANN model improved the mean squared error (MSE) of the MLR model by 59% and R^2 by 15% that the ANN represents a valuable tool for predicting of lamb fat tail weight from birth, weaning and finishing weights.

KEY WORDS artificial neural network, fat tail, multiple linear regression, sheep.

INTRODUCTION

The sheep industry is the largest enterprise of animal agriculture in Iran. The total number of sheep in Iran is estimated to be about 52 million, which accounts for nearly 42% of the available total animal units. Fat tail breeds are an important class of sheep breeds and these breeds are commonly found in a wide range of countries in Asia, especially the Middle East and Iran (Davidson, 2006). The fat-tail is regarded as an adaptive response of animals to a hazardous environment and is a valuable reserve for the animal during migration and winter (Kashan *et al.* 2005). Until recently, it had additional value because it was used to preserve cooked meat for longer periods of time and also as an energy reserve during times of drought and famine. Therefore the climatic variation as well as the associated re-

quirements of humans led to artificial selection for higher fat tail weight across generations. Nowadays, in intensive and semi-intensive systems most of the advantages of a large fat tail have reduced their importance and therefore, a decrease in fat tail size is often desirable for producers and consumers. Fat deposition requires more energy than the deposition of lean tissue (Moradi *et al.* 2012). Also, ruminant edible fats are particularly rich in saturated fatty acids due to the extensive microbial hydrogenation of dietary polyunsaturated fatty acids (PUFA) in the rumen and in many countries, edible fat is usually an unpopular part of meat for consumers, being considered unhealthy and is desirable to select against large fat-tails (Zamiri and Izadifar, 1997). This has been practiced by crossing fat-tailed breeds with lean-tailed breeds or selection for lower fat tail weight across generations (Kashan *et al.* 2005). In selection strategies, live weight, average daily gain, and fat tail weight are important components influencing the profitability of sheep. Live weight and average daily gain were measured on live sheep, but for measuring of fat tail weight, the animal should be slaughtered. To overcome this problem, fat tail measurements (length, width and circumference) were performed on live animals and used as a measure of tail weight in breeding programs (Vatankhah and Talebi, 2008).

However, a few studies reported low correlation between tail fat length, width, circumference measurements and its weight (Zamiri and Izadifar, 1997; Safdarian *et al.* 2008).

Artificial neural networks (ANN) are new analytical tools based on the models of neurological structures and processing function in the brain. The main advantage of ANNs in prediction is that a priori assumptions about the relations between independent and dependent variables are not necessary. However, those relations learned by an the ANNs are hidden in its neural architecture and cannot be expressed in traditional mathematical terms. The comparative advantage of the ANNs over more conventional econometric models, such as multiple linear regression (MLR) is that they can model complex, possibly non-linear relationships without any prior assumptions about the underlying data-generating process. They are able to learn and to generalize relations between input and output data from examples presented to the network. The strength of ANNs is pattern recognition and pattern classification, but these programs can also be used for predictive purposes (Dayhoff, 1990). Because of these features of ANNs, there has been an increasing tendency to apply the ANNs in biological science (Marengo *et al.* 2006; Alp and Cigizoglu, 2007; Norouzian and Asadpour, 2012).

In the present study, we compared the performance of the classic approach, the multiple linear regression and ANNs for estimating of fat tail weight from empirical data that were obtained from farm experiment. This study was undertaken to obtain prediction models for estimating fat tail weight of weight of Balouchi lambs from birth, weaning and finishing weights for breed characterization and selection for genetic improvement. In this study, artificial neural network (ANN) is employed to investigate the relationship between fat tail and live weights of lambs.

MATERIALS AND METHODS

Animal data
The current study was conducted on a sheep farm with approximately 200 lambs per year Mashhad, Iran (latitude 36 °20', longitude 54 °11', and altitude 1830 m). The climate is semi-arid, with a mean annual precipitation of 236 mm and a mean annual temperature of 33.9 °C. A total of 69 Balouchi lambswere used in this study.

At birth, lamb identification number, sex and birth type were recorded for each lamb. In addition to the ewe's milk, the lambs were offered alfalfa hay *ad libitum* and a concentrate mixture including 40% corn, 20% soybean meal, 20% beet pulp and 20% wheat bran (300 g per lamb per day) until weaning. The lambs were weaned at 60 days of age and body weights were recorded. After weaning, lambs were maintained under uniform feeding, fattening diet (Table 1), for 12 weeks and performance were determined.

Table 1 Ingredients and chemical composition of fattening diet

Item	% DM
Ingredients	
Corn	43
Soybean meal	10
Alfalfa hay	25
Beet pulp	21
Limestone meal	0.4
White Salt	0.2
Vitamin-mineral premix[1]	0.4
Metabolizable energy and chemical composition	
Metabolizable energy (Mcal/kg)	2.66
Crude protein (%)	14.8
Ether extract (%)	2.54
Neutral detergent fiber (%)	24.5
Acid detergent fiber (%)	14.9
Calcium (%)	0.68
Phosphor (%)	0.39

[1] Each kg of supplement contained: vitamin A: 50000 IU; vitamin D$_3$: 10000 IU; vitamin E: 0.1 g; Calcium: 196 g; Phosphorus: 96 g; Sodium: 71 g; Magnesium: 19 g; Iron: 3 g; Copper: 0.3 g; Manganese: 2 g; Zinc: 3 g; Cobalt: 0.1 g; Iodine: 0.1 g and Selenium: 0.001 g.

At the end of finishing period, the animals were slaughtered for determination of fat tail weight. Lambs were fasted for 12 hours and weighed before slaughter. Dressed carcass weight and weight of tail fat were recorded in kilograms.

Development of artificial neural network models
The data were randomly divided into two subsets. The first subset is the training set (n=52), which is used for building the model. The second subset is the testing set (n=17), which is not used during the training and is used to evaluate performance of different models.

A 2-layer feed-forward network formed by 1 input neuron, 1 output layer, and a number of hidden units fully connected to both input and output neurons were adopted in this study. The most used learning procedure is based on the back propagation algorithm in which the network reads inputs and corresponding outputs from a proper data set (training set) and iteratively adjusts weights and biases in order to minimize the error in prediction. In this study, training gradient descent with Levenberg Marquardt algorithm is applied and the performance function was the mean square error (MSE), the average squared error between the network outputs and the actual output.

Development of multiple regression models

To compare the effectiveness of the ANN for the prediction of body weight, the MLR model was developed using the three body measurements, birth, weaning and finishing weight as input variables to predict the fat tail weight.

The multiple regression procedure will estimate b_0, $b_1,..., b_q$ parameters of the linear equation:

$$y= b_0 + b_1x_1 + ... + b_qx_q$$

Where:

b_0, $b_1,..., b_q$: independent contributions of each independent variable $x_1,..., x_q$ to the prediction of the dependent variable y.

The global statistical significance of the relationship between y with the independent variables is analyzed by means of an analysis of variance to ensure the validity of the model in a quantified manner. The same training data set was used to develop the regression equations, and the effectiveness of prediction from the MLR model a tested using test data set. The Neural Network Toolbox of MATLAB 8.3 was employed to construct ANN models. For comparison, MLR models were generated using the training and test dataset by MATLAB 8.3 Statistics Toolbox.

Models evaluation

The following parameters were calculated to evaluate the performance and predictive ability of the model: R^2 (correlation coefficient between predicted and observed values) and MSE (mean squared error). The R^2 and MSE value between predicted and observed data is calculated by the following equations:

$$MSE = \frac{\sum_{t=1}^{n} \left| y_t - \hat{y}_t \right|^2}{n}$$

$$R^2 = \frac{SSReg}{SST} = 1 - \frac{SSE}{SST}$$

$$SSReg = \sum_{t=1}^{n} (\hat{y}_t - \overline{y})^2 \quad SST = \sum_{t=1}^{n} (y_t - \overline{y})^2$$

$$SSE = \sum_{t=1}^{n} (y_t - \hat{y}_t)^2$$

Where:

y_t: observed value.

$\hat{y}_t$: estimated value.

n: number of observations.

SSReg: sum of square of regression model.

SST: sum of square of the total.

SSE: sum of square of residuals.

To compare the predicted values with the results of laboratory assays, t-student test was used.

RESULTS AND DISCUSSION

Conformation of artificial neural network model

Architecture, specification and statistic information of the neural network were listed in Table 2.

Table 2 Architecture, specification and statistical information of the neural network model

Measurements	Value
No. layers	2
No. nodes in the hidden layer	25
No. nodes in the output layer	1
Training algorithm	Levenberg-Marquardt
Epoch	1000
Hidden layer transfer function	Sigmoid
Output layer transfer function	Pureline

Selecting inputs and outputs, the number of layers, number of neurons in each layer and number of hidden layer nodes of the ANNs can affect the benefits and abilities of them, significantly. A previous study (Cybenko, 1989) showed that one hidden layer neural network was enough to approximate any function, if enough hidden nodes were presented. The topology of the network, along with the neuron processing function, determines the accuracy and degree of representation of the model developed to correctly represent the system behavior. Therefore, the first aim was to determine the optimal number of hidden layer nodes. There are no rigorous theoretical principles for determining this. However, there are many empirical rules (Berry and Linoff, 1997). For example, the number of neurons in the hidden layer can be confirmed by the formula:

$$m = \text{Log}_2 (n) + \alpha$$

Where:

m: number of neurons in the hidden layer.

n: number of input variables.

α: integer between 0 and 10 (Berry and Linoff, 1997).

A series of neural networks with different numbers of hidden layer nodes were trained. According to its generalization ability of the testing set, MSE was calculated on different numbers of the hidden layer nodes. The model which gave the lowest value of MSE was chosen as the final ANN model. The best number of hidden layer nodes was 25 for prediction of fat tail weight (Table2).

For ANN, the training was stopped after 1000 epochs because the error increased. The linear transfer for output layer and the sigmoid transfer function for input and hidden layer are used in the ANN.

This transfer function gives an appropriate response for many applications with respect to linear transfer function.

Comparison between ANN and MLR models

The statistical values of empirical and predicted values by ANN and MLR models, as well as residues (difference between predicted and observed values) and relative residues are listed in Table 3 and Figure 1.

Also, scatter plot comparing observed and estimated fat tail weights for the MLR and ANN and scatter plot comparing estimated fat tail weight and residues (observed minus estimated values) are shown in Figures 2 and 3.

Comparison between actual and predicted fat tail weights using ANN and MLR models revealed statistically significant differences (Table 3).

Compared with the MLR model, the ANN model gave a better prediction.

The ANN predictions gave higher R^2 values with lower MSE in comparison with MLR (Figure 1). The R^2 and MSE of the MLR model were 0.81 and 1.24, respectively. However, the ANN model for the same dataset produced much improved results with $R^2 = 0.93$ and MSE$= 0.51$, respectively. The ANN model improved the MSE of the MLR model by 59% and R^2 by 15%. In ANN model, the regression between observed and estimated fat tail weights showed a slope very close to one and a low dispersion around the regression line (Figures 2 and 3). On the other hand, estimated fat tail weights versus residues (observed values minus estimated values) showed a slope very close to zero and homogeneous deviations around this value.

Table 3 Mean, maximum, minimum and standard deviation (SD) of empirical and predicted data, as well as residues

Item	Empirical	Predicted		Residuals		Relative error*	
		MLR	ANNs	MLR	ANNs	MLR	ANNs
Maximum	11.27	7.72	11.11	4.8	2	42.6	28.2
Minimum	2.19	3.61	2.82	-2.8	-4.21	127.9	61.0
Means	5.99	5.92	6.11	-0.06	-0.12	4.2a	2.9b
SD	1.33	0.88	1.34	1.12	0.71	23.1	12.2

* Relative error= (((predicted-observed)/observed)×100)
The means within the same row with at least one common letter, do not have significant difference (P>0.01).
ANNs: artificial neural networks and MLR: multiple linear regression.
SD: standard deviation.

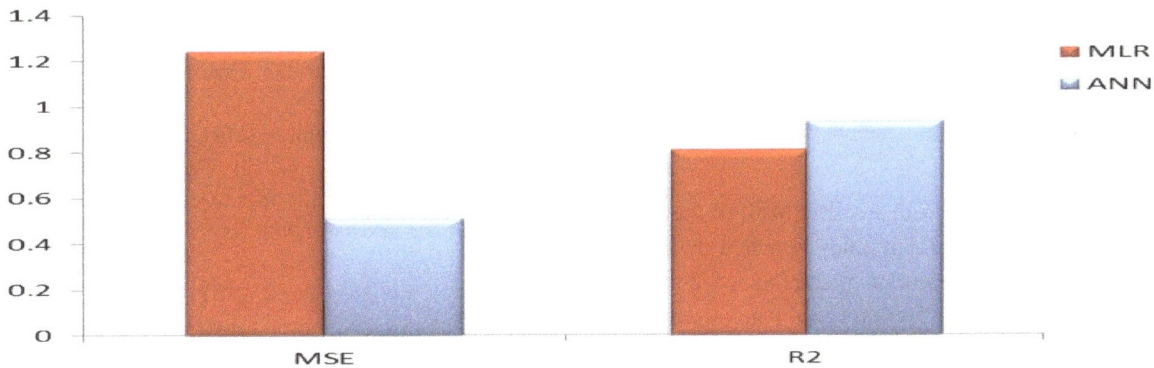

Figure 1 Performance comparison of ANN and MLR prediction models

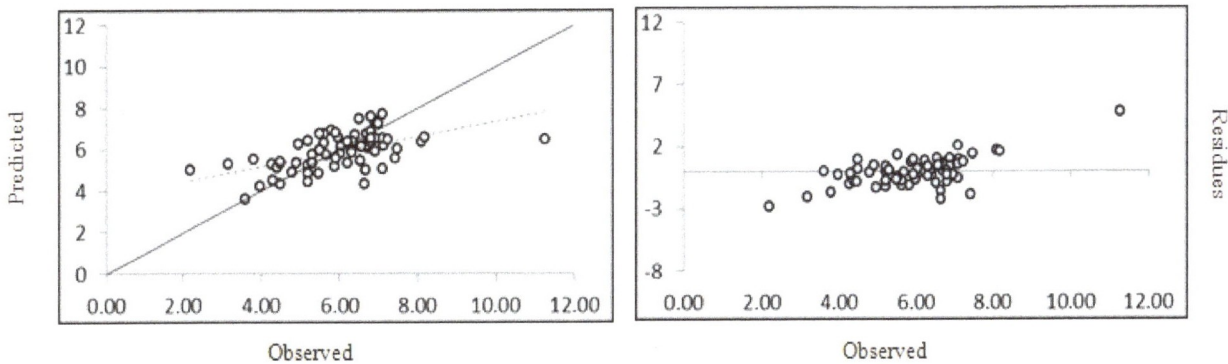

Figure 2 Scatter plot comparing observed and estimated fat tail weight for the multiple regression and scatter plot comparing estimated fat tail weight and residues (observed minus estimated values)

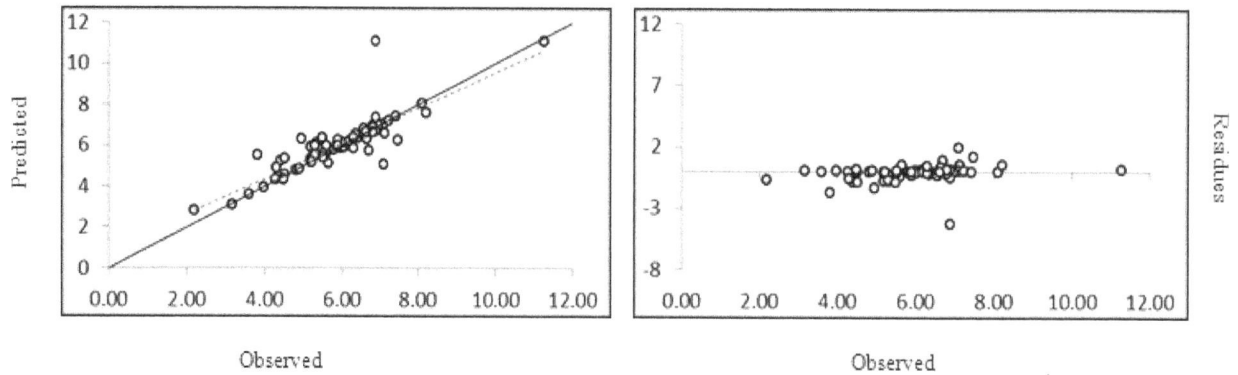

Figure 3 Scatter plot comparing observed and estimated fat tail weight for the artificial neural networks and scatter plot comparing estimated fat tail weight and residues (observed minus estimated values)

However, the MLR provides worse results than an ANN model. The mean relative error was 4.2 and 2.9 for MLR and ANN prediction, respectively, and was significantly (P<0.01) lower for ANN than MLR (Table 3).

As far as we know, there is no literature about ANN modeling for fat tail weight prediction in sheep. However, some ANN models have been used in dairy science. Fernandez *et al.* (2007) used a three layer feed-forward ANN to model and predict the weekly milk prediction on dairy goats. These authors demonstrated that artificial ANN is a suitable tool for the prediction of next week's milk yield from goat factors recorded on a farm and present milk yield.

CONCLUSION

Based on maximum R^2 value with smallest MSE of models the results obtained in the present study revealed that the ANN model gave a more accurate prediction of fat tail weight than MLR models. This suggests that the ANN method may be a promising tool for the rapid estimation of fat tail weight from body measurements in the sheep industry. However, further efforts with larger data sets are required to better determine the feasibility of rapidly predicting fat tail weight by using ANN methods.

REFERENCES

Alp M. and Cigizoglu H.K. (2007). Suspended sediment load simulation by two artificial neural network methods using hydromete- orological data. *Environ. Model Softw.* **22,** 2-13.

Berry M.J.A. and Linoff G. (1997). Data Mining Techniques. John Wiley and Sons, New York.

Cybenko G.C. (1989). Approximations by super positions of a sigmoidal function. *Math. Cont. Sig. Sys.* **2(3),** 303-314.

Davidson A. (2006). The Oxford Companion to Food. Oxford University Press, USA.

Dayhoff J.E. (1990). Neural network architectures. Van Nostrand Reinhold, New York.

Fernandez C., Soria S., Sanchez-Seiquer E.P., Gomez-Chova S., Magdalen E.R., Martin-Guerrero J.D., Navarro M.D. and Serrano A.J. (2007). Weekly milk prediction on dairy goats using neural networks. *Neural Comput. Appl.* **16,** 373-381.

Kashan N.E.J., Manafi Azar G.H., Afzalzadeh A. and Salehi A. (2005). Growth performance and carcass quality of fattening lambs from fat-tailed and tailed sheep breeds. *Small Rumin. Res.* **60,** 267-271.

Marengo E., Bobba M., Robotti E. and Liparota M.C. (2006). Modeling of the polluting emissions from a cement production plant by partial least squares, principal component regression, and artificial neural networks. *Environ. Sci. Technol.* **40,** 272-280.

Moradi M.H., Nejati-Javaremi A., Moradi-Shahrbabak M., Dodds K.G. and McEwan J.C. (2012). Genomic scan of selective sweeps in thin and fat tail sheep breeds for identifying of candidate regions associated with fat deposition. *BMC Genet.* **13,** 10-12.

Norouzian M.A. and Asadpour S. (2012). Prediction of feed abrasive value by artificial neural networks and multiple linear regression. *Neural Comput. Appl.* **21,** 905-909.

Safdarian M.M., Zamiri M.I., Hashemi M. and Noorolahi H. (2008). Relationships of fat-tail dimensions with fat-tail weight and carcass characteristics at different slaughter weights of Torki-Ghashghaii sheep. *Meat Sci.* **80,** 686-689.

Vatankhah M. and Talebi M.A. (2008). Heritability estimates and correlations between production and reproductive traits in Lori-Bakhtiari sheep in Iran. *South African J. Anim. Sci.* **38(2),** 110-118.

Zamiri M.J. and Izadifard J. (1997). Relationships of fat-tail weight with fat-tail measurements and carcass characteristics of Mehraban and Ghezel rams. *Small Rumin. Res.* **15,** 261-266.

Performance Hematology and Correlation between Economical Traits under the Effects of Dietary Lysine and Methionine in Broilers

M. Bouyeh[1*] and O.K. Gevorgyan[2]

[1] Department of Animal Science, Rasht Branch, Islamic Azad University, Rasht, Iran
[2] Department of Veterinary and Animal science, Armenia National Agrarian University, Yerevan, Armenia

*Correspondence E-mail: mbouyeh@gmail.com

ABSTRACT

Lysine (Lys) and Methionine (Met) as two primary essential amino acids and precursors of carnitine bio-synthesis are involved in most of economical traits function in domestic animals. We assessed the impact of dietary Lys and Met on the performance, lipid redistribution, intramuscular fat, carcass quality and especially phenotypic correlations among some studied parameters in broiler chickens. 300 day-old male Ross 308 chicks were randomly divided among 5 treatments, with 4 replicates per treatment. There were 15 chicks in each replicate in a completely randomized design. Same basal diet was supplemented with 5 levels of synthetic Lys and Met in amount of 0, 10, 20, 30 or 40% higher than National Research Council (NRC), 1994 recommendation for starter and grower periods. The collected data were analyzed and determined the correlation coefficient by SAS software and Duncan's test was used to compare the means on a value of ($P<0.05$). The results indicated that the two highest levels of Lys and Met treatments (30 and 40% more than NRC recommendation) led to significant increase in carcass efficiency, European Production Efficiency Factor (EPEF), blood albumin, breast muscle, heart, spleen, lymphocytes and liver weight ($P<0.05$), whereas feed conversion ratio (FCR), crude fat contents of breast and thigh muscles and plasma triglyceride were the least in these two treatment groups ($P<0.05$). Statistical analysis also showed many numbers of significant (at levels of $P<0.01$ or $P<0.05$) positive or negative correlations between the studied traits. For example highly positive correlations between carcass efficiency with heart, liver, spleen and breast weights and negative correlations with FCR, abdominal fat, plasma triglyceride, heterophyles breast and thigh fat was observed. As a conclusion of present study, dietary Lys and Met in higher levels of NRC recommendation could influence the parameters relate to performance, fat distribution, carcass quality and immune system in broilers.

KEY WORDS correlation, economical traits, lysine, methionine.

INTRODUCTION

The most important goal of broiler breeding is to increase profitability of broiler meat production. Until the last few decades, most birds were sold whole, but there has been a dramatic increase in the proportion of birds being grown for portioning and further processing (Ewart, 1993). Poultry production and processing technologies have become rap-idly accessible and are being implemented on a worldwide basis, which will allow continued expansion and competitiveness in this meat sector (Aho, 2001). So, the success of poultry meat production has been strongly related to improvements in growth factors, liveability and carcass quality together with each other especially by increasing breast muscle proportion and reducing abdominal fat pad. Abdominal and subcutaneous fat are being regarded as the main sources of waste in the slaughterhouse. Because abdominal fat is highly correlated (r=0.6 to 0.9) with total carcass lipids, it is used as the main criteria reflecting excessive fat deposition in broilers (Chambers, 1990). Havenstein *et al.* (2003) described that fat in broiler (at 43 d of age) accounts for as much as 10 to 15% of the total carcass weight. Therefore, there is substantial potential to improve feed efficiency and carcass quality by further reducing fatness.

There are a number of studies have been conducted to determine the influences of lysine and methionine as the first two limiting amino acids in practical corn-soybean based diets for broiler chicks. Some researches have suggested that levels of lysine and methionine in excess of NRC (1994) recommendations may result in enhanced performance, especially in regard to breast meat yield (Si *et al.* 2004; Schutte and Pack, 1995; Hicking *et al.* 1990; Moran and Bilgili, 1990), weight gain and feed conversion ratio (Si *et al.* 2001; Gorman and Balnave, 1995) and abdominal fat (Bouyeh and Gevorgyan, 2011a). Some studies else that have been conducted to evaluate the effects of these amino acids in excess of NRC recommendations on laying hens performance, confirmed it effect on egg production, feed conversion ratio, egg weight, egg mass and livability specially in low protein diets (Bouyeh and Gevorgyan, 2011b). Murray *et al.* (1998) found that addition of synthetic amino acids like lysine and methionine at high levels to the diet can stimulate insulin secretion from pancreas and aggregate in plasma which in turn releases amino acids and fatty acids (Sturkie, 1986) from the bodily saved sources and leads to protein synthesis. Moreover, some reports have shown the positive effect of adding more lysine to the diet than required on the chickens suffering different stresses (Ayupov, 1985).

On the other hand, lysine and methionine as precursors of L-carnitine (Borum, 1983) can play important roles in lipid and energy metabolism in poultry. L-carnitine is a natural, vitamin-like substance that acts in the cells as a receptor molecule for activated fatty acids. The major metabolic role of it appears to be the transport of long-chain fatty acids into the mitochondria for B-oxidation (Coulter, 1995). A short age of this substance results primary in impaired energy metabolism and membrane function (Harmeyer, 2002). In this regard, some researches indicated that carnitine supplementation of diets can be used to augment carnitine supply for use in metabolism, thereby facilitating fatty acid oxidation and reducing the amount of long-chain fatty acids available for storage in adipose tissue (Golzar Adabi *et al.* 2006; Kidd *et al.* 2009). Improvement in weight gain, feed conversion ratio, carcass characteristics or decrease in serum triglyceride in birds fed supplemented L-carnitine reported by researchers such as Vonlettner *et al.*

(1992) and Xu *et al.* (2003). In this regard, determining the quantities and qualities of relationship among the important traits can help to improve both main section of poultry industry: breeding and production (Gorgani Firozjah *et al.* 2015). This study aimed to estimate phenotypic correlations between some performance, fat and immune related traits as the most economical parameters in broiler chicks under the influence of different levels of dietary lysine and methionine.

MATERIALS AND METHODS

This experiment was conducted at the broiler farm belonged to Islamic Azad University, Rasht branch, using three hundred day-old male broiler chickens (Ross 308) that were selected very carefully in aspects of uniformity in body weight, good appearance, motility, etc., so that the body weight deviation of mean (46 g) was only 0.5 g. The chicks allotted to five experiment groups, each of which included four replicates of 15 birds, performed in a completely randomized design. Same basal diet was supplemented with 5 levels of synthetic lysine (as Lys-HCl) and methionine (DL-methionine) in amount of T1= 0 (control), T2= 10, T3= 20, T4= 30 and T5= 40% higher than NRC (1994) recommendation, regarding with lysine and total sulfur amino acids (TSAA) for broilers. Diets were fed from 1 to 42 d and included starter (1 to 21 d) and grower (22 to 42 d). Nutrient levels of the basal diets were based on the NRC (1994) recommendations. In order to buffer the excess chloride provided by L-Lys HCl, there was added 0.1% $NaHCO_3$ to both basal diets including starter and grower that were supplied in mash physical form (Table 1). The broiler chickens were maintained in 2×1 m pens, equipped with bell drinkers and hanging tube feeders, feed and water were available *ad libitum*, light schedule, temperature and general management were performed according to Ross 308 (2007). During 42 d experimental period, body weight gain, feed consumption, mortality, feed conversion ratio and European Production Efficiency Factor (EPEF), were recorded weekly, birds were checked twice a day for mortality; dead birds were weighed and the weight was used to adjust feed conversion ratio (FCR) (total feed consumed divided by weight of live birds plus dead birds weight). At 21 and 42 day of age three birds from each pen that were within one-half standard deviation of the overall pen body weights mean and free from visible defects were randomly chosen for blood sampling which collected into a syringe from wing vein and placed into proper tubes. These blood samples were urgently sent to determine triglyceride, cholesterol, low density lipoprotein (LDL), high density lipoprotein (HDL), uric acid, alkaline phosphatase, lymphocytes, heterophyles and glucose.

At the end of the experiment, after blood sampling, feed but not water was withheld 6 hr. prior to slaughter and then, those three birds of each replicate were processed for carcass characteristics.

Table 1 Composition (g/kg) of basal diets

Ingredient	Starter (0-21 d)	Grower (22-42 d)
Yellow corn	550	620
Dehulled soybean meal	380	320
Corn oil	21.5	14
Dicalcium phosphate	22	18
Oyster shell	14	15.8
Sodium chloride	2	2
Vitamin premix[1]	3	3
Trace mineral mix[2]	3	3
L-lysine-HCl	1.2	1.1
DL-methionine (98%)	2.3	2.1
Sodium bicarbonate	1	1
Total	1000	1000
Nutrient		
ME (kcal/kg)	3000	3000
CP (%)	21.90	19.8
Lysine (%)	1.10	1.00
Methionine (%)	0.50	0.40
TSAA (%)	0.90	0.75

[1] Provides per kg of diets: vitamin A: 17500 IU; Cholecalciferol 5000 IU; vitamin E: 25 IU; B_{12}: 0.03 mg; Riboflavin: 15 mg; Niacin: 75 mg; Choline: 700 mg; Folic acid: 1.5 mg; Pyridoxine: 6.25 mg; Biotin: 0.127 mg and Thiamine: 3.05 mg.
[2] Provides per kg of diet: Zinc: 100 mg; Manganese: 120 mg; Copper: 10 mg; Iron: 75 mg; Iodine: 2.5 mg; Selenium: 0.15 mg and Calcium: 130 mg.

After weighing the carcass pieces, thigh (biceps femoris) and breast (pectoralis major) muscles, without skin were taken, chopped, ground and frozen at -20 °C until further analyses. After thawing, tissues were extracted with 2:1 chloroform: methanol. Total lipids were extracted as described by Folch *et al.* (1957) and cholesterol content of these tissues was determined enzymatically by the method of Allain *et al.* (1974), as modified by Sale *et al.* (1984). For evaluation the fatty acids profile of the muscles, it was used a gas chromatograph (not shown it results in this paper). The weight of breast and thigh (with leg) muscles, calculated as carcass weight percentage. At the end, data were analyzed by software, the partial correlation coefficients among the traits were estimated, using the software SAS, version 6.12 (SAS, 1999), fitting the same values of lysine and methionine. Path analysis was used by expanding the matrix of partial correlation in coefficients which give the direct influence of one trait on another, regardless the effect of the other traits.

RESULTS AND DISCUSSION

Main effects of treatments on the studied traits

Table 2, shows statistical comparison between the means of traits.

The effects of different dietary levels of lysine and methionine on the most number of parameters were significant (P<0.01 or P<0.05). For example, live body weight (at d 42), absolute weight gain, EPEF, carcass weight, (as a percentage of carcass weight) and blood albumin attained the highest value in T4.

There was a linear reduction of abdominal fat with increasing dietary lysine and methionine and also decrease in some other fat related parameters, so that reduce about 45% in plasma triglyceride, 50% in abdominal fat, 35% in breast fat content and 27% in thigh muscle fat in the highest level of lysine and methionine group in comparison with the control group was observed (not shown in the Table). Whereas the means of breast muscle and blood albumin were significantly higher in T4 and T5 experiment groups (Table 2). The trends like this, were observed in regard to bursa (bursa of fabricius), Lymphocytes, heterophyles, heterophyles/lymphocytes ratio (known as a stress criteria) which showed the better result by the two highest levels means T4 and T5 experiment groups (not shown in the Table). This result can be caused by two separate effects of lysine and methionine: 1) as two amino acids in high levels can tend to release amino acids from bodily save sources following stimulate pancreas for further secretion insulin into blood and so stimulate tissues synthesis and 2) adding of these two amino acids to diets as precursors of L-carnitine, could be used to augment carnitine supply for use in metabolism, thereby facilitating fatty acid oxidation and so reducing the amount of long-chain fatty acids available for storage in adipose tissues. There is a positive correlation between consumption lysine and methionine with the level of serum carnitine (Krajkovikova, 2000). Beside, methionine participates in protein synthesis as an essential amino acid and is also as a glutathione precursor that helps to protect cells from oxidative stress, and is required for the synthesis of polyamines (spermine and spermidine), which take part in nucleus and cell division processes and also, methionine is the most important methyl group donor for methylation reaction of DNA and other molecules (Jankowski *et al.* 2014).

On the other hand lysine is also an essential amino acid that is necessary to produce proteins like antibodies, so adequate dietary levels of these amino acids are needed to support optimum performance of immune system. Some poultry nutritionists use the level recommended by NRC as a guideline in establishing their own amino acid requirements regardless of location, health or environmental conditions. There are few research works relate to the effect of lysine and methionine on immune system. Several studies demonstrated that methionine and lysine constructively affect the immune system improving both cellular and humeral immune response.

Table 2 Effects of lysine and methionine on some performance related parameters of the broilers

Variable	Amount of dietary lysine and methionine (based on TSAA) relative to NRC recommendation					Significant
	Groups					
	T1-control (NRC)	T2 (1.1 NRC)	T3 (1.2 NRC)	T4 (1.3 NRC)	T5 (1.4 NRC)	
Live body weight (day 42)	2989.6±65.80[a]	2949.2±70.65[a]	2878.5±66.78[ab]	2999.7±62.87[a]	2757.3±57.07[b]	*
Absolute gain of live weight (g)	2943.6±65.12[a]	2902.7±69.90[a]	2833.0±66.12[ab]	2953.2±62.25[a]	2711.3±56.50[b]	*
EPEF	388±6.45[b]	385±7.12[b]	372±7.10[b]	440±6.25[a]	389±8.15[b]	**
Carcass weight (g)	2116.5±47.32[b]	2131.5±45.35[b]	2109.0±47.32[b]	2316.5±49.78[a]	2110.0±41.65[b]	**
Breast muscle weight (%) (with bone)	34.67±0.11[b]	36.65±0.12[b]	35.60±0.23[b]	38.12±0.16[a]	39.10±0.15[a]	*
Thigh muscle weight (%) (with bone)	37.22±3.23	40.82±2.95	36.95±3.25	37.65±3.87	33.67±3.13	NS
Abdominal fat (%)	0.91±0.01[a]	0.85±0.01[ab]	0.84±0.01[b]	0.67±0.01[c]	0.44±0.01[d]	**
Blood alkaline phosphatase (IU/L)	1125±48.56	1098±62.37	1112±89.47	1178±75.19	1169±87.21	NS
Blood albumin (g/dL)	1.64±0.14[b]	1.58±0.11[b]	1.62±0.14[b]	1.82±0.16[a]	1.79±0.12[a]	*
Heart weight (%)	0.605±0.003[c]	0.617±0.004[c]	0.710±0.03[b]	0.810±0.04[a]	0.827±0.03[a]	**

EPEF: European Production Efficiency Factor and TSAA: total sulfur amino acids.
[*] (P<0.05) and [**] (P<0.01).
The means within the same row with at least one common letter, do not have significant difference (P>0.05).
NS: non significant.

It was reported that methionine and lysine requirements for optimal immunity are higher than for optimal growth (Tsiagbe et al. 1987; Swain and Johri, 2000; Shini et al. 2005; Khalil et al. 2010).

Also it is reported that restriction of sulfur amino acids (SAA) results in severe lymphocyte depletion in intestinal tissues (Swain and Johri, 2000).

Correlation coefficients between the studied traits
Table 3 and 4 show the correlation coefficients between some of the studied traits which were more important or had more significant correlations with the other traits. As it is shown in the tables, some of correlation data are positive and others are negative, and also some of them are significant at statistical level of 0.01 or 0.05 and others nonesignificant. These correlation coefficients which are separated here into three groups include performance, lipid and immune related traits, indicate the quality and quantity of relationship between the traits under the influences of dietary lysine and methionine as experimental treatments.

Performance related traits
There was some significant correlations between the traits relate to performance with each other or with other traits (Table 3 and 4) which are classified into two following groups including positive and negative correlations at level of (P<0.05 or P<0.01).

Positive correlations
As it is shown in Table 3 and 4, there were some significant positive correlations between following parameters:
The correlations between feed conversion ratio (FCR) and some traits such as breast fat (r=0.683), thigh fat (r=0.526), plasma triglyceride (r=0.585) and abdominal fat pad (r=0.620).

These results show that the higher amount of breast and thigh fat, plasma triglyceride or abdominal fat pad tends to the higher FCR and so, the lower feed efficiency. This result can be acceptable because production of fat in the body is usually in companion with the more metabolic costs than other products such as protein to synthesis the body tissues.

With regard carcass efficiency, it was observed positive correlation with spleen weight (r=0.899), liver weight (r=0.800) and heart weight (r=0.915). It can be concluded that increasing in mentioned above organ weights tend to higher carcass efficiency. This result may be due to the positive effect of stronger heart, liver and spleen on health and performance of the bird.

Negative correlations
Negative correlations between FCR and heart weight (r=-0.660), shows that the higher amount of heart weight (as a percentage of the carcass weight) could decrease FCR. In regard with carcass efficiency, it observed significant negative correlations with some traits such as breast fat (r=-0.634), thigh fat (r=-0.641), plasma triglyceride (r=-0.680), and abdominal fat (r=-0.763). This result emphasize the negative effect of fat content on carcass efficiency of the broilers.

Lipid related traits
It was observed some significant correlations between the traits relate to lipids with each other or with other traits (Table 3 and 4) which are classified into two following groups including positive and negative correlations.

Positive correlations
Table 3 and 4 shows some significant positive correlations between breast fat with some traits such as thigh fat (r=0.823) and plasma triglyceride (r=0.880).

Table 3 Correlations between the studied traits

Traits	Breast fat	Thigh fat	Breast cholesterol	Thigh cholesterol	Plasma triglyceride	Lymphocytes (%)	Heterophyles (%)	Spleen weight	Bursa weight
Breast fat	1	0.823**	-0.398	-0.454*	0.880**	-0.741**	0.838**	-0.669**	-0.348
Thigh fat	0.823**	1	-0.359	-0.364	0.899**	-0.573**	0.748**	-0.736**	-0.234
Breast cholesterol	-0.398	-0.359	1	0.951*	-0.447*	0.241	-0.451*	0.499*	0.032
Thigh cholesterol	-0.454*	-0.364	0.951*	1	-0.447*	0.355	-0.495*	0.416	0.084
Plasma triglyceride	0.880**	0.899**	-0.447*	-0.447*	1	-0.638**	0.683**	-0.792**	-0.269
Lymphocytes (%)	-0.741**	-0.573**	0.241	0.355	-0.638**	1	-0.654**	0.246	0.360
Heterophyles (%)	0.838**	0.748**	-0.451*	-0.495*	0.683**	-0.654**	1	-0.495*	-0.376
Spleen weight	-0.669**	-0.736**	0.499*	0.416	-0.792**	0.246	-0.495*	1	0.075
Bursa weight	-0.348	-0.234	0.032	0.084	-0.269	0.360	-0.376	0.075	1
Live body weight	0.306	0.474*	0.021	0.020	0.021	-0.126	0.212	-0.276	-0.208
FCR	0.683**	0.526*	-0.468*	-0.462*	0.585**	-0.432	0.484*	-0.685**	-0.017
Carcass efficiency	-0.684**	-0.641*	0.429	0.315	-0.680**	0.135	-0.469*	0.899**	-0.033
Breast weight	-0.604**	-0.694**	0.256	0.228	-0.711**	0.317	-0.638**	0.657**	0.258
Thigh weight	0.250	0.363	-0.299	-0.263	0.376	-0.146	0.001	-0.278	0.172
Abdominal fat	0.781**	0.863**	-0.388	-0.335	0.938**	-0.479*	0.546*	-0.885**	-0.229
Liver weight	-0.417	-0.539*	0.326	0.189	-0.652**	-0.038	-0.148	0.796**	-0.118
Heart weight	-0.543*	-0.595**	0.441	0.307	-0.700**	0.117	-0.313	0.893**	-0.057
Glucose	-0.215	-0.271	0.058	-0.070	-0.364	-0.060	-0.084	0.653**	-0.030

FCR: feed conversion ratio.
* $(P<0.05)$ and ** $(P<0.01)$.

Table 4 Correlations between the studied traits (continue)

Traits	Live body weight	FCR	Carcass efficiency	Breast weight	Thigh weight	Abdominal fat	Liver weight	Heart weight	Glucose
Breast fat	0.306	0.683**	-0.684**	-0.604**	0.250	0.781**	-0.417	-0.543*	-0.215
Thigh fat	0.474*	0.526*	-0.641**	-0.694**	0.363	0.863**	-0.539*	-0.595**	-0.271
Breast cholesterol	0.021	-0.468*	0.429	0.256	-0.299	-0.388	0.326	0.441	0.058
Thigh cholesterol	0.020	-0.462*	0.315	0.228	-0.263	-0.335	0.189	0.307	-0.070
Plasma triglyceride	0.509*	0.585**	-0.680**	-0.711**	0.376	0.938**	-0.652**	-0.700**	-0.364
Lymphocytes (%)	-0.126	-0.432	0.135	0.317	0.146	-0.479	-0.038	0.117	-0.060
Heterophyles (%)	0.212	0.484*	-0.469*	-0.638**	0.001	0.546**	-0.148	-0.313	-0.084
Spleen weight	-0.276	-0.685**	0.899**	0.657**	-0.278	-0.885**	0.796**	0.893**	0.653**
Bursa weight	-0.208	-0.017	-0.033	0.258	0.172	-0.229	-0.118	-0.057	-0.030
Live body weight	1	-0.012	-0.231	-0.293	0.286	0.509*	-0.337	-0.266	-0.145
FCR	-0.012	1	-0.730**	-0.280	0.203	0.620**	-0.423	-0.660**	-0.287
Carcass efficiency	-0.231	-0.730**	1	0.600**	-0.276	-0.763**	0.800**	0.915**	0.634**
Breast weight	-0.293	-0.280	0.600**	1	-0.203	-0.684**	0.598**	0.554*	0.429
Thigh weight	0.286	0.203	-0.276	-0.203	1	0.398	-0.514*	-0.278	-0.120
Abdominal fat	0.509*	0.620**	-0.763**	-0.684**	0.398	1	-0.737**	-0.767**	-0.505*
Liver weight	-0.337	-0.423	0.800**	0.598**	-0.514*	-0.737**	1	0.870**	0.655**
Heart weight	0.266	-0.660**	0.915**	0.554*	-0.278	-0.767**	0.870**	1	0.610**
Glucose	-0.145	-0.278	0.634*	0.429	-0.120	-0.505*	0.655**	0.610**	1

FCR: feed conversion ratio.
* $(P<0.05)$ and ** $(P<0.01)$.

In regard with breast muscle cholesterol, there were some significant positive correlations with thigh muscle cholesterol (r=0.951) and spleen weight (r=0.499). Correlation between abdominal fat with breast fat (r=0.781), plasma triglyceride (r=0.938), live body weight (r=0.509) was also observed.

Negative correlations
Negative significant correlations between breast fat with some traits including Lymphocytes (r=-0.741), spleen weight (r=-0.669), breast weight (r=-0.604) and heart weight (r=-

0.543) indicates the negative effects of excess body fat contents on these traits. Breast cholesterol had also negative correlations with heterophyles (r=-0.451) and FCR (r=-0.468) which may evaluate the negative effect of high content of tissues cholesterol on immune system.

Immune related traits
Some significant correlation coefficients was observed between the traits relate to immune system of the broilers with each other or with other traits (Table 3 and 4) which can be

classified into two following groups including positive and negative correlations.

Positive correlations

As it is shown in table 3 and 4, there were some significant positive correlations between heterophyles (%) with some traits such as breast fat (r=0.838), thigh fat (r=0.748), plasma triglyceride (r=0.683) and abdominal fat (r=0.546). In regard with spleen weight, it was observed positive correlations with carcass efficiency (r=0.899), breast weight (r=0.675) and blood glucose (r=0.653).

Negative correlations

Here were some significant negative correlations between lymphocytes (%) with some traits such as breast fat (r=-0.741), and plasma triglyceride (r=-0.638). Negative correlation between heterophyles and spleen weight (r=-0.495) was also observed. Spleen weight with breast fat, thigh fat, plasma triglyceride, FCR and abdominal fat were also significant (Table 3 and 4) which may due to negative effects of high content of tissues and plasma fat on the broiler immune system.

CONCLUSION

The results obtained from this study implicate that increasing lysine and methionine could reduce abdominal fat content, breast and thigh crude fat and plasma triglyceride and improve feed conversion ratio, breast muscle yield, carcass efficiency (as the most important economical traits in broiler chicks) and some immune relate traits. Investigation on the correlations indicates a close relationship between the lipid content of studied body organs and plasma (especially breast muscle fat, thigh muscle fat, plasma triglyceride and abdominal fat pad) with the most numbers of studied traits relate to performance, carcass characteristics and also immune system of the broilers, so that for example, increasing in fat content of the body tend to suppressing those traits (negative correlation) under the effects of higher levels of lysine and methionine (more than NRC recommendations), and also results reported here support the hypothesis that it is possible to produce poultry meat with different fat content by supplementation lysine and methionine in excess of ordinary levels. So, it is suggested that amount of lysine and methionine in higher levels of NRC (1994) recommendations may result in enhance economical trait performance in broilers.

ACKNOWLEDGEMENT

The authors thank Islamic Azad University, Rasht Branch for financial support of this study. Appreciation is also extended to Veterinary and Animal Science Faculty of National Agrarian University of Armenia for technical and laboratory support.

REFERENCES

Aho P. (2001). Subject: Poultry Elite. Watt Poultry, USA. Available at: http://www.wattnet.coarchives/docs/501wp20.pdf. Accessed May. 2001.

Allain C.C., Poon L.S., Chan C.S.G., Richmond W. and Fu P.C. (1974). Enzymatic determination of total serum cholesterol. *Clin. Chem.* **20,** 470-475.

Ayupov F.G. (1985). Effect of supplementary lysine and aspartic acid on anabolic process in hens under stress. *Sb. Nauchn. Tr.* **31,** 106-109.

Borum P.R. (1983). Carnitine. *Annu. Rev. Nutr.* **3,** 233-259.

Bouyeh M. and Gevorgyan O.K. (2011a). Influence of excess lysine and methionine on cholesterol, fat and performance of broiler chicks. *J. Anim. Vet. Adv.* **10(12),** 1546-1550.

Bouyeh M. and Gevorgyan O.K. (2011b). Influence of different levels of lysine, methionine and protein on the performance of laying hens after peak. *J. Anim. Vet. Adv.* **10(4),** 532-537.

Chambers J.R. (1990). Genetics of growth and meat production in chickens. Pp. 599-643 in Quantitative Genetics and Selection. R.D. Crawford, Ed. Poultry Breeding and Genetics. Elsevier, Amsterdam.

Coulter D.L. (1995). Carnitine deficiency in epilepsy-risk factors and treatment. *J. Child. Neurol.* **10(2),** 2532-2539.

Ewart J. (1993). Evaluation of genetic selection techniques and their application in the next decade. *Br. Poult. Sci.* **34,** 3-10.

Folch J., Lees M. and Sloane G. (1957). A simple method for the isolation and purification of total lipids from animal tissues. *J. Biol. Chem.* **226,** 497-509.

Golzar-Adabi S., Moghaddam G., Taghizadeh A., Nematollahi A. and Farahvash T. (2006). Effect of L-carnitine and vegetable fat on broiler breeder fertility, hatchability, egg yolk and serum cholesterol and triglyceride. *Int. J. Poult. Sci.* **5(10),** 970-974.

Gorgani Firozjah N., Atashi H. and Zare A. (2015). Estimation of genetic parameters for economic traits in Mazandaran native chickens. *J. Anim. Poult. Sci.* **4(2),** 20-26.

Gorman I. and Balnave D. (1995). The effect of dietary lysine and methionine concentrations on the growth characteristics and breast meat yields of Australian broiler chickens. *Australian J. Agric. Rev.* **46(8),** 1569-1577.

Harmeyer J. (2002). The physiological role of L-carnitine. *Lohmann Inform.* **27,** 15-21.

Havenstein G.B., Ferket P.R. and Qureshi M.A. (2003). Carcass composition and yield of 1957 versus 2001 broilers when fed representative 1957 and 2001 broiler diets. *Poult. Sci.* **82,** 1509-1518.

Hicking D., Guenter W. and Jackson M. (1990). The effect of dietary lysine and methionine on broiler chicken performance and breast meat yield. *Canadian J. Anim. Sci.* **70,** 673-678.

Jankowski J., Kubińskal M. and Zduńczyk Z. (2014). Nutritional and immune modulatory function of methionine in poultry diets. *Ann. Anim. Sci.* **14(1),** 17-31.

Khalil R.H., Saad T.T. and Derballa A.E. (2010). Effect of lysine and methionine deficiency on immunity in fresh water fish. *J. Arab. Aquacult. Soc.* **5(1),** 65-78.

Kidd M.T., Gilbert J., Corzo A., Page C., Virden W.S. and Woodworth J.C. (2009). Dietary L-carnitine influences broiler thigh yield. *Asian-australas. J. Anim. Sci.* **22,** 681-685.

Krajkovikova M. (2000). Correlation of carnitine levels to methionine and lysine intake. *Physiol. Res.* **44(3),** 399-402.

Moran E.T. and Bilgili S.F. (1990). Processing losses, carcass quality and meat yields of broiler chickens receiving diets marginally deficient or adequate in lysine prior to marketing. *Poult. Sci.* **69,** 702-710.

Murray R.K., Granner D.K., Mayes P.A. and Rodwell V.W. (1998). Harper's Biochemistry. Appleton and Lana, Norwalk, Connecticut.

NRC. (1994). Nutrient Requirements of Poultry, 9th Rev. Ed. National Academy Press, Washington, DC., USA.

Ross 308. (2007). Ross 308 Broiler: Nutrition Specification. Available at: www.aviagen.com.

Sale F.O., Marchesini S., Fishman P.H. and Berr B. (1984). A sensitive enzymatic assay for determination of cholesterol in lipid extracts. *Anal. Biochem.* **142,** 347-350.

SAS Institute. (1999). SAS®/STAT Software, Release 6.12. SAS Institute, Inc., Cary, NC. USA.

Schutte J.B. and Pack M. (1995). Sulfur amino acid requirement of broiler chicks from fourteen to thirty-eight days of age. 1. Performance and carcass yield. *Poult. Sci.* **74,** 480-487.

Shini S., Li X. and Bryden W.L. (2005). Methionine requirement and cell-mediated immunity in chicks. *Br. J. Nutr.* **94,** 746-752.

Si J., Fritts C.A., Burnham D.J. and Waldroup P.W. (2001). Relationship of dietry lysine level to the concentration of all essential amino acids in broiler diets. *Poult. Sci.* **80,** 1472-1479.

Sturkie P.D. (1986). Avian Physiology. Springer-Verlag, New York.

Swain B.K. and Johri T.S. (2000). Effect of supplemental methionine, choline and their combinations on the performance and immune response of broilers. *Br. Poult. Sci.* **41,** 83-88.

Tsiagbe V.K., Cook M.E., Harper A.D. and Sunde M.I. (1987). Enhanced immune responses in broiler chick fed methionine supplemented diets. *Poult. Sci.* **66,** 1147-1154.

Vonlettner F., Zollitsh W. and Halbmayer E. (1992). Use of L-carnitine in the broiler ration. *Bodenkultur.* **43,** 161-167.

Xu Z.R., Wang M.Q., Mao H.X., Zhan X.A. and Hu C.H. (2003). Effects of L-carnitine on growth performance, carcass composition and metabolism of lipids in male broilers. *Poult. Sci.* **82,** 408-441.

AMELX and AMELY Structure and Application for Sex Determination of Iranian Maral deer (*Cervus elaphus maral*)

T. Farahvash[1], R. Vaez Torshizi[1*], A.A. Masoudi[1], H.R. Rezaei[2] and M. Tavallaei[3]

[1] Department of Animal Science, Faculty of Agriculture, Tarbiat Modares University, Tehran, Iran
[2] Department of Environmental Science, Faculty of Fisheries and Environmental Science, Gorgan University of Agricultural Science and Natural Recourses, Gorgan, Iran
[3] Human Genetic Research Center, Baqiyatallah University of Medical Science, Tehran, Iran

*Correspondence E-mail: rasoult@modares.ac.ir

ABSTRACT

In order to have a good perspective of wild animals, it is necessary to determine their population and genetic structure. It provides an opportunity to decide on better conservation managements. In the wilderness, due to the escapable nature and sometimes not having the distinguishable bisexual appearance, sex identification could be difficult by observing animals. The X- and Y- chromosome linked amelogenin (AMELX and AMELY) due to its independent and different evolution on both chromosomes could play an important role in sex determining of wild animals. To determine the sex ratio and also the genetic structure of AMELX and AMELY in Maral deer (*Cervus elaphus maral*), 37 samples were collected from populations were located in north parts of Iran. Results showed that in female deer, the amelogenin gene had one banding patterns (231 bp, for X chromosome) and the male deer had two banding pattern (231 bp and 180 bp for X and Y chromosomes, respectively). The AMELY of Maral had in/del mutation (54 bp). The genetic distance (D) of AMELX from Maral deer and Red deer was 0.12 ± 0.02, it was calculated zero for AMELY. The phylogenetic analysis of AMELX and AMELY of different deer species, showed no distance for AMELY and the D was 0.048 ± 0.009 for AMELX. It is recommended that sex determination of wild animals, especially mammalian populations using amelogenin gene would be a useful and simple method which could provide further information for genetic conservation strategies.

KEY WORDS amelogenin, *Cervus elaphus maral*, sex determination, wilderness.

INTRODUCTION

In order to genetic conservation of a population, it is necessary to have information on the structure and genetic diversity of the population (Yamazaki *et al.* 2011). Sex determination of the wild animal populations is an effective technique to evaluate the population structure and to decide on a useful conservation management and keep the population dynamic (Shaw *et al.* 2003). The main sample which is available from the wild animals is fecal that is collecting without knowing the sex of the animal. In this situation, genetic markers could be useful means for achieving genetic information of the populations. The result would help in gathering statistical and evolutional information to make the best conservation management decision (Carranza *et al.* 2009). Numerous molecular techniques have been improved in mammals for sex determination that some are based on polymerase chain reaction. For example, SRY locus on the Y chromosome had widely been applied in this manner (Matsubara *et al.* 2001). The main problem with applying this marker is that a male individual would be distinguished only when the SRY locus was not amplified. However, this condition may also happen due to the experimental errors. So to solve the problem, another gene (mostly Cytb or an autosomal microsatellite marker) should be included in the experiment (Barbosa et al. 2009). The application of two pairs of primers would raise the cost and also make difficulty, since the annealing temperature and PCR protocol should be the same as the SRY gene (Takahashi et al. 1998). Considering this fact, a simple method that is able to recognize both X and Y chromosome at the same time is of great importance (Pilgrim et al. 2005). Amelogenin gene, in mammalians, is both X- and Y-chromosome linked (AMELX and AMELY, respectively) and this gene controls the development of the enamel. The conserved structure of this gene turns it into a useful marker for sex determination. This gene is conserved and has independent different evolution of X and Y chromosomes (Royo et al. 2007; Sullivan et al. 1993). Because of the in/del mutation in AMELY, two distinguishable bands with different sizes would be amplified on agarose gel (Pfeiffer and Brenig, 2005; Babo et al. 2002). Amelogenin was first used to sex identification of cow that was reported two different patterns of amplification: classI for X chromosome with 280 bp length in female cow and class I and class II for X and Y chromosome with 280 bp and 217 bp length in male animals, respectively (Ennis and Gallagher, 1994). The similar pattern of one and two amplified bands is reported for sheep (Pfeiffer and Brenig, 2005). The Cervidae has escapable nature of life style which results in difficulty in sex determination of deer from the appearance of animal, so the amelogenin gene could be an informative marker for this manner. Maral deer are a big game animal of Iran, which is suffering from decreased population size, abundance of natural habitats and illegal hunting that expose these animals to decline genetic diversity. Determining sex ratio of maral populations would provide additional information to decide on conservation management of these populations. This study has been conducted to evaluate the structure of amelogenin gene from maral deer, also to determine the sex ratio of some captive Maral deer populations of Iran using AMELX and AMELY.

MATERIALS AND METHODS

Sample collection and DNA extraction

A total of 37 samples, included tissue, fecal and blood samples were collected from East Azerbaijan (Aynali), Qazvin (Ziyaran and Barajin), Guilan, Gorghan (Ghorogh), Semnan (Parvar) and Mazandaran naturally reserved Maral populations. DNA was extracted by using Bioneer Dynabio Blood/Tissue DNA Extraction Mini Kit (Bioneer, South

Corea) and AccuPrep Stool DNA Extraction Kit (Bioneer, South Corea).

Primers

The primers were the same as described by Ennis and Gallagher (1994):
SE47:5'-CAGCCAAACCTCCCTCTGC-3' and
SE48:5'-CCCGCTTGGTCTTGTCTGTTGC-3'.

Polymerase chain reaction

PCR reaction was carried out in a 25 µL mixture containing 12.5 µL Taq DNA polymerase 2X mix red Amplicon master mix, 1 µl of each external primers (5 pmol/µL) and 0.5 µL DNE template (5 ng/µL). Cycling was carried out under the following conditions, 95 °C for 15 min followed by 35 cycles of 95 °C for 30sec, 57 °C for 40 s, 72 °C for 30 s and the final extension of 5 min at 72 °C.

Sex determination of amplified samples

PCR products were run on 2% agarose gel and the sex of the animals was determined using the one and two banding patterns.

The Sequencing

20 µL of PCR products were sequenced (Macrogene Company, South Korea). The results were blast using blastn procedure of NCBI (http://www.ncbi.nlm.nih.gov/BLAST). The AMELX and AMELY sequences were trimmed with SEQSCAPE2.6. The genetic distance (D) was calculated by MEGA.6 software (Tamura et al. 2013) software. The polymorphic and parsimony informative sites were determined using DNAsp.51001 software. In order to phylogenetic analyzing of the data, the AMELX and AMELY sequences of Red deer, Sika deer, Follow deer, Roe deer and cow were obtained from NCBI (Table 1). The model parameters were calculated by the model test 2.1.10 software and phylogenetic analysis was carried out for AMELX and AMELY sequences using maximum likelihood method for MEGA.6.

RESULTS AND DISCUSSION

Sex determination using amelogenin amplification

The amplification of all the samples was successful. The sex of animals was determined by using 2% agarose gel by the following pattern: female animals: 1 band, 231 bp length and male animals: 2 bands: first, 231 bp and the second, 180 bp length (Figure 1).

Figure 1 shows that AMELY has two bands pattern. It is the consequence of an in/del mutation in this gene so it has two bands with different sizes, one with the very same of the X chromosome and the other with a shorter length.

Table 1 The AMELX and AMELY sequences of Red deer, Sika deer, follow deer, Roe deer and cow

No	Species	Name	Acc. No.	References
1	*Cervus elaphus*	Red deer	AY453391	Pfiffer and Brenig (2005)
2	*Cervus nippon*	Sika deer	AB028027	Yamauchi *et al.* (2000)
3	*Dama dama*	Follow deer	KJ542361	Nichols and Spong (2014)
4	*Capreolus capreolus*	Roe deer	KJ542360	Nichols and Spong (2014)
5	*Bos taurus*	Cow	EU569299	Pursak and Grzybowski (2008)

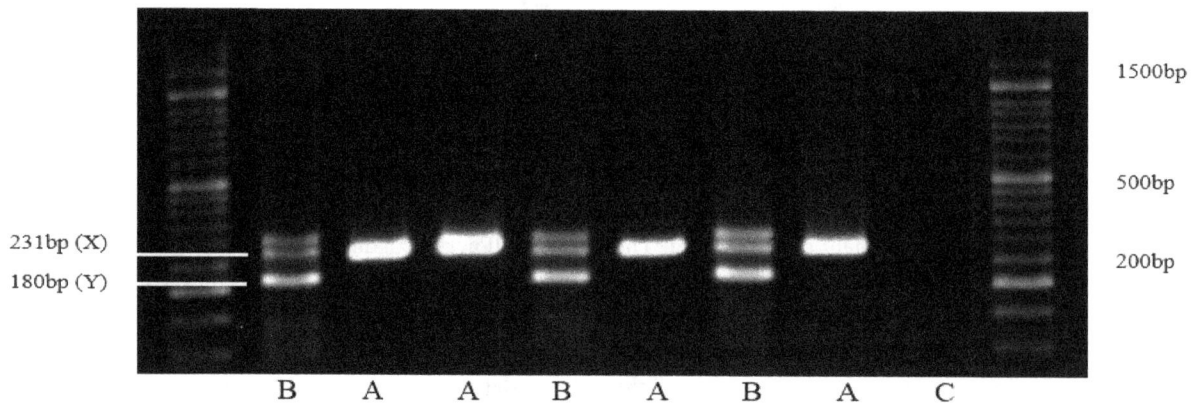

Figure 1 The results of amplification of amelogenin gene were used to determine the sex of maral deer
Female animals had 1 band (231 bp) and male animals had 2bands (231 bp for AMELX and 180 bp for AMELY) A: female animal; B: male animal and C: negative control

This pattern has been reported by other researchers in cow (Ennis and Gallagher, 1994), sheep (Pfeiffer and Brenig, 2005), Red deer (Gurgul *et al.* 2010; Pajares *et al.* 2007; Pfeiffer and Brenig, 2005) and Sika deer (Yamazaki *et al.* 2011; Yamauchi *et al.* 2000). This is the most important advantage of amelogenin gene for sex determination of wilderness. Because of this fact, there is a possibility to amplify two primers at the same tube and get reliable results with no need to test more primers. This method could be done in all no toothless mammalian species (Royo *et al.* 2007).

It should be noted that there is a third band in the male animals but it does not have influence on the sex determination.

Most researchers have been reported this third band and some suggested that it is likely due to poor amplification of poor samples especially fecal samples (Pfeiffer *et al.* 2005; Yamauchi *et al.* 2000). The results of sex determination of Maral deer naturally reserved populations are shown in Table 2.

Table 2 The results of sex determination of Iranian Maral deer populations

Location	Captive populations	No.	Male	Female
East Azerbaijan	Aynali	7	3	4
Qazvin	Ziyaran	5	-	5
Qazvin	Barajin	10	3	7
Guilan	-	5	2	3
Gorghan	Ghorogh	4	3	1
Semnan	Parvar	3	2	1
Mazandaran	-	3	-	3

The sequence results

The sequences of Maral AMELX and AMELY were as followed:

***Cervus elaphus maral* AMELX**
CAGCCCTTCCAGGCCCAGCCCATCCAGCCACAGCC
TCACCAACCCCTACAGCCCCAGTCACCTGTGCACC
CCATCCAGCCCTTGCCACCCCTGCAGCCCCTGTCA
CCTGTGCACCCCATCCAGCCCTTGCCCCCACAGCC
ACCTCTGCCTCCGATATTCCCCATGCAGCCTTTGCC
CCCTGTGCTTCCTGACCTGCCTCTGGAAGCTTGG-
CCAGCAACAGACAAGACCAAG

***Cervus elaphus maral* AMELY**
CAGCCCTTCCAGGCCCAGCCCATCCAGCCACAGCC
TCACCAACCCCTACAGCCCCAGTCACCTGTGCACC
CCATCCAGCCCTTGCCACCTCTGCCTCCGATATTCC
CCATGCAGCCTTTGCCCCCTGTGCTTCCTGACCTGC
CTCTGGAAGCTTGGCCAGCAACAGACAAGAC-
CAAGCGG

The nucleotide composition and protein sequences of AMELX and AMELY were calculated (Table 3). Sequences were blasted and they had 96%, 89%, 86%, 84% and 82% X homogeny and 92%, 82%, 83%, 70% and 83% Y homogeny with Red deer, Sika deer, Follow deer, Roe deer, sheep and cow, respectively. However, Pfeiffer and Brenig (2005) reported 97 and 96% X homogeny and 90 and 86% Y homogeny for sheep and Red deer with the original sequence of the cow, respectively.

Table 3 Nucleotide composition and protein sequences of Maral AMELX and AMELY

	Nucleotide composition (bp)				Protein sequence
AMELX	T(U) 16.9	C 48.1	A 18.6	G 16.5	QPFQAQPIQPQPHQPLQPQSPVHPIQPLPPLQPLSPVHPIQPLPPQPPLPPI FPMQPLPPVLPDLPLEAWPATDKTK
AMELY	T(U) 17.2	C 45.6	A 19.4	G 17.8	QPFQAQPIQPQPHQPLQPQSPVHPIQPLPPLPPIFPMQPLPPVLP-DLPLEAWPATDKTKR

Other study mentioned 91% and 87% AMELX and AMELY similarity between red deer and cow, respectively (Gurgul *et al.* 2010).

The length of AMELX and AMELY of Maral deer was determined and compared with the same sequences of other deer species and the original sequence of the cow. Results are summarized in Table 4. Maral deer had the same length of AMELX and AMELY with Red deer. The Sika deer had the shortest sequence of AMELX and AMELY. The comparison of AMELX and AMELY sequences from Maral deer showed that the Y had shorter sequence (54 bp) (Figure 4). This is the consequence of in/del mutation in Y-chromosome amelogenin (from site 90 to site 143) and it is the reason why there are two different banding patterns. Pfeiffer and Brenig (2005) mentioned 51 bp in/del in AMELY from Red deer, whereas Gurgul *et al.* (2010) reported 49 bp. It is reported 54 bp in Sika deer (Yamauchi *et al.* 2000) and the in/del mutation of the AMELY of sheep and cow was reported to be 68 bp and 72 bp, respectively (Pfeiffer and Brenig, 2005; Ennis and Gallagher, 1994).

The phylogenetic analyses

In order to compare amelogenin sequences of Maral deer and Red deer, the AMELX and AMELY sequences from Red deer were downloaded from NCBI and aligned with the same sequences of Maral deer. Calculated genetic distance (D) was 0.12 0.02 and 0.00 ± 0.00 for AMELX and AMELY, respectively. These amounts indicated that the diversity of amelogenin sequence of Maral deer and Red deer was low.

It confirmed that the amelogenin gene is has a conserved sequence. AMELX sequence was aligned between Maral deer, Red deer, Sika deer, Follow deer and Roe deer. There were 25 polymorphic sites with no parsimony informative sites. The protein sequences of AMELX from these groups were aligned and no specific differences were seen. The calculated D was 0.048 ± 0.009.

Sika deer had the shortest sequence (214 bp) in comparison with other deer species. The phylogenetic tree of AMELX was illustrated using maximum likelihood method (Figure 5).

Table 4 The length of AMELX and AMELY of Maral deer, Red deer, Sika deer, Follow deer, Roe deer and the original sequence of cow

Species	Name	AMELX (bp)	AMELY (bp)	References
Cervus elaphus maral	Maral deer	231	180	-
Cervus elaphus	Red deer	231	180	Pfiffer and Brenig (2005)
		255	205	Pfiffer and Brenig (2005)
Cervus nippon	Sika deer	221	167	Yamauchi *et al.* (2000)
		219	165	Yamazaki *et al.* (2011)
Dama dama	Follow deer	607	607	Nichols and Spong (2014)
Capreolus capreolus	Roe deer	608	608	Nichols and Spong (2014)
Bos taurus	Cow	280	217	Pursak and Grzybowski (2008)

```
Cervus_elaphus_maralX CAG CCC TTC CAG GCC CAG CCC ATC CAG CCA CAG CCT CAC CAA CCC CTA CAG CCC CAG TCA CCT GTG CAC CCC ATC CAG  [ 78]
Cervus_elaphus_maralY CAG CCC TTC CAG GCC CAG CCC ATC CAG CCA CAG CCT CAC CAA CCC CTA CAG CCC CAG TCA CCT GTG CAC CCC ATC CAG  [ 78]

Cervus_elaphus_maralX CCC TTG CCA CCC CTG CAG CCC CTG TCA CCT GTG CAC CCC ATC CAG CCC TTG CCC CCA CAG CCA CCT CTG CCT CCG ATA  [156]
Cervus_elaphus_maralY CCC TTG CCA CC- --- --- --- --- --- --- --- --- --- --- --- --- --- --- --- --- --- --- --- --T CTG CCT CCG ATA  [156]

Cervus_elaphus_maralX TTC CCC ATG CAG CCT TTG CCC CCT GTG CTT CCT GAC CTG CCT CTG GAA GCT TGG CCA GCA ACA GAC AAG ACC AAG ---  [234]
Cervus_elaphus_maralY TTC CCC ATG CAG CCT TTG CCC CCT GTG CTT CCT GAC CTG CCT CTG GAA GCT TGG CCA GCA ACA GAC AAG ACC AAG CGG  [234]
```

Figure 4 The in/del position (54 bp) of AMELY in comparison with AMELX sequence

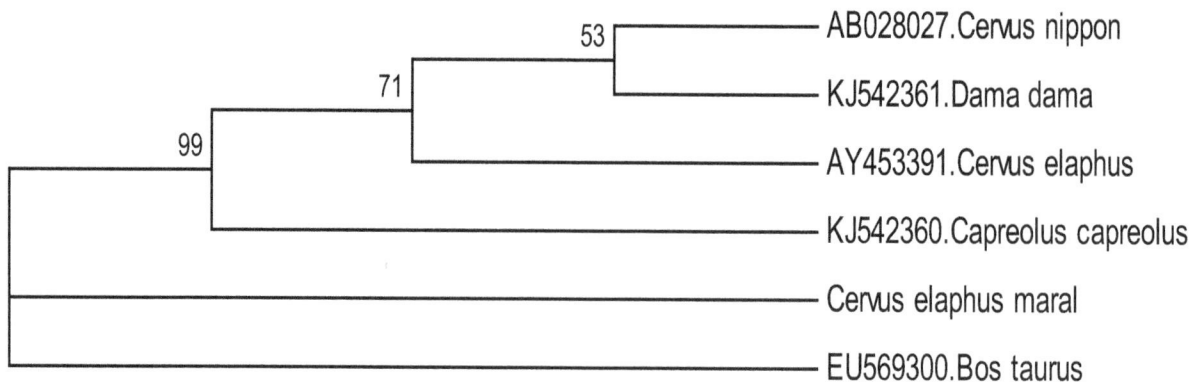

Figure 5 The phylogenetic tree of the AMELX. The phylogeny has been analyzed using Maral deer (*Cervus elaphus maral*), Red deer (*Cervus elaphus*), Sika deer (*Cervus nippon*), Follow deer (*Dama dama*) and Roe deer (*Capreolus capreolus*) AMELX sequences. *Bos taurus* was an out group and the tree has been analyzed using maximum likelihood method with HKY model and 1500 bootstrap

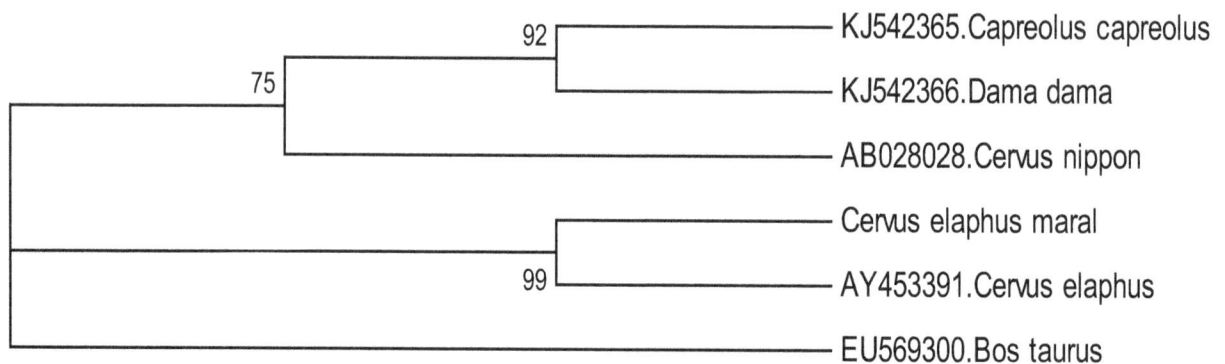

Figure 6 The phylogenetic tree for the AMELY. The phylogeny has been analyzed using Maral deer (*Cervus elaphus maral*), Red deer (*Cervus elaphus*), Sika deer (*Cervus nippon*), Follow deer (*Dama dama*) and Roe deer (*Capreolus capreolus*) sequences. *Bos taurus* was an out group and the tree has been analyzed using maximum likelihood method with Tajima-Nei model and 1500 bootstrap

The alignment of AMELY sequence from Maral deer, Red deer, Sika deer, Follow deer and Roe deer showed there were no any polymorphic sites. The protein coded by these sequences had no significant difference with the original sequence of the cow. Estimated D was 0.00. Figure 6 shows the phylogenetic relationship of AMELY sequences of deer populations. The results of phylogenetic analysis confirmed this fact that X- and Y- chromosome linked amelogenin have independent and different evolution.

CONCLUSION

Sex determination of wild animals is a useful method that would help to have a better conservation management of wilderness. Amelogenin gene due to its structure and different evolution of X- and Y- chromosomes linked amelogenin, could be a reliable molecular technique in sex identification and phylogenetic study of mammalian populations. The results of this study confirmed that AMELX and AMELY could be easily applied to determine the sex ratio of Iranian deer, especially Maral deer.

ACKNOWLEDGEMENT

This study has been conducted with the financial support of Tarbiat Modares University. The authors want to thank Department of Environment Islamic Republic of Iran, for the support and samples they were provided. We also very much welcome the facilities and technical support from Noor Human Genetic Research Center, Baqiyatallah University of Medical Science.

REFERENCES

Babo O., Takahashi N., Terashima T., Li W., Denbesten P.K. and Takano Y. (2002). Expression of alternatively spliced RNA transcripts of amelogenin gene exons 8 and 9 and its end products in the rat incisor. *J. Histochem. Cytochem.* **50,** 1229-1236.

Barbosa A.M., Fernandez-Garcia J.L. and Carranza J. (2009). A new marker for rapid sex identification of red deer (*Cervus elaphus*). *Hystrix It. J. Mamm.* **20(2),** 169-172.

Carranza J., Pérez-González J., Mateos C. and Fernández-García J.L. (2009). Parents' genetic dissimilarity and offspring sex in a polygynous mammal. *Mol. Ecol.* **18,** 4964-4973.

Ennis S. and Gallagher T.F. (1994). A PCR based sex-determination assay in cattle based on the bovine amelogenin locus. *Anim. Genet.* **25**, 425-427.

Gurgul A., Radko A. and Slota E. (2010). Characteristics of X- and Y- chromosome specific regions if the amelogenin gene and a PCR-based method for sex identification in red deer (*Cervus elaphus*). *Mol. Biol. Rep.* **37**, 2915-2918.

Matsubara K., Ishibashi Y., Ohdachi S. and Matsuda Y. (2001). A new primer set for sex identification in the genus *Sorex* (Soricidae, Insectivora). *Mol. Ecol. Notes.* **1**, 241-242.

Nichols R.V. and Spong G. (2014). An eDNA-based SNP assay for ungulate species and sex identification. Ph D. Thesis. Swedish Univ., Skogsmarksgrand.

Pajares G., Alvarez I., Fernandez I., Perez-Paravol L., Goyache F. and Royo L.I. (2007). A sexing protocol for wild ruminants based on PCR amplification of amelogenin gene AMELX and AMELY. *Arch. Tierz. Dummerstorf.* **50(5)**, 442-446.

Pfeiffer I. and Brenig B. (2005). X- and Y- chromosome specific variants of the amelogenin gene allow sex determination in sheep (*Ovis aries*) and European red deer (*Cervus elaphus*). *BMC Genet.* **6**, 16.

Pilgrim K.L., Mckelvey K.S., Riddle A.E. and Schwartz M.K. (2005). Felid sex identification based on noninvasive genetic samples. *Mol. Ecol. Notes.* **5**, 60-61.

Prusak B. and Grzybowski G. (2008). Amelogenin X Gene sequence in the Polish Red Cattle Population Kept. Institute of Genetics and Animal Breeding Publication, Polish Academy of Sciences, Poland.

Royo L.J., Pajares G., Alvarez I., Fernandez I. and Goyache F. (2007). Genetic viability and differentiation in the Spanish roe deer (*Capreolus capreolus*) characterized via mitochondrial DNA and microsatellite markers: a phylogeographic reassessment in the European framework. *Mol. Phyl. Evol.* **42**, 747-761.

Shaw C.N., Wilson P.J. and White B.N. (2003). A reliable molecular method of gender determination for mammals. *J. Mammal.* **84**, 123-128.

Sullivan K.M., Mannucci A., Kimpton C.P. and Gill P. (1993). A rapid and quantitative DNA sex test: fluorescence-based PCR analysis of X-Y homologous gene amelogenin. *Biotech.* **15**, 636- 641.

Takahashi M., Masuda R., Uno H., Yokoyama M., Suzuki M., Yoshida M.C. and Ohtaishi N. (1998). Sexing of carcass remains of the sika deer (*Cervus nippon*) using PCR amplification of the SRY gene. *J. Vet. Med. Sci.* **60**, 713-716.

Tamura K., Stecher G., Peterson D., Filipski A. and Kumar S. (2013). MEGA6: molecular evolutionary genetics analysis. *Mol. Bio. Evol.* **30**, 2725-2729.

Yamauchi K., Hamasaki S., Miyazaki K. and Kikusui T. (2000). Sex determination based on fecal DNA analysis of the amelogenin gene in sika deer (*Cervus nippon*). *J. Vet. Med. Sci.* **62**, 669-671.

Yamazaki S., Motoi Y., Nagai K., Ishinazaki T., Asani M. and Suzuki M. (2011). Sex determination of Sika deer (*Cervus nippon yesoensis*) using nested PCR from feces collected in the field. *J. Vet. Med. Sci.* **73(12)**, 1611-1616.

15

Economic Value and Produced Milk Quality in Holstein Lactating Cows in Organic System

M. Sharifi[1*], R. Pahlavan[2] and A. Aghaei[3]

[1] Department of Animal Science, Faculty of Agriculture, University of Tabriz, Tabriz, Iran
[2] Department of Animal Science, Faculty of Agriculture and Natural Resources, University of Tehran, Karaj, Iran
[3] Department of Management and Accounting, Farabi Compus University of Tehran, Qom, Iran

*Correspondence E-mail: majidsharifi@tabrizu.ac.ir

ABSTRACT

In the past decade, a global demand for products from organic agriculture has increased rapidly. Milk quality is of major interest for all parties. Therefore, the objective of this study was to compare cow performance and product quality in conventional and organic system. Twenty Holstein dairy cows were allotted to one of 2 diet groups, which including: a conventional diet (CON), and an organic system with high forage (OHF). Multiparous cows (3rd and 4th parity) were randomly assigned to the treatments. Range forages were used as part of diets and cows were offered concentrate and silage two times a day. Dry matter intake (DMI) and milk yield were measured across 200 d. Furthermore, somatic cell count, feed cost and feed efficiency were determined at 20 day intervals. The milk yield was different for cows that treated with the OHF (22.5 kg/d) and CON (28.9 kg/d) systems, respectively. Body weights were not affected by treatments; however, differences in body condition scores (P<0.05) were observed. Although energy corrected milk, milk urea nitrogen, cortisol and β-hydroxybutyrate acid were higher in cows fed CON system; milk fat, phytanic acid, hippuric acid and profit to cost ratio were higher (P<0.05) in cows fed organic system. Additionally, lower feed efficiency, feed cost and blood urea nitrogen were observed in cows fed organic diets (P<0.05).

KEY WORDS blood metabolite, economic, milk composition, organic system.

INTRODUCTION

In recent years the market for organic products has grown considerably and along with this the consumer's awareness of the production process. Therefore, organic farming using domestic livestock has recently become widespread around the world. Clearly, crucial prerequisites in order to produce high quality milk are healthy cows fed with feed free from unusual feeds. Organic farming defines clear rules for feeding of livestock, health management, and housing of animals (Mullen *et al.* 2015). With the transition from conventional to organic dairy farming, milk yield and its composition change drastically (Ellis *et al.* 2006; Prandini

et al. 2009; Slots *et al.* 2009; Butler *et al.* 2011). There is a growing body of research comparing organic and conventional farming systems. In a critical review, Mullen *et al.* (2015) demonstrated that the benefits of organic systems are more influenced by specific farm management policies than by production system itself. Although organic systems may reduce milk yield and growth rates (Slots *et al.* 2009; Butler *et al.* 2011), organic production methods may improve animal health and welfare, human health and improve the environment (Ellis *et al.* 2006; Slots *et al.* 2009; Prandini *et al.* 2009; Mullen *et al.* 2015). Milk yield on organic dairy farm is lower than milk yield on conventional farms (Adler *et al.* 2013). The reasons for lower milk yield in organic dairy herds may be due to differences in genetics, management, feeding practices, and increased subclinical mastitis (Mullen *et al.* 2015). In addition to differences in feeding and milk yield, reproductive efficiencies may also be different in organic method. Organic milk is produced in rural and nomadic breeders of Iran for many years, and organic dairy products are available in villages and cities humid climate area (without a dry season and with temperate summers and winter) in Iran (Sharifi *et al.* 2015). However, the high demand for organic milk in recent years asks for production on a larger scale. No published studies have evaluated the effect of organic systems on the performance of dairy cows in Iran. Therefore, the objective of this study was to evaluate the use of organic systems in different levels of forage and their effects on body weight, body condition score, milk production, and milk quality of Holstein dairy cows in Iran.

MATERIALS AND METHODS

Animals, diets and experimental design

The study was conducted at Valfajr Agricultural Research Center farm, located in the central Alborz range lands of Noshahr region of Mazandaran Province, Iran. All activities were performed under the guidelines approved by the Standard Committee of the Ministry of Agriculture and Veterinary Organization of Iran. Of 20 selected Holstein dairy cows for this study the average (Mean±SD) initial body weight (BW) was 495.5 ± 39.6 kg and the previous year daily milk yield (MY) was 27.5 ± 1.1 kg/d. Cows (as a group) were held at all times freely in the farm and pasture except when being milked. A completely randomized block design with 2 treatments was used and cows were randomly distributed into groups, and blocked by average body weight (BW), MY and parity (3rd and 4th parity). The two treatments during the study were: a conventional system (CON) and an organic system with high forage (OHF). Daily amounts of diet in CON were offered at morning (05:00), mid-day (13:00) and evening (21:00) and the non-consumed feed were collected and subtracted from the provided amount of feed. Organic cows were fed according to the rules of Bystrom *et al.* (2002) and offered *ad libitum* forage. For the OHF the requirements were calculated from a predicted forage intake (2.0-2.25 kg DM/100 kg live weight) and predicted milk yield, while, in conventional system CON the requirements were calculated based on milk yield. All cows received concentrate, minerals and vitamins in relation to the expected nutritional needs for MY. Cows were milked 3-time a day at 04:00, 12:00 and 20:00 hours and had *ad libitum* access to fresh water. The concentrate component and amount in diets are in Table 2.

The rangeland forages (in spring, summer and autumn) used in diets contained mainly *Hordeum bulbosum* (35%), *Lolium perenne* (27%), *Prangos ferulacea* (22%), *Poa pratensis* (10%) and others species (6%).

Range management

The rangeland area was 20.33 hectares of grassland from the Noshahr region of Mazandaran Province, Iran. Furthermore, 8.92 hectares of agricultural land was used for production of conventional feed (Table 1).

Table 1 Tillable land and major crops in organic and conventional farm

Item	Treatments	
	OHF	CON
Extent of cultivation (Hectare, Ha)		
Rangeland forage	6.08	2.12
Alfalfa hay	1.80	1.20
Maize forage	1.00	0.60
Barley grain	8.65	3.00
Corn grain	2.80	2.00
Total farm area (Ha)	20.33	8.92
Rangeland forage (kg DM/Ha)		
Spring	1100	-
Summer	1500	-
Autumn	900	-
Winter	200	-
Alfalfa (kg DM/Ha)		
Spring	2500	3500
Summer	3500	4500
Autumn	2700	3500
Winter	1800	3000
Maize forage (kg DM/Ha)	30000	45000
Barley grain (kg DM/Ha)	2500	5000
Corn grain (kg DM/Ha)	3100	5300

DM: dry matter; OHF: organic system with high forage and CON: conventional system.

For grazing in rangeland, a paddock was divided into 4 sections and forage was grazed every 50 days. The range areas were not irrigated and the yearly average rainfall amounted to about 1205 mm with 16.7 °C of average yearly temperature. The soil structures were classified as loam and/or sandy-loam, with a neutral pH (6.83-7.19) and 2.99%-4.75% organic matter. During each 50-day, botanical characterisation was performed using the method of Sharifi *et al.* (2016b) and the percentage of each species in the samples were recorded. Pasture samples were also collected at the same time and used for chemical analysis.

Data collection

Forages and other feeds were evaluated according to the association of official analytical chemists (AOAC, 1991) method. Samples of sun-dried forages were packaged and sent to the laboratory for analysis of dry matter, crude protein, crude fiber, ether extract, Ash and neutral detergent fiber measurements.

Estimation of intake for Holstein cows (n=10 per group)

Item	Treatments	
	OHF	CON
Corn silage 40% grain (kg per day)		
Early lactation	3.98	2.36
Mid lactation	3.85	2.31
Concentrate intake (kg per day)		
Early lactation	4.16	13.55
Mid lactation	4.03	13.23
Concentrated components (g/kg)		
Wheat bran	130.4	47.60
Barley grain	347.8	317.5
Corn grain	87.00	273.0
Canola	387.0	311.1
Carbonate calcium	15.20	22.5
Di-calcium phosphate	10.90	7.10
Limestone	10.90	14.4
Mineral and vitamin mix[1]	10.90	6.70

[1] Mineral composition: Ca: 180 g/kg; P: 60 g/kg; Mg: 50 g/kg; Na: 50 g/kg; Cu: 1.3 g/kg; Zn: 6.0 g/kg; Mn: 3.5 g/kg; I: 0.06 g/kg; Co: 0.032 g/kg; Se: 0.02 g/kg; vitamin A: 600000 IU/kg; vitamin D_3: 120000 IU/kg and vitamin E: 1300 IU/kg. OHF: organic system with high forage and CON: conventional system.

The BW of cows was recorded at 20-day intervals prior to the morning feed allotment. Cows were weighed when leaving the milking parlor using a digital scale. A scale of 1 to 9 for body condition score (BCS) (Khadem *et al.* 2009) was estimated for cows by the same person throughout experiment. The scoring scale ranged from 1 for very thin to 9 for very fat. The BCS values were recorded 4 times which included the dry period (pre-calving), post-calving, and the early and mid lactation. Data for daily dry matter intake (DMI) was recorded from the beginning to the end of the experiment. With data spanning 200 days, cows had DMI records for 10 periods (d 1 to d 20, d 21 to d 40, d 41 to d60, etc.). Also, milk yield for each period was recorded as the average milk produced by cows in each day of the 20-day periods. Additionally, milk fat and protein contents were measured once a week and somatic cell count (SCC) and milk urea nitrogen (MUN) were analyzed every 20 days. The fat content of milk was measured using the Smart-Trac rapid fat analyzer (CEM, Matthews, NC.). The Lacti-Check ultrasound milk analyzer (P&P International Ltd., Hopkinton, MA) was used to measure the protein contents, while the SCC was determined with Fossomatic 90 (Foss electric). For the MUN measurement, samples were collected every 20 days. A concentration of MUN was determined in all trials using the same diacetyl monoxime colorimetric assay adapted to a continuous flow analyzer (Khadem *et al.* 2009). Hippuric and phytanic acid were determined with a gas chromatograph (GC-2010, Shimadzu Co., Japan) equipped with a 100 m capillary column (0.25 mm i.d., 0.20 mm film thickness) and a flame ionization detector. Furthermore, feed efficiency (FE) was estimated by dividing energy-corrected milk (ECM) by the daily DMI of cows (Sharifi *et al.* 2016a).

For blood metabolite and urea nitrogen, blood samples (20 mL) were collected from the tail vein of cows at 20-day farm visits using evacuated tubes containing EDTA at a level of 1.8 g/L of blood. Samples were kept on ice for 15 min after collection and then centrifuged at 1000 × g for 20 min. Plasma was harvested and stored frozen in plastic tubes at −20 °C until further analysis. The plasma urea concentration was determined using the method described by Chaney and Marbach (1962).

β-hydroxybutyrate acid and non esterified fatty acids were assayed by colorimetric method (Ranbut®, Ireland). Cortisol of serum was measured by hormonal cortisol kit using Gama counter (Kon. Pron.) system.

Statistical analyses

For statistical analysis, the dependent variables were BW, BCS, MY, milk components, DMI, feed efficiency (FE), MUN, blood urea nitrogen (BUN) and feed cost, the fixed effects were dietary treatment, parity and 20-d period nested within dietary treatment.

The MIXED linear model procedure of SAS (SAS, 2004), in which cow was the random variable and sample sequence was the repeated measures, was used. The autoregressive covariance [AR (1)] structure was used because it resulted in the lowest Akaike's information criterion (Littell *et al.* 1998). The GLM PROC model was also used when necessary. Results are presented as least square means and statistical differences were considered significant at (P<0.05). Trends towards significance were considered at (0.05≤P<0.10).

RESULTS AND DISCUSSION

Body weight and body condition score

The change of BW over time is shown in Table 3. There were no statistically significant differences for BW between treatment groups; however, the CON cows lost slightly more BW than the organic cows during experiment because of higher energy density. Pre-calving BCS were not different between treatment groups (Table 3); but, significant differences were observed between treatments after calving (P<0.05) and early (P<0.001) and mid-lactation (P<0.01). Although trends for BW change in CON cows was higher than organic system, results showed that organic system can improve body condition in cows. Across all periods CON cows had higher BCS than organic system except in early lactation.

Also, dry cow BCS was similar for all treatment groups. Similar results were reported by Roesch *et al.* (2005) but their finding that BCS was not different between cows fed organic compared to conventional system is in contrast with our results.

Table 3 Estimation of body condition score, body weight, dry matter intake and milk yield of cows fed organic and conventional systems in Holstein lactating

Item	Treatments		SEM	P-value
	OHF	CON		
Body condition score				
Dry cow	5.95	5.94	0.03	NS
Post calving	4.90	5.08	0.03	*
Early lactation	4.68	4.59	0.03	***
Mid lactation	5.23	5.53	0.03	**
Body weight (kg)	483	472	8.47	NS
Early lactation	484	475	8.50	NS
Mid lactation	483	469	8.50	NS
Dry matter intake (kg/d)	17.8	21.2	0.13	**
Early lactation	18.1	21.5	0.15	**
Mid lactation	17.5	21.0	0.15	**
Milk yeild (kg/d)	22.5	30.9	0.09	**
Early lactation	23.3	30.1	0.13	***
Mid lactation	21.7	27.8	0.13	**
Feed efficiency (ECM/DMI)	1.33	1.41	0.006	*
Early lactation	1.34	1.45	0.008	**
Mid lactation	1.32	1.37	0.008	*

OHF: organic system with high forage; CON: conventional system; ECM: energy-corrected milk and DMI: Dry matter intake.
* (P<0.05); ** (P<0.01) and *** (P<0.001).
SEM: standard error of the means.

Bystrom *et al.* (2002) found that cows fed organic system weighed less at the end of lactation compared to cows fed conventional system although BW and BCS change was greater for conventional cows compared to organic cows. Furthermore, Trachsel *et al.* (2000) reported changes in body condition score of cows fed organic systems was lower than cows fed conventional system.

Dry matter intake and feed efficiency
Table 3 show the DMI intake for treatment groups over the 200 day of lactation. For the mean of all periods, cows fed CON system had a higher (P<0.05) DMI than cows fed OHF system (21.2 kg/d *vs.* 17.8 kg/d, respectively). Normally, DMI reduction can be a result of bulkier feed intake and fewer digestibilities in OHF system. Table 3 show the change in FE during the trial period. Cows fed CON system had higher (P<0.01) FE than cows fed OHF system (1.41 *vs.* 1.33, respectively). It may be difficult to determine how cows in different treatments partitioned consumed energy into alternative components of production (i.e., body maintenance and growth or restoration of body reserves); however, it is possible to determine how treatment groups utilized energy from consumed feed to restore of body reserves. Our observation of relatively lower DMI in organic system is in line with many reports in the literature (Prandini *et al.* 2009; Butler *et al.* 2011; Adler *et al.* 2013; Stiglbauer *et al.* 2013). However, organic cows in the current study consumed more forage than conventional cows (Table 2); but conventional cows consumed more concentrates than organic cows.

In contrast to our study, Bystrom *et al.* (2002) found that feed efficiency was not different between animals fed conventional *vs.* organic systems. In contrast, Thomassen *et al.* (2008) reported that feed efficiency was greater for cows fed organic systems compared to conventional system, because of the higher forage consumption of organic cows compared to conventional cows.

Milk yield and composition
In the Table 3 means and standard error of means for MY for each period and across to the 200 days is given. Changes in DMI and MY of cows were similar from beginning to the end of the experiment. Also, the changes of milk fat and protein, ECM, SCC, hippuric and phytanic acid in the study periods are shown in Table 4. Milk yield was different among treatment groups, especially early lactation periods (P<0.001).

Across the lactation period, cows fed CON system had higher (P<0.01) MY than cows fed OHF system (28.9 kg *vs.* 22.5 kg/d, respectively).

Fat percentage was higher for cows fed OHF system compared to the CON system (3.73% *vs.* 3.56%, respectively). For protein content, cows fed CON diets had higher (P<0.05) value than OHF diet (3.51% *vs.* 3.36%, respectively).

Furthermore, Table 4 shows that CON system was associated with significantly (P<0.05) higher SCC than other system at early and mid-lactation periods with a tendency for CON system to have fairly higher (P<0.10) SCC than organic system for mean of periods.

Table 4 Estimation of milk composition of cows fed organic and conventional systems in Holstein lactating

Item	Treatments		SEM	P-value
	OHF	CON		
Energy corrected milk (kg/d)	23.7	30.5	0.09	**
Early lactation	24.3	31.2	0.14	**
Mid lactation	23.1	28.9	0.14	***
Milk fat percentage	3.73	3.56	0.006	**
Early lactation	3.68	3.56	0.007	*
Mid lactation	3.79	3.57	0.007	**
Milk protein percentage	3.36	3.51	0.007	*
Early lactation	3.33	3.50	0.008	**
Mid lactation	3.39	3.51	0.008	*
Somatic cell count (10^3 cells/mL)	228.0	238.4	42.2	†
Early lactation	238.7	243.6	45.1	*
Mid lactation	217.3	233.3	45.1	*
Milk urea nitrogen (mg/dL)	15.0	19.0	0.06	**
Early lactation	14.6	19.6	0.11	***
Mid lactation	15.5	18.5	0.11	**
Blood urea nitrogen (mmol/L)	4.28	4.79	0.01	*
Early lactation	4.30	4.78	0.03	**
Mid lactation	4.27	4.81	0.03	*
Hippuric acid (mg/L)	31.3	18.9	0.10	***
Early lactation	29.3	17.7	0.15	***
Mid lactation	33.4	20.1	0.15	***
Phytanic acid (mg/100 g fat)	215	139	3.29	**
Early lactation	191	123	4.11	**
Mid lactation	240	155	4.11	**

OHF: organic system with high forage and CON: conventional system.
† ($P<0.10$) * ($P<0.05$); ** ($P<0.01$) and *** ($P<0.001$).
SEM: standard error of the means.

The milk yield and milk composition of the current study are in agreement with finding of others Roesch *et al.* (2005), Prandini *et al.* (2009), Butler *et al.* (2011), Stiglbauer *et al.* (2013) and Kuhnen *et al.* (2015) which all reported that MY was higher for conventional cows than organic cows. While in contrast with our results Thomassen *et al.* (2008) reported that milk yield was not significantly different for cows fed organic or conventional systems. No difference could be related to the same rate of forage or energy density in both treatments. Roesch *et al.* (2005), Butler *et al.* (2011) and Kuhnen *et al.* (2015) reported that percentages of milk protein was equal between groups, but a higher fat percentage was seen with organic feeding compared to the conventional feeding because of more forage intake. Conversely, Adler *et al.* (2013) showed that conventionally kept cows had higher milk, fat and protein content in milk than organically managed cows. Muller and Helga (2010) reported that SCC was not significantly different between an organic *vs.* conventional feeding. In contrast, Kuhnen *et al.* (2015) reported that average individual cow SCC levels were significantly higher in organic herds compared to conventional except in winter. Moreover, Table 4 indicated that consumption of organic system did statically ($P<0.01$) higher phytanic acid than CON system.

Similarly, content of hippuric acid was increased ($P<0.001$) by OHF system than CON system (31.3 *vs.* 18.9 mg/L). Cows on organic system consume more forage than CON system.

Therefore, they consume more phytol (part of chlorophyll) which is broken down in ruminant's stomachs to phytanic acid. In agreement to our study, Larsen *et al.* (2010) and Adler *et al.* (2013) founds that, on average, organic milk had double the phytanic acid levels than conventional.

Dietary intake of phytanic acid have been suggested to be involved in both health and disease promoting processes, thus some researchers have suggested that it can prevent diabetes and metabolic diseases, while others have suggested that it promotes development of prostate cancer (Werner *et al.* 2011).

The potential health-promoting properties is based on the fact that animal and *in vitro* studies have shown that phytanic acid might have preventive effects on metabolic dysfunctions, since in animal studies it increases expression of genes involved in fatty acid oxidation, enhances glucose uptake and metabolism in hepatocyte and potentially reduce metabolic efficacy through increased differentiation of brown adipocyte differentiation and expression of uncoupling protein-1 (Werner *et al.* 2011).

Hippuric acid such as phytanic acid could be a marker to distinguish organic milk from cows fed on different feeding system. However, it is necessary to check if the hippuric acid content comes from fresh forage or from organic handling.

Milk urea nitrogen and blood urea nitrogen

Across the 200 days, CON cows had higher (P<0.01) MUN than OHF cows (19.0 vs. 15.0 mg/dL) groups respectively (Table 4). Similarly, CON cows had higher (P<0.05) BUN (Table 4) than OHF cows (4.79 vs. 4.28 mmol/L, respectively). This finding is in accordance to the works of Dystrom et al. (2002), Roesch et al. (2005) and Adler et al. (2013) who found that high forage diets compared to high concentrate diets appear to significantly decrease MUN levels. Furthermore, results of this study are similar to those reported by Trachsel et al. (2000) and Roesch et al. (2005) who reported BUN levels were greater in CON cows compared to organic cows. BUN reduction in the OHF cows can be due to the lower crude protein in diets. It can lead to a decreased ammonia production in the rumen and consequently decreased urea production in the liver. Our study provide support to other study that the participating systems adequately reflect the performance level for the two management systems and are not due to biased selection of systems. Consequently, MUN a fraction of milk protein that is derived from BUN may be one the useful tools (for quality of milk) that may help monitoring of any change required in the feeding and management of a herd.

Blood metabolite

Regarding the blood serum parameters, non esterified fatty acids (NEFA) showed the highest levels at 10 days of parturition for organic system (Figure 1), whereas CON cows had higher (P<0.01) NEFA concentration on days 60 and 90. Blood serum β-hydroxybutyrate acid (BHBA) increased in CON cows after 20 days of parturition above 0.4 mmol/L and reached maximum levels at 20-day in OHF (Figure 1). NEFA serum concentration is an indicator of the lipid mobilization degree from reserve adipose tissue and, in conclusion, of the negative energy balance in ruminants (Ellis et al. 2006). Altogether, studies demonstrate that organic system is not more prone to develop a negative energy balance than conventional system (Roesch et al. 2005; Muller and Sauerwein, 2010; Stiglbauer et al. 2013).

Contrary to our results, Fall et al. (2008) showed that the profiles of all tested metabolic variables NEFA, and BHBA were very similar between organic and conventional systems. Roesch et al. (2005) compared NEFA and BHBA at 30-day post-partum without discovering any differences. But, there have been concerns that the high energy demands

of early lactation cannot be satisfied in organic management.

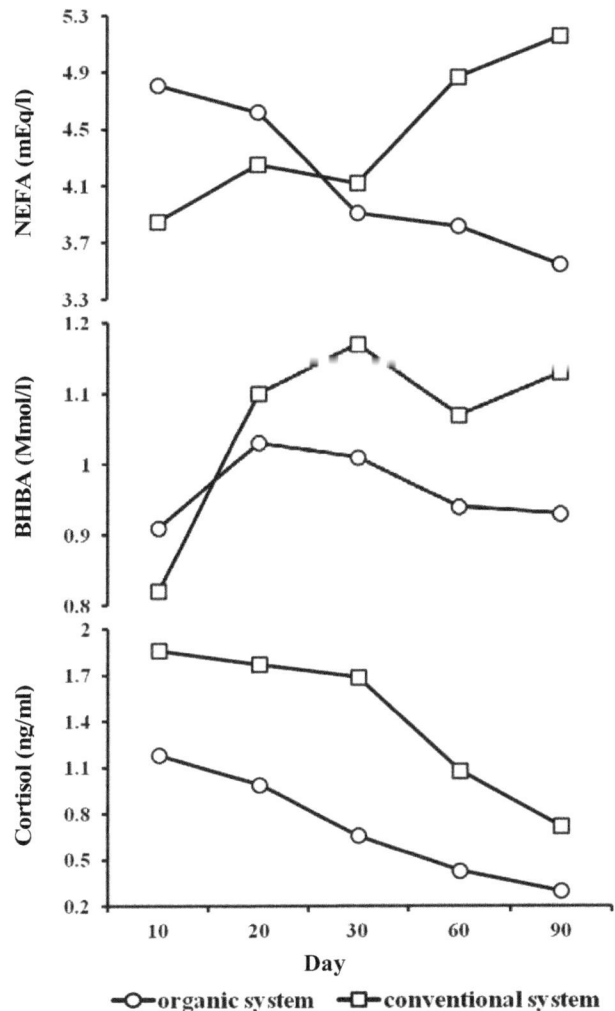

Figure 1 Time courses of non esterified fatty acids (NEFA), β-hidroxy butyrate acid (BHBA), and cortisol in an organic system with high forage (OHF) and conventional system (CON) during the observation periods. Means of NEFA, BHBA and cortisol were significantly diferent (P<0.05) detect among treatments for all periods. Overall SEM for NEFA= 0.08, BHBA= 0.03 and cortisol= 0.06

The results showed that the cortisol level on days 10 was higher in CON cows than OHF cows (1.86 vs. 1.18 ng/mL, respectively, Figure 1) and that the cortisol levels on days 20, 30, 60 and 90 were significantly different for all treatments.

Cortisol, often referred to as the "stress hormone", is a glucocorticoid secreted by the hypothalamic-pituitary-adrenal axis (Katharina et al. 2015).

Levels of cortisol are known to increase in response to physical or psychological stress. Cortisol plays an important role in the body and primarily affects metabolic and immune function (Katharina et al. 2015).

Improved animal welfare is considered one of the key benefits of converting to the organic system (Mullen *et al.* 2015) which logically should equate to reduce cow stress.

Feed cost (FC) and income over feed cost (IOFC)

Across the 200 days, cows fed CON system had significantly higher FC than cows fed OHF systems (7.81 *vs.* 6.09 \$/day, respectively). Cows fed CON system had higher IOFC compared to cows fed OHF system (Table 5). On the other hand, the average cost of production for each of the 2 treatments with organic and conventional systems is listed in Table 5. Production costs were higher in the CON system

than other group because of the higher use of transport, chopping silage, worker, harvesting and electricity costs. The results of this study indicate that the highest profit to cost ratio (PCR=total revenue/total expenses) was related to the dairy cow of OHF system. Economy of animal production is closely associated with the efficiency of breeding. It is generally understood as the company's ability to change the material inputs (expressed as costs) into the marketable product under the common production conditions (Michalickova *et al.* 2014). High level of milk yield which finally reduced the unit cost per kg of milk was the main determinant of difference in this case.

Table 5 Estimation of feed cost, income over feed cost (IOFC) and profit to cost ratio of cows fed organic and conventional systems in Holstein lactating

Item	Treatments		SEM	P-value
	OHF	CON		
Feed cost (\$/d)	6.09	7.81	0.07	*
Early lactation	6.19	7.92	0.09	**
Mid lactation	5.99	7.74	0.09	*
IOFC (%)	54.1	60.0	0.39	**
Early lactation	53.1	58.5	0.44	*
Mid lactation	55.1	61.8	0.44	**
Income[1]				
Total milk (kg)	45000.0	57800.0	-	-
Total milk (kg/per cow)	7500.0	9633.3	-	-
Price of total milk (\$)	22499.3	26010.0	-	-
Price of total milk (\$/per cow)	3749.8	4335.0	-	-
Gross revenue(\$/farm)[2]	8848.3	8826.8	-	-
Profit to cost ratio[3]	1.65	1.51	0.03	*
Feed cost				
Total feed Consumption (kg DM)	35600.0	42400.0	-	-
Gain consumption (kg DM)	8188.0	26712.0	-	-
Forage consumption (kg DM)	27412.0	15688.0	-	-
Feed intake cost (\$)	12178.7	15628.7	-	-
Grain intake cost (\$)	4503.3	12020.3	-	-
Forage intake cost (\$)	7675.3	3608.2	-	-
Fixed cost (\$)[4]	1472.2	1554.5	-	-
Production cost				
Field preparation (\$)	4276.0	8275.0	-	-
Seeds (\$)	521.0	926.3	-	-
fertilizer and pesticides (\$)	0.0	2065.7	-	-
Machinery (\$)	2067.5	4105.5	-	-
Storage feed (\$)	1687.7	1177.5	-	-
Harvesting (\$)	2501.8	1940.5	-	-
Transport (\$)	1315.5	1220.3	-	-
Electricity and fuel (\$)	328.7	226.8	-	-
Chopping (\$)	310.2	172.5	-	-
Worker (\$)	2020.5	1551.8	-	-
Laboratory (\$)	832.0	862.2	-	-
Animal health (\$)	594.2	1379.5	-	-

[1] Income, minus income from calf production has been reported.
[2] Aggregation of income from the sales of farm outputs (GR=$\sum$N RxiYi).
[3] Profit to cost ratio (PCR): PCR= total revenue/total expenses.
[4] Fixed costs included: maintenance costs of farm.
OHF: organic system with high forage and CON: conventional system.
* (P<0.05) and ** (P<0.01).
SEM: standard error of the means.

Michalickova *et al*. (2014) noted comparable value for the profit to cost ratio in milk production in spite of extremely low milk yield per cows reared in mountain and foothill regions.

CONCLUSION

By focusing on different levels of evaluation, we conclude that Holstein cows fed on organic system differ from conventional system. DMI and MY were higher in cows fed CON system compared to those in cows fed the other system. However, profitability is likely to be higher for cows fed OHF system, because cows fed CON system had significantly higher FC and IOFC than OHF system. Therefore, despite the lower MY in OHF system, this system is deemed more appropriate due to the lower cost with the same profit to cost ratio compared with conventional system. Feed efficiency was the greatest for CON cows compared to organic cows. Furthermore, the results indicate that the maximum BUN and MUN in CON system can be effective in reducing milk quality. Moreover, cows of OHF system had higher phytanic and hippuric acids contents than cows of CON system during experiment.

ACKNOWLEDGEMENT

We gratefully thank the team at the experimental dairy farm at the company PAK for financial support. Moreover, University of Tehran is grateful for the support of research.

REFERENCES

Adler S.A., Jensen S.K., Govasmark E. and Steinshamn H. (2013). Effect of short term versus long term grassland management and seasonal variation in organic and conventional dairy farming on the composition of bulk tank milk. *J. Dairy Sci*. **96**, 5793-5810.

AOAC. (1991). Official Methods of Analysis. Vol. I. 15th Ed. Association of Official Analytical Chemists, Arlington, VA, USA.

Butler G., Stergiadis S., Seal C., Eyre M. and Leifert C. (2011). Fat composition of organic and conventional retail milk in northeast England. *J. Dairy Sci*. **94**, 24-36.

Bystrom S., Jonsson S. and Martinsson K. (2002). Organic versus conventional dairy farming studies from the Ojebyn project. Pp. 179-184 in Proc. Orga. Res. Conf. Aberystwyth, UK.

Chaney A.L. and Marback E.P. (1962). Modified reagents for determination of urea and ammonia. *Clin. Chem*. **8**, 130-132.

Ellis K.A., Innocent G., Grove-White D., Cripps P., McLean W.G., Howard C.V. and Mihm M. (2006). Comparing the fatty acid composition of organic and conventional milk. *J. Dairy Sci*. **89**, 1938-1950.

Fall N., Grohn Y., Forslund K., Essen-Gustavsson B., Niskanen R. and Emanuelson U. (2008). An observational study on early lactation metabolic profiles in Swedish organically and conventionally managed dairy cows. *J. Dairy Sci*. **91**, 3983-3992.

Katharina G., Sigl T., Meyer H.H.D. and Wiedemann S. (2015). Cortisol levels in skimmed milk during the first 22 weeks of lactation and response to short-term metabolic stress and lameness in dairy cows. *J. Anim. Sci. Biotechnol*. **6**, 31-38.

Khadem A.A., Sharifi M., Afzalzadeh A. and Rezaeian M. (2009). Effects of alfalfa hay or barley flour mixed alfalfa silage contained diets on feeding behavior, productivity, rumen fermentation and blood metabolites of lactating cows. *Anim. Sci. J*. **80**, 403-410.

Kuhnen S., Stibuski R.B., Honorato L.A. and Machado Filho L.C.P. (2015). Farm management in organic and conventional dairy production systems based on pasture in Southern Brazil and its consequences on production and milk quality. *Anim*. **5**, 479-494.

Larsen M.K., Nielsen J.H., Butler G., Leifert C., Slots T., Kristiansen G.H. and Gustafsson A.H. (2010). Milk quality as affected by feeding regimens in a country with climatic variation. *J. Dairy Sci*. **93**, 2863-2873.

Littell R.C., Henry P.R. and Ammerman C.B. (1998). Statistical analysis of repeated measures data using SAS procedures. *J. Anim. Sci*. **76**, 1216-1231.

Michalickova M., Krupova Z. and Krupa E. (2014). Determination of economic efficiency in dairy cattle and sheep. *Slovak J. Anim. Sci*. **47**, 39-50.

Mullen K.A.E., Dings E.H.A., Kearns R.R. and Washburn S.P. (2015). A comparison of production, reproduction and animal health for pastured dairy cows managed either conventionally or with use of organic principles. *Prof. Anim. Sci*. **31**, 167-174.

Muller U. and Sauerwein H. (2010). A comparison of somatic cell count between organic and conventional dairy cow herds in West Germany stressing dry period related changes. *Live. Sci*. **127**, 30-37.

Prandini A., Sigolo S. and Piva G. (2009). Conjugated linoleic acid (CLA) and fatty acid composition of milk, curd and Grana Padano cheese in conventional and organic farming systems. *J. Dairy Res*. **76**, 278-282.

Roesch M., Doherr M.G. and Blum J.W. (2005). Performance of dairy cows on Swiss farms with organic and integrated production. *J. Dairy Sci*. **88**, 2462-2475.

SAS Institute. (2004). SAS®/STAT Software, Release 9.1. SAS Institute, Inc., Cary, NC. USA.

Sharifi M., Hosseinkhani A., Sofizade M. and Mosavi J. (2016a). Effects of fat supplementation and chop length on milk composition and ruminal fermentation of cows fed diets containing Alfalfa silage. *Iranian J. Appl. Anim. Sci*. **6**, 293-301.

Sharifi M., Khadem A.A., Heins B.J., Pahlavan R. and Safdari M. (2015). The effect of weaning age on performance and economics of Holstein calves reared under organic farming system. *Iranian J. Appl. Anim. Sci*. **5**, 29-33.

Sharifi M., Taghizade A. and Hosseinkhani A. (2016b). Organic Production in Livestock. Education and Agricultural Promotion Press, Tehran, Iran.

Slots T., Butler G., Leifert C., Kristensen T., Skibsted L.H. and Nielsen J.H. (2009). Potentials to differentiate milk composition by different feeding strategies. *J. Dairy Sci.* **92,** 2057-2066.

Stiglbauer K.E., Cicconi-Hogan K.M., Richert R., Schukken Y.H., Ruegg P.L. and Gamroth M. (2013). Assessment of herd management on organic and conventional dairy farms in the United States. *J. Dairy Sci.* **96,** 1290-1300.

Thomassen M.A., Van Calker K.J., Smits M.C.J., Iepema G.L. and De Boer I.J.M. (2008). Life cycle assessment of conventional and organic milk production in the Netherlands. *Agric. Syst.* **96,** 95-107.

Trachsel P., Busato A. and Blum J.W. (2000). Body condition scores of dairy cattle in organic farms. *J. Anim. Physiol. Nutr.* **84,** 112-124.

Werner L.B., Hellgren L.I., Raff M., Jensen S.K., Petersen R.A., Drachmann T. and Tholstrup T. (2011). Effect of dairy fat on plasma phytanic acid in healthy volunteers a randomized controlled study. *Lipids. Health. Disease. J.* **10,** 95-102.

16

Association between *MTNR1A* and *CYP19* Genes Polymorphisms and Economic Traits in Kurdi Sheep

Z. Davari Varanlou[1], S. Hassani[1*], M. Ahani Azari[1], F. Samadi[1],
J. Fakhraie[2] and A. R. Khan Ahmadi[3]

[1] Department of Animal Science, Gorgan University of Agricultural Science and Natural Resources, Golestan, Iran
[2] Department of Animal Science and Veterinary, Khorasan Razavi Agricultural and Natural Resources Research and
  Education Center, AREEO, Mashhad, Iran
[3] Department of Animal Science, Faculty of Agricultural and Natural Resources, Gonbad University, Gonbad, Iran

*Correspondence E-mail: hasani@gau.ac.ir

ABSTRACT

The ovine melatonin receptor 1A (*MTNR1A*) and aromatase (*CYP19*) genes were structurally characterized and the association between their variants and reproductive and growth traits was studied in Kurdi sheep at Kurdi sheep breeding station located in Shirvan, Iran. The genomic DNA was extracted by guanidine thio-cyanate-silica gel method. Polymerase chain reaction was carried out to amplify 824 bp fragment of exon 2 of *MTNR1A* and 140 bp fragment of the exon 3 of the ovine *CYP19* genes. The PCR products were digested with restriction endonucleases RsaI for *MTNR1A* and *BstMBI* for *CYP19* genes and checked by poly-acrylamide gel electrophoresis for the presence of restriction sites. Two alleles were found for all the loci investigated, which were named as A and B for *CYP19*, and R and r for *MTRN1A*. Allelic frequencies for *MTRN1A* were 0.49 and 0.51 for R and r alleles, while in the case of *CYP19* gene, frequencies were 0.475 and 0.525 for A and B alleles, respectively. Association analysis did not show any significant relations between *MTNR1A* gene polymorphisms and litter size (LS), age at first lambing (AFL) and lambing interval (LI). Moreover, *CYP19* gene polymorphism did not affect birth weight (BW), weaning weight (WW), 6, 9 and 12 months (YW) body weights, age at first lambing (AFL) and lambing interval (LI).

KEY WORDS *CYP19* gene, growth traits, *MTNR1A* gene, polymorphism, reproductive traits.

INTRODUCTION

Aromatase is a cytochrome P450 enzyme complex that is encoded by the *CYP19* gene and catalyzes a critical reaction for estrogen biosynthesis involving the formation of aromatic C18 estrogens from C19 androgens. The cytochrome P450 aromatase (P450aro, CYP19) is a microsomal member of the cytochrome P450 superfamily (Nelson *et al.* 1993). The aromatase cytochrome P450 is necessary for the biosynthesis of estrogens in several tissues, most importantly ovaries, adipose tissue and brain. Estrogens play fundamental roles including endocrine, paracrine and autocrine activities involved in there gulation of male and female reproduction also in metabolic processes like fat deposition and growth (Heine *et al.* 2000; Jones *et al.* 2000; Simpson *et al.* 2000). The *CYP19* gene has been mapped to bands q24-q31 of chromosome 7 in sheep (Payen *et al.* 1995; Goldammer *et al.* 1999). In codon 69 which is located in exon 3, a silent C/T transition in several animals was found (Vanselow *et al.* 1999). Lôbo *et al.* (2009) have reported that in Brazilian sheep breeds, this polymorphism makes the differences in performance traits including litter weight, lambing interval, lambing age, reproductive and maternal ability. The melatonin pineal hormone (N-acetyl-5-

methoxytryptamine) occurs only during the hours of darkness which regulates circadian rhythms and reproduction changes in mammals with seasonally reproductive function (Reppert *et al.* 1994). The MLT (melatonin) can also be produced by extra-pineal sites like the retina, the gastrointestinal tract and the innate immune system (Jaworek *et al.* 2005). In mammals, two specific receptors sub types i.e. MT1 and MT2, encoded by the *MTNR1A* and *MTNR1B* genes, respectively. The MT1 and MT2 receptors are involved in the melatonin secretion, of which, only the melatonin receptor subtype 1A (*MNTR1A*) gene is considered to be a candidate gene and seems to play a key role in the control of photoperiod-induced seasonality mediated by the circadian concentrations of melatonin (Dubocovich *et al.* 1988; Weaver *et al.* 1996). The *MTNR1A* gene has been mapped to ovine chromosome 26, consists of two exons divided by a large intron (Reppert *et al.* 1994; Messer *et al.* 1997). Exon II of the gene encoding the MT1 receptor in sheep has two restriction fragment length polymorphism (RFLP) sites, one for *MnlI* and the second for *RsaI* enzyme (Messer *et al.* 1997). In sheep, the MT1 receptor encoded by exon 2 of the *MTNR1A* gene and this exon has two restriction fragment length polymorphism (RFLP) sites, one for *MnlI* and the second for *RsaI* enzyme (Messer *et al.* 1997). The structure and polymorphism of exon 2 of the *MTNR1A* gene using the *RsaI* restriction enzyme has been evaluated in several sheep breeds (Chu *et al.* 2003; Notter *et al.* 2003; Mateescu *et al.* 2009; Hristova *et al.* 2012; Martínez-Royo *et al.* 2009; Moradi *et al.* 2014). Melatonin acts as a natural inhibitor of the aromatase activity and expression by regulating the gene expression of specific aromatase promoter regions (Martınez-Campa *et al.* 2012). The geographic origins of the animals and photoperiod, with the intermediary activity of melatonin are important factors regarding the sheep reproductive activity through effecting on the aromatase activity (Mora *et al.* 2014). The objectives of the present study were first to detect the PCR-RFLP polymorphism of *MTNR1A* and *CYP19* genes and secondly to investigate the associations between *MTNR1A* and *CYP19* genes and growth and reproductive traits in Kurdi sheep.

MATERIALS AND METHODS

Genotyping

In this study, venous jugular blood samples (5 mL per ewe) were collected from 120 pure bred Kurdi ewes from Kurdi sheep breeding station located in Shirvan, Iran and transferred into vacutainer tubes containing 0.5 molar ethylene diamine tetracetic acid (EDTA) as anticoagulant and frozen at -20 °C. Genomic DNA was extracted from whole bloodusing a commercial kit (Diatom DNA Prep100, ISO

Gene, Moscow) following the manufacturer's protocol.

The quantity and quality of the isolated DNA were determined using spectrophotometry and agarose gel electrophoresis. Polymerase chain reactions (PCR) were carried-out using Personal Cycler™ thermocycler (Biometra, Germany) and PCR Master Kit (Cinnaclon Inc., Iran). Master Mix consisted of 0.04 U/μL of TaqDNA polymerase, 10X PCR buffer, 3 mM MgCl$_2$ and 0.04 mM dNTPs (each). Each reaction mixture consisted of12.5 μL of the master mix, 1 μL of the DNA solution (50 to 100 ng/μL), 1 μL of each primer (5 pmol/μL) and some deionized water making up a final volume of 25 μL.

Amplification of a 140 bp fragment of the exon 3 of the ovine *CYP19* gene was carried out using primers (synthesized by CinnaGen, Iran) described by Vanselow *et al.* (1999), in agreement with the sequence deposited in GenBank (AJ012153):

CYP19-F (5'-CCA GCT ACT TTC TGG GAA TT-3')
CYP19-R (5'-AAT AAG GGT TTC CTC TCC ACA-3')

The amplification program consisted of an initial denaturation at 94 °C for 5 min followed by 35 cycles of denaturation at 94 °C for 30 sec, annealing at 55 °C for 30 sec, extension at 72 °C for 30 sec and a final extension at 72 °C for 5 min. For amplifying an 824 bp fragment of the main part of the exon 2 of the ovine *MTNR1A* gene with specific primers (synthesized by CinnaGen, Iran) as described by Messer *et al.* (1997), in agreement with the sequence deposited in GenBank (U14109):

MTNR1A-F(5'-TGTGTTTGTGGTGAGCCTGG-3')
MTNR1A-R(5'-ATGGAGAGGGTTTGCGTTTA-3')

The amplification reaction was carried-out under the following conditions: an initial denaturation step at 94 °C for 5 min followed by 35 cycles of denaturation at 94 °C for 1 min, annealing at 58.5 °C for 1 min and extension at 72 °C for 2 min and a final extension of 72 °C for 5 min. Then, products of amplification were analyzed by 1.5% agarose gel electrophoresis. The gels were stained with ethidium bromide and visualized under ultraviolet light. A 10 μL of PCR products were incubated for 14h at 37 °C with 1 μL (10 units) of *BstMBI* and *RsaI* enzymes for Cyp19 and *MTNR1A* genes, respectively. The digestion products were also electrophoresed on 8% acrylamide gel and visualized in parallel with a 50 bp DNA marker.

Statistical analysis

Determination of genotypic and allelic frequencies and Hardy-Weinberg (H-W) equilibrium test were carried out using Pop-Gene software (V 1.31) (Yeh *et al.* 1997).

In order to test the association of different conformational patterns with the studied traits, statistical analysis was performed using general linear model (GLM) procedure of the SAS program and least squares means of the banding patterns were compared using the Tukey-Kramer test at 5 percent probability level (SAS, 2000).

Studied traits were growth and reproductive traits including birth weight (BW), weaning weight (WW), 6, 9 and 12 (YW) month weights, age at first lambing (AFL) and lambing interval (LI).

The Following models were used for growth and reproductive traits, respectively:

$$y_{ijklm} = \mu + G_i + A_j + B_k + T_l + e_{ijklm}$$
$$y_{ijklmn} = \mu + G_i + YC_j + MC_k + A_l + YB_m + e_{ijklmn}$$

Where:

y_{ijklm} and y_{ijklmn}: growth and reproductive traits, respectively.

μ: overall mean.

G_i: fixed effect of the ith banding patterns (i=1,2,3).

A_j: fixed effect of the jth dam age (j=1,...,8).

B_k: fixed effect of the Kth year.

T_l: fixed effect of the lth birth type.

YC_j: fixed effect of the jth lambing year.

MC_k: effect of the kth lambing season.

YB_m: effect of mth birth year.

e_{ijklm} and e_{ijklmn}: random residual errors.

RESULTS AND DISCUSSION

A 140 bp fragment of the ovine cyp19 gene from exon 3 was amplified successfully. *BstMBI* restriction enzyme was used to digest the PCR products. The PCR digestion products of 120 samples showed only two genotypes: AB and BB. AB genotype exhibited 140, 82 and 58 bpfragments and BB genotype had only one fragment, 140 bp (Figure 1) which was in agreement with Vanselow *et al.* (1999) and Lôbo *et al.* (2009) reports. The frequencies of individual alleles and genotypes in the present study are shown in Table 1.

The frequency of B (0.525) was higher than allele A (0.475) and frequency of AB (0.95) was higher than BB genotype. No significant relation (P>0.05) was found between cyp19 conformational patterns and all the studied traits in the population (Table 2). In study of Vanselow *et al.* (1999) on five breed groups of European sheep, the allele frequencies were 0.74 for allele A and 0.26 for allele B in Hungarian Merino sheep (n=38), in Awassi, Tsigaja, Brith Milk sheep (0.6 for allele A and 0.4 for allele B; n=5) and for Lacaune breed, a frequency of 1.0 for allele A and zero for allele B (n=5).

In another study on several breed groups, a greater frequency of allele B was observed in the Brazilian Somali (1, n=13), Poll Dorset (0.61, n=9) Santa Inês (0.6, n=71) and 1/2 Dorper (0.8, n=18).

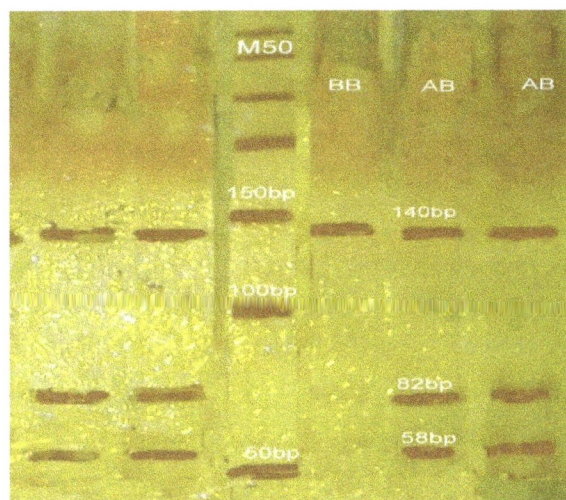

Figure 1 Analysis of RFLP polymorphism of aromatase gene (*Cyp19*) in Kurdi sheep. Non-digested PCR products are 140 bp in size (allele B). In the case of allele A, there were two fragments of 82 bp and 58 bp, respectively
PM: 50 bp molecular weight ladder
AB and BB: deduced genotypes

Table 1 Observed alleles and genotypic frequencies for *CYP19* gene in Kurdi sheep

Allele frequencies		Genotype frequencies		
A	B	AA	AB	BB
0.475	0.525	0	0.95	0.05

In the studied population, AA genotype was not observed and results indicated a relation between the genotypes and some growth and reproductive traits, so that, lower age at first lambing in all 1/2 Dorper BB and lower lambing interval in Santa Inês BB ewes and higher litter weight at weaning for AB ewes (in same genetic groups) were observed (Lôbo *et al.* 2009). Mora *et al.* (2014), investigated C242T polymorphism at the Cyp19 gene in four breed groups composite of Texel, Dorper, White Dorper and Santa Inês and three distinct genotypes: AA, AB and BB were observed. In their study, the Texel sheep group with European origin had highest frequencies for allele A but highest frequencies of allele B was observed in White Dorper sheep originated from tropical countries. Results suggested a relation between the higher frequency of alleles A and B with the ancestral geographic origin of the sheep. In agreement with the results of the present investigation, allele B frequency in Brazilian Somali, Poll Dorset, Santa Inês and 1/2 Dorper sheep (Lôbo *et al.* 2009) and Dorper , White Dorper and Santa Inês (Mora *et al.* 2014) was higher than allele A.

Table 2 Least square means of studied traits for the *CYP19* gene

Trait	Genotypes		F-statistic	P-value
	AB	BB		
BW (kg)	4.49±0.07	4.86±0.42	1.06	0.35
WW (kg)	26.78±0.36	25.08±1.93	0.98	0.37
6MW (kg)	29.59±0.41	28.60±2.35	0.10	0.90
9MW (kg)	41.73±8.12	43.43±49.91	0.05	0.95
YW (kg)	41.41±0.53	43.25±3.05	0.49	0.61
AFL (d)	676.82±10.58	662.36±30.14	0.07	0.93
LI (d)	353.94±3.92	357.54±11.08	0.35	0.70

BW: birth weight; WW: weaning weight; 6MW: 6-month weight; 9MW: 9-month weight; YW: yearling weight; AFL: age at first lambing and LI: lambing interval.

But in Texel sheep (Mora *et al*. 2014) and Hungarian Merino, Awassi, Tsigaja, Brith Milk and Lacaune sheep (Vanselow *et al*. 1999) higher frequency of allele A was found. Also, there were no animals with AA genotype in our study, probably due to the low frequency of allele A which was in agreement with the results obtained by Vanselow *et al*. (1999) and Lôbo *et al*. (2009), but disagree with the results found by Mora *et al*. (2014). Different genotypes for the cype19 gene among sheep produced a differences in some reproductive and growth traits (first lambing and lambing interval, weight at birth and at weaning, and daily weight gain) (Lôbo *et al*. 2009).

In the present study, no significant association was found between genotypes and the studied traits in Kurdi sheep. Apart from above, this locus did not show Hardy-Weinberg equilibrium. This approves that factors leading to disequilibrium, especially selection, may influence the genetic structure of the population. Exon 2 of *MTNR1A* gene with 824 bp length was amplified. *RsaI* restriction enzyme recognizes and cuts the PCR products. For *RsaI*, four cleavage sites (53 bp, 267 bp, 23 bp, 411 bp and 70 bp) within the amplification fragment was found but only one fragment was polymorphic (Chu *et al*. 2003).

This site was at 604 positionin the reference sequence (Reppert *et al*. 1994). Digestion of 120 samples with *RsaI* revealed three genotypes i.e. RR (411 bp/267 bp), Rr (411/290 bp/267 bp) and rr (411 bp/290 bp) in Kurdi sheep (Figure 2).

These results were consistent with those of Notter *et al*. (2003), Chu *et al*. (2006) and Martínez-Royo *et al*. (2009), while the rr genotype was not found in local Karnobatska breed (Hristova *et al*. 2012). Furthermore, for Chios, White Karaman and Awassi breeds, only one genotype (rr) was detected and no polymorphism at the *RsaI* cleavage sites was founding three sheep breeds (Şeker *et al*. 2011). Frequencies of individual alleles and genotypes in the present study are shown in Table 3. In the present study, frequencies of RR, Rr and rr genotypes were 0.275, 0.5 and 0.275, respectively which were similar to those recorded in German Mutton Merino ewes (0.24 RR, 0.48 Rr and 0.28 rr) by Chu *et al*. (2006).

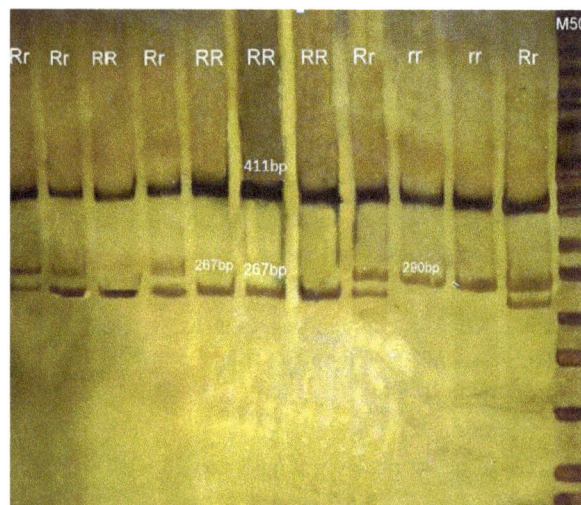

Figure 2 Analysis of RFLP polymorphism exon 2 of the *MTNR1A* gene in Kurdi sheep. Three genotypes: RR (411 bp/267 bp), Rr (410/290 bp/267 bp) and rr (411 bp/290 bp) were detected
PM: 50 bp molecular weight ladder

Table 3 Observed alleles and genotypic frequencies for *MTNR1A* gene in Kurdi sheep

Allele frequencies		Genotype frequencies		
R	r	RR	Rr	rr
0.49	0.51	0.275	0.45	0.275

Chu *et al*. (2006), determined allele and genotype frequencies of *MTNR1A* gene in non-seasonal estrous breeds (Small Tail Han, Hu ewes) and in seasonal estrous breeds (Suffolk, Dorset and German Mutton Merino ewes). A frequency of RR genotypes was higher, and frequency of rr genotype was lower in non-seasonal estrous sheep breeds than in seasonal estrous sheep breeds. Moreover, they detected a relation between rr genotype and seasonal estrus in ewes and association between RR genotypes and non-seasonal estrus in ewes, while in the Rasa Aragonesa breed rallele of SNP606/*RsaI* of *MNTR1A* gene was associated with a higher percentage of oestrous cyclic ewes. These findings, indicated that other genes closely linked or regulatory sequences of the *MNTR1A* gene could be influencing the ability to breed out of season (Martínez-Royo *et al*. 2009).

Table 4 Least square means of studied traits for the *MTNR1A* gene

Trait	Genotypes			F-statistic	P-value
	RR	Rr	rr		
AFL (d)	678.5±15.43	675.96±16.33	678.58±16.18	1	0.39
LI (d)	351.94.5±5.94	352.74.5±5.80	355.50±6.06	0.55	0.65
LS (d)	1.13±0.14	1.16±0.14	1.07±0.14	1.08	0.35

AFL: age at first lambing; LI: lambing interval and LS: litter size.

In the present study, no significant relation (P>0.05) was found between *MTNR1A* conformational patterns and the studied traits in Kordi sheep (Table 4). Similar to these findings, in the study of Notter *et al.* (2003), genotypic effects on litter size were small and not significant, while Chu *et al.* (2003) identified a relation between the *MTNR1A* gene and litter size of ewes at second lambing seasonal and highly prolific Han sheep. The local populations of Bulgarian sheep breeds, Starozagorska, Karnobatska, Breznishka and Sofiiska (Elin-Pelinska) were characterized by frequency of the R allele: 0.302, 0.729, 0.520, 0.526 and r allele: 0.698, 0.271, 0.480, 0.474 and respectively. These findings confirmed the importance of MTNR1A gene as a potential DNA marker in marker – assisted selection (Hristova *et al.* 2012).

The present study should be considered as preliminary investigation and further research is needed to provide better distinguishing function of *MTRN1A* and *CYP19* genes and determination of their effects on economic traits of Kurdi sheep.

CONCLUSION

Genetic polymorphism was approved for *MTNR1A* and *CYP19* genes in Kurdi sheep. No significant association between the polymorphisms of these genes with reproductive and growth traits was found. Further researches with more number of observations are needed for more reliable association study.

ACKNOWLEDGEMENT

This work was supported by by Gorgan University of Agricultural Science and Natural Resources, (Golestan, I.R. Iran). The authors are also thankful to the staff of Hoseianabad sheep breeding station, Shirvan, Iran, for their great help to provide blood samples and data.

REFERENCES

Chu M., Cheng D., Liu1 W., Fang L. and Ye S. (2006). Association between melatonin receptor 1A gene and expression of reproductive seasonality in sheep. *Asian-Australas J. Anim. Sci.* **19**, 1079-1084.

Chu M.X., Ji C.L. and Chen G.H. (2003). Association between PCR-RFLP of melatonin receptor 1a gene and high prolificacy in Small Tail Han sheep. *Asian-Australas J. Anim. Sci.* **16**, 1701-1704.

Dubocovich M.L., Yun K., Al Ghoul W.M., Benloucif S. and Masana M.I. (1998). Selective MT2 melatonin receptor antagonists block melatonin mediated phase advances of circadian rhythms. *Faseb. J.* **12**, 1211-1220.

Goldammer T., Brunner R.M., Vanselow J., Zsolnai A., Fürbass R. and Schwerin M. (1999). Assignment of the bovine aromatase encoding gene CYP19 to 10q26 in goat and 7q24-q31 in sheep. *Cytogenet. Cell. Genet.* **85**, 258-259.

Heine P.A., Taylor J.A., Iwamoto G.A., Lubahn D.B. and Cooke P.S. (2000). Increased adipose tissue in male and female estrogen receptor-alpha knockout mice. *Proc. Natl. Acad. Sci. USA.* **97**, 12729-12734.

Hristova D., Georgieva S., Yablanski T., Tanchev S., Slavov R. and Bonev G. (2012). Genetic polymorphism of the melatonin receptor MT1 gene in four Bulgarian sheep breeds. *J. Agric. Sci. Technol.* **4**, 187-192.

Jaworek J., Brzozowski T. and Konturek S.J. (2005). Melatonin as an organoprotector in the stomach and the pancreas. *J. Pineal. Res.* **38**, 73-83.

Jones M.E., Thorburn A.W., Britt K.L., Hewitt K.N., Wreford N.G., Proietto J., Oz O.K., Leury B.J., Robertson K.M. and Yao S. (2000). Aromatase-deficient (ArKO) mice have a phenotype of increased adiposity. *Proc. Natl. Acad. Sci.* **97**, 12735-12740.

Lôbo A.M., Lôbo R.N. and Paiva S.R. (2009). Aromatase gene and its effects on growth, reproductive and maternal ability traits in a multi breed sheep population from Brazil. *Genet. Mol. Biol.* **32**, 484-490.

Martínez-Campa C., González A., Mediavilla M.D., Alonso-González C., Alvarez-García V., Sánchez-Barceló E.J. and Cos S. (2009). Melatonin inhibits aromatase promoter expression by regulating cyclooxygenases expression and activity in breast cancer cells. *Br. J. Cancer.* **101**, 1613-1619.

Martinez-Royo A., Lahoz B., Alabart J.L., Folch J. and Calvo J.H. (2012). Characterisation of the melatonin receptor 1A (*MTNR1A*) gene in the Rasa Aragonesa sheep breed: association with reproductive seasonality. *Anim. Reprod. Sci.* **133**, 169-175.

Mateescu R., Lunsford A. and Thonney M. (2009). Association between melatonin receptor 1A gene polymorphism and reproductive performance in Dorset ewes. *J. Anim. Sci.* **87**, 2485-2488.

Messer L.A., Wang L., Tuggle C.K., Yerle M., Chardon P., Pomp D., Womack J.E., Barendse W., Crawford A.M., Notter D.R. and Rothschild M.F. (1997). Mapping of the melatonin receptor 1a (*MTNR1A*) gene in pigs, sheep and cattle. *Mamm. Gen.* **8**, 368-370.

Mora N.H., Silva S.C., Tanamati F., Schuroff G.P., Macedo F.A. and Gasparino E. (2014). Polymorphism C242T in the Cyp19 gene in meat sheep. *Brazilian J. Biol.* **76,** 205-208.

Moradi N., RahimiMianji G., Nazifi N. and Nourbakhsh A. (2014). Polymorphism of the melatonin receptor 1A gene and its association with litter size in Zel and Naeinisheep breeds. *Iranian J. Appl. Anim. Sci.* **4,** 79-87.

Nelson D.R., Kamataki T., Waxman D.J., Guengerich F.P., Estabrook R.W., Feyereisen R., Gonzalez F.J., Coon M.J., Gunsalus I.C., Gotoh K. and Nebert D.W. (1993). The P450 superfamily: update on new sequences, gene mapping, accession numbers, early trivial names of enzymes and nomenclature. *DNA. Cell. Biol.* **12,** 1-51.

Notter D.R., Cockett N.E. and Hadfield T.S. (2003). Evaluation of melatonin receptor 1a as a candidate gene influencing reproduction in an autumn-lambing sheep flock. *J. Anim. Sci.* **81,** 912-917.

Payen E., Saidi-Mehtar N., Pailhoux E. and Cotinot C. (1995). Sheep gene mapping: assignment of ALDOB, CYP19, WT and SOX2 by somatic cell hybrid analysis. *Anim. Genet.* **26,** 331-333.

Reppert S.M., Weaver D.R. and Ebisawa T. (1994). Cloning and characterization of a mammalian melatonin receptor that mediates reproductive and circadian responses. *Neuron.* **13,** 1177-1185.

SAS Institute. (2000). SAS®/STAT Software, Release 8.1. SAS Institute, Inc., Cary, NC. USA.

Şeker İ., Özmen Ö., Çinarkul B. and Ertugrul O. (2011). Polymorphism in melatonin receptor 1A (*MTRN1A*) gene in chios, White Karaman and Awassi sheep breeds. *Kafkas. Univ. Vet. Fak. Derg.* **17,** 865-868.

Simpson E., Rubin G., Clyne C., Robertson K., O'Donnell L., Jones M. and Davis S. (2000). The role of local estrogen biosynthesis in males and females. *Trends. Endocrinol. Metab.* **11,** 184-188.

Vanselow J., Kühn C., Fürbass R. and Schwerin M. (1999). Three PCR/RFLPs identified in the promoter region 1.1 of the bovine aromatase gene (CYP19). *Anim. Genet.* **30,** 232-233.

Weaver D.R., Liu C. and Reppert S.M. (1996). Nature's knockout: the Mel1b receptor is not necessary for reproductive and circadian responses to melatonin in Siberian hamsters. *Mol. Endocrinol.* **10,** 1478-1487.

Yeh F.C., Yang R.C. and Boyle T. (1999). POPGENE: Microsoft Windows Based Freeware for Population Genetic Analysis. Molecular Biology and Technology Center, Unversity of Alberta. Canada.

Validation of Reference Genes for Real Time PCR Normalization in Milk Somatic Cells of Holstein Dairy Cattle

M. Muhaghegh-Dolatabady[1*], H. Hossainy-Dolatabady[1], E. Heidari Arjlo[2]
and K. Mahmoudi[?]

[1] Department of Animal Science, Faculty of Agriculture, Yasouj University, Yasouj, Iran
[2] Cellular and Molecular Research Center, Yasouj University of Medical Science, Yasouj, Iran

*Correspondence E-mail: mmuhaghegh@yu.ac.ir

ABSTRACT

Real time-qPCR is the most reliable method for evaluation of mRNA expression levels. However, to obtain accurate results, selection of suitable reference genes is necessary for normalizing the real-time qPCR data. The aim of this research was to validate the expression stability of three potential reference genes (*ACTB*, *GAPDH* and *UXT*) in milk somatic cells of Holstein dairy cattle under different lactation stages. For this purpose two types of milk samples from eighteen healthy cows at three lactation stages (early, middle and late of lactation cycle) and four mastitic cows were included in this experiment. Total RNA was extracted from the milk somatic cells and then cDNA was synthesized. Real-time polymerase chain reaction (PCR) performed for *ACTB*, *GAPDH* and *UXT* genes as candidate reference genes. Then, the real-time PCR results were analyzed with BestKeeper program. The evaluation of selected genes by real-time PCR revealed that all genes were expressed in the healthy and mastitic dairy cows. In addition, the *UXT* and *GADPH* genes displayed the lowest and highest values of expression level, respectively. The *ACTB* gene was considered as the most suitable internal controls as it was stably expressed in milk somatic cells regardless of dairy cows conditions. Taken together, our results could help to select suitable reference gene for the normalization of expression levels in milk somatic cells of dairy cattle.

KEY WORDS BestKeeper, dairy heifers, milk somatic cell, reference gene.

INTRODUCTION

Quantitative real-time PCR (qPCR) technique is considered to be the most accurate and reliable method for gene expression analysis. It has the advantages of sensitivity, real time detection of reaction progress, speed of analysis and precise quantification of the material in the sample (Gachon *et al.* 2004). The qPCR is a multistage process and the accuracy of obtained results depends on several factors including the quality, stability and input of RNA, the efficiency of reverse transcription, primer performance, reference genes, PCR steps and method chosen for data analysis

(Bustin, 2002; Bustin and Nolan, 2004; Pfaffl, 2001; Skern *et al.* 2005; Fleige and Pfaffl, 2006; Derveaux *et al.* 2010). Among them, the choice of suitable reference genes to normalize data is a great importance to obtain accurate results. A suitable reference gene should be expressed at a constant level among samples, and its expression is assumed to be unaffected by the experimental conditions (Bustin, 2002). The use of unsuitable reference genes may lead to errors in quantification and, then, the expression data may lead to misinterpretation. Reference gene validation was carried out in different organs of dairy and beef cattle, such as adipose tissue (Saremi *et al.* 2012), liver, kidney, pituitary and thyroid (Lisowski *et al.* 2008), milk somatic cells (Varshney *et al.* 2012; Verbeke *et al.* 2015), mammary gland (Bionaz and Loor, 2007; Bougarn *et al.* 2011), oocyte (Macabelli *et al.* 2014; Mahdipour *et al.* 2015) and whole blood samples of cows (Devrim *et al.* 2012; Kishore *et al.* 2013; Kizaki *et al.* 2013).

Several statistical procedures or software packages have been reported to evaluate the stability expression in candidate reference genes, such as geNorm (Vandesompele *et al.* 2002), NormFinder (Andersen *et al.* 2004), BestKeeper (Pfaffl *et al.* 2004) and Stability index (Brunner *et al.* 2004), with the ranking of candidate reference genes depending upon the selected software. Up to now, based on our knowledge, there is no any report for validation of reference genes in milk somatic cells of Holstein dairy heifers. Therefore, the aim of this study was to evaluate the stability of β-actin, related to cell structure (*ACTB*), glyceraldehyde-3-phosphate dehydrogenese, related to carbohydrate metabolism (*GAPDH*) and ubiquitously-expressed transcript, related to activation of transcriptional activation (*UXT*) genes in milk somatic cells of Holstein dairy cows at first lactation under different lactation stages using BestKeeper program.

MATERIALS AND METHODS

Eighteen healthy Holstein dairy cows at first lactation were classified according to their lactation stages (6 at 7-10, 6 at 140-150 and 6 at 290-295 days after parturition). The selection criteria was somatic cell count (SCC) less than 350000/mL milk for early lactation stage and SCC < 100000/mL for middle and late lactation stages. In addition of healthy cows, four dairy cows at first lactation with clinical mastitis were also included in this experiment. In healthy cows, one liter of milk sample representing all four quarters was collected in sterile tubes. The milk samples from cows with mastitis were collected from the quarter with clinical mastitis immediately after the onset of clinical signs and before drug treatment. Then, the milk sample was centrifuged for 20 min at 1500 g at 4 °C. The cell pellet was washed in phosphate-buffered saline (PBS) pH 7.4 twice and centrifuged for 20 min at 4 °C and 220 g according to Liebe (1996). The pellets were resuspended with 500 μL phosphate-buffered saline (PBS)- ethylenediaminetetraacetic acid (EDTA) and kept at -40 °C until RNA isolation.

Total RNA was extracted using Denazist kit according to the manufacturer's protocol. All samples were DNase I (Cinnagen) treated to eliminate DNA genomic contamination. RNA quality was assessed by electrophoresis on 1% agarose gel stained with ethidium bromide. All RNA samples were reverse transcribed using *AccuPower® Rocket-Script*™ *RT PreMix* kit (Bioneer) and random hexamer

primers (Takapozist) according to the manufacture instructions. The final volume was adjusted to 50 μL with RNase free water. The amplified cDNA samples were then stored at -20 °C until use in real-time PCR. The primers sequences and their characteristics are shown in Table 1. Real-time PCR using CFX96 (BIORAD, USA) was performed on 3 candidate endogenous control genes for each of the 22 bovine milk cell samples using the HotTaq EvaGreen qPCR kit (Cinnagen). A PCR mix (10 μL) was prepared to give the end concentrations: 5 μL water, 1 μL each of the forward and reverse primers (10 pm), 1 μL of cDNA and 2 μL of *HotTaq EvaGreen* qPCR master mix. All reactions were performed in duplicate. The amplification conditions were based on reference papers (Table 1). In each reaction of real-time PCR, the cycle number at which the fluorescence rises appreciably above the background fluorescence is determined as crossing point (CP). Subsequently, a melting step was performed, consisting of 95 °C for 5 s. 65 °C for 5 s. and slow heating at a rate of 0.5 °C per 5 second up to 95 °C, with continuous fluorescence measurement, and finally followed by cooling down to 25 °C.

Expression stability of potential references genes are evaluated by BestKeeper program. To identify the most stable reference gene, this program use raw CP values and amplification efficiencies. The stability estimation of reference genes expression is based on the inspection of calculated variations (standard deviation (SD) and coefficient of variation (CV) values). According to the variability observed, reference genes can be ordered from the most stably expressed, exhibiting the lowest variation, to the least stable one, exhibiting the highest variation. Any studied gene with the SD higher than 1 can be considered inconsistent (Pfaffl *et al.* 2004). In addition, pearson correlations between each individual gene and the BestKeeper index (geometric mean between CP values of stable genes) were calculated as the BestKeeper correlation coefficient. Genes with the highest BestKeeper correlation coefficient were considered the most stably expressed (Bonefeld *et al.* 2008; Mehta *et al.* 2010). For each primer, the efficiency of qPCR and R^2 were estimated using a standard curve obtained from a pooled cDNA of all samples serially diluted 10-fold over 6 measuring points with two replications.

RESULTS AND DISCUSSION

The screening of three potential references genes by real-time PCR showed that all genes were expressed in the healthy (at different lactation stages) and mastitic Holstein dairy cows. Gene expression levels of the candidate reference gene (expressed in CP values) are displayed in Table 2. In this study, the expression levels of *UXT* and *GADPH* genes revealed lowest and highest values, respectively.

Table 1 Primers characteristics of 3 potential reference genes in real time PCR

Gene	Primer	Sequence (5'-3')	Length bp	Accession	Reference
ACTB	β-actin.f38	CCTTTTACAACGAGCTGCGTGTG	391	AH00130	(Lee et al. 2006)
	β-actin.r428	ACGTAGCAGAGCTTCTCCTTGATG			
GADPH	GADPH.463f	GGCGTGAACCACGAGAAGTATAA	120	AF022183	(Leutenegger et al. 2000)
	GADPH582r	CCCTCCACGATGCCAAAGT			
UXT	UXT.323F	TGTGGCCCTTGGATATGGTT	101	BQ676558	(Bionaz and Loor, 2007)
	UXT.423R	GGTTGTCGCTGAGCTCTGTG			

Gene expression variation was calculated for all three candidate reference genes based on CP-values and displayed as the standard deviation (SD) and coefficient of variation (CV). BestKeeper highlighted ACTB as the reference gene with the least overall variation from the three candidate genes with an SD of 0.96 (Table 2), which represents an acceptable 1.95 fold change in expression. The variation in expression of UXT and GADPH genes was greater than two-fold (SD greater than 1.0). Among the examined three candidate reference genes, UXT lies in middle with respect to its stability (Figure 1). Subsequently, pair-wise correlation between genes and also correlation between each gene and the BestKeeper index were calculated (Table 3). Correlations between the three genes ranged from -0.05 for UXT/ACTB to 0.44 for GADPH/ACTB. The highest correlation between candidate genes and the BestKeeper index was obtained for GADPH gene (0.88) that was as the least stable expressed reference gene. However, the ACTB gene also displayed significant correlation with the BestKeeper index (P<0.01).

A number of authors have studied expression profiles of housekeeping genes in milk somatic cells of cattle, goat and yak using qRT-PCR, but based on our knowledge; this is the first report that the validation of housekeeping gene was investigated in Holstein dairy cows at first lactation in three lactation stages. The three candidate housekeeping genes that were evaluated in this experiment were selected from three studies investigating gene expression in milk somatic cells of Holstein dairy cattle (Table 1). PCR amplification products were obtained for three housekeeping genes in all samples but UXT was displayed consistently high CP values (greater than 35) in both sample groups and suggesting it is not expressed in sufficient quantity to be used as an effective housekeeping gene in milk somatic cells (Table 2). Very low level of amplification was also observed for UXT gene in milk somatic cell of Sahiwal dairy cattle through lactation (Varshney et al. 2012). In similar study, GAPDH showed the highest expression, whereas the expression of UXT was the lowest in mammary epithelial cells of buffalo during all stages of lactation (Yadav et al. 2012).

In the present study, ACTB was the most stable reference genes in Holstein dairy heifers at different conditions, making it as a suitable reference gene for normalization of real-time PCR data in milk somatic cells.

Actins are the main structural protein of cytoplasm and play important role in cell secretion, motility, cytoplasm flow and cytoskeleton maintenance (Hunter and Garrels, 1977). Verbeke et al. (2015) identified ubiquitin C (UBC), ribosomal protein S15a (RPS15A) and ACTB as the most stable genes based on their expression in bovine milk somatic cells. The expression stability of ribosomal protein L4 (RPL4), elongation factor 1 alpha (EEF1A1), GAPDH and ACTB genes were also reported in mammary epithelial cells across different lactation stages of Indian cows (Jatav et al. 2016). Nine candidate reference genes including UXT, GAPDH and ACTB were assessed in milk somatic cells of Sahiwal dairy cattle and results revealed that PPP1R11 (Protein phosphatase 1, regulatory (inhibitor) subunit 11), ACTB, UBC and GAPDH were stably expressed genes among all candidate reference genes (Varsheny et al. 2012). Jarczak et al. (2014) evaluated six potential reference genes and found that peptidylprolyl isomerase A (PPIA) and ribosomal phosphoprotein P0 (RPLP0) are the most suitable internal controls as they were stably expressed in goat milk somatic cells regardless of disease status. In addition, the expression stability of ten commonly used reference genes such as ACTB and GAPDH were examined in milk somatic cells from goats in mammary gland challenged with Staphylococcus aureus and in milk somatic cells from healthy controls. The Glucose 6-phosphate dehydrogenase (G6PD), Tyrosine 3-monooxygenase/tryptophan 5-monooxygenase activation protein, zeta polypeptide (YWHAZ), and ACTB gens were recommended as reference genes to normalize the qPCR data (Modesto et al. 2013). Bai et al. (2014) investigated the transcriptional stability of 10 candidate reference genes in milk somatic cells of lactating yak, including the genes of our study. Four genes, ribosomal protein S9 (RPS9), PPP1R11, UXT, and mitochondrial ribosomal protein L39 (MRPL39), were identified as being the most stable genes in milk somatic cells of lactating yak.

Expression evaluation of nine candidate reference genes including ACTB, GAPDH and UXT was investigated in bovine mammary gland during the lactation cycle. UXT, RPS9 and RPS15 displayed the most expression stability across cow and time (Bionaz and Loor, 2007). Bonnet et al. (2013) studied eight candidate reference genes including UXT in different bovine and / or caprine tissues.

Table 2 Descriptive statistics of the 3 potential reference genes

Gene*	UXT	GADPH	ACTB
N	22	22	22
GM (CP)	37.04	28.88	32.26
AM (CP)	37.12	28.95	32.29
Min (CP)	32.04	24.03	29.71
Max (CP)	40.76	32.59	33.87
SD (±CP)	1.85	2.00	0.96
CV (% CP)	4.98	6.9	2.99
Min (x-fold)	-32.01	28.85	-5.87
Max (x-fold)	13.45	13.08	3.05
SD (±x-fold)	3.6	4.00	1.95
Ranking	2	3	1
r^2	0.97	0.95	0.98
E (%)	92.3	94.7	97.23

GM: the geometric mean and AM: arithmetic mean.
Min: minimal value and Max: maximal value.
SD: standard deviation and CV: coefficient of variance.
Min (x-fold) and Max (x-fold): the extreme values of expression levels expressed as an absolute x-fold over- or under-regulation coefficient.
SD [± x-fold]: standard deviation of the absolute regulation coefficient.
r2: coefficient of correlation estimated by qPCR.
E: polymerase chain reaction (PCR) efficiency.

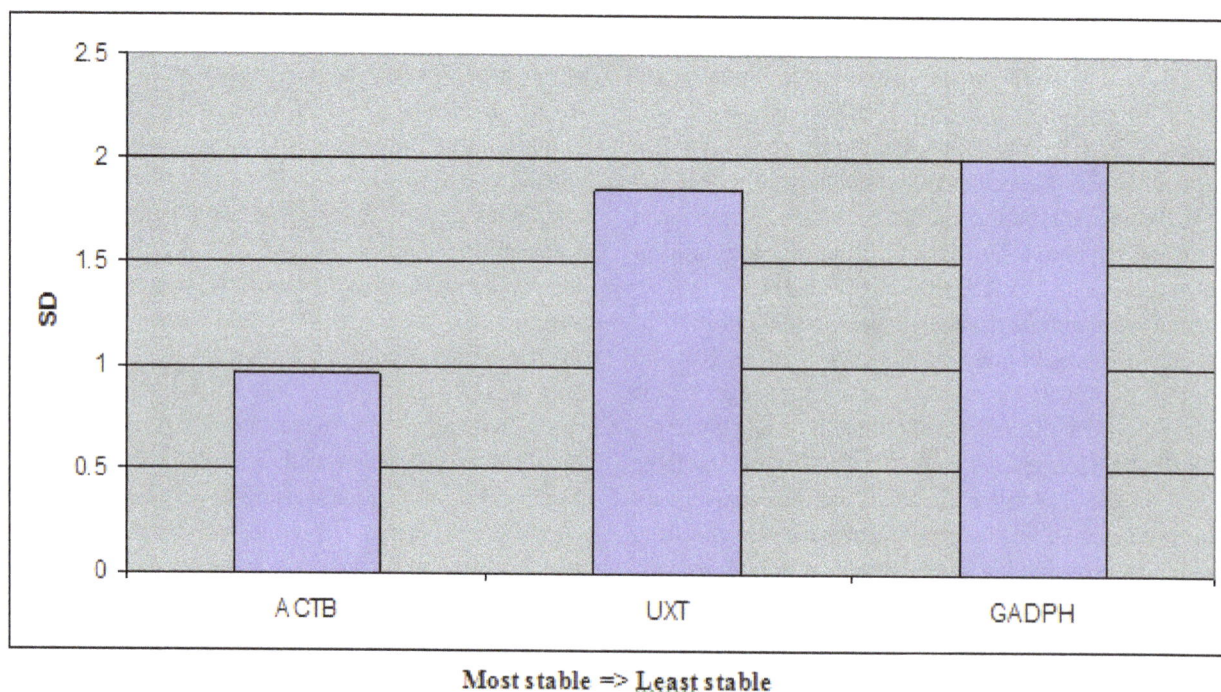

Figure 1 Expression stability of the 3 candidate reference genes in milk somatic cells of dairy heifers

Table 3 Repeated pair-wise correlation analysis of three candidate reference genes and correlation analysis reference genes versus BestKeeper index

Gene	UXT	GADPH	ACTB
GADPH	0.30	-	-
P-value	0.17	-	-
ACTB	-0.05	0.44	-
P-value	0.83	0.05	-
BestKeeper vs.	UXT	GADPH	ACTB
r^1	0.57	0.88	0.81
P-value	0.001	0.001	0.001

r: coefficient of correlation.

In bovine, *UXT*, initiation factor 3 subunit K (*EIF3K*) and *TBP* were the most stable genes in mammary gland while in caprine, genes with the highest stability in the mammary gland were *UXT*, *PPIA* and *MRPL39*. The most stable genes in bovine and caprine mammary gland were *UXT*, *EIF3K* and ceroid- lipofuscinosis neuronal 3 (*CLN3*). Finot *et al.* (2011) reported that among six potential reference gene such as *ACTB* and *GAPDH*, the genes encoding for ribosomal proteins, 18S rRNA and *RPLP0* presented the best expression stability in caprine mammary gland. Evaluation of appropriate housekeeping genes in the bovine mammary gland tissue samples and epithelial cell revealed that *UXT* and *GAPDH* were the most stable reference genes (Jedrzejczak and Szatkowska, 2014).

The results of our study showed unstable expression with respect to the one of the most commonly used reference gene, i.e. *GAPDH*. Historically, *GAPDH* gene has been used quite frequently as single endogenous control gene in the most of studies on bovine gene expression. In the other hand, the expression of this reference gene has been shown to highly unstable in bovine mammary gland (Kadegowda *et al.* 2009). Therefore, to correct interpretation of qPCR results, evaluation of *GAPDH* as reference gene in any tissue of interest is mandatory.

CONCLUSION

In conclusion, this study identified that *ACTB* gene can be used as reference genes in genes expression studies on bovine milk somatic cells as it demonstrated stable expression under different experimental conditions.

ACKNOWLEDGEMENT

This research was supported by Yasouj University.

REFERENCES

Andersen C.L., Jensen J.L. and Orntoft T.F. (2004). Normalization of real-time quantitative reverse transcription-PCR data: a model-based variance estimation approach to identify genes suited for normalization, applied to bladder and colon cancer data sets. *Cancer. Res.* **64**, 5245-5250.

Bai W.L., Yin R.H., Zhao S.J., Jiang W.Q., Yin R.L., Mas Z.J., Wang Z.Y., Zhu Y.B., Luo G.B., Yang R.J. and Zhao Z.H. (2014). Technical note: selection of suitable reference genes for studying gene expression in milk somatic cell of yak (*Bos grunniens*) during the lactation cycle. *J. Dairy Sci.* **97**, 902-910.

Bionaz M. and Loor J. (2007). Identification of reference genes for quantitative real-time PCR in the bovine mammary gland during the lactation cycle. *Physiol. Genomics.* **29**, 312-319.

Bonefeld B.E., Elfving B. and Wegener G. (2008). Reference genes for normalization: a study of rat brain tissue. *Synapse.* **62**, 302-309.

Bonnet M., Bernard L., Bes S. and Leroux C. (2013). Selection of reference genes for quantitative real-time PCR normalisation in adipose tissue, muscle, liver and mammary gland from ruminants. *Animal.* **7**, 1344-1353.

Bougarn S., Cunha P., Gilbert F.B., Meurens F. and Rainard P. (2011). Technical note: validation of candidate reference genes for normalization of quantitative PCR in bovine mammary epithelial cells responding to inflammatory stimuli. *J. Dairy Sci.* **94**, 2425-2430.

Brunner A.M., Yakovlev I.A. and Strauss S.H. (2004). Validating internal controls for quantitative plant gene expression studies. *BMC Plant. Biol.* **4**, 14-21.

Bustin S.A. (2002). Quantification of mRNA using real-time reverse transcription PCR (RT-PCR): trends and problems. *J. Mol. Endocrinol.* **29**, 23-29.

Bustin S.A. and Nolan T. (2004). Pitfalls of quantitative real-time reverse-transcription polymerase chain reaction. *J. Biomol. Tech.* **15**, 155-166.

Derveaux S., Vandesompele J. and Hellemens J. (2010). How to do successful gene expression analysis using real-time PCR. *Methods.* **50**, 227-230.

Devrim A.K., Sozmen M., Yigitarslan K., Sudagidan M., Kankavi O. and Atabay H.I. (2012). Assessment of TNF-alpha and leptin gene expression by RT-PCR in blood of cows with left abomasal displacement. *Rev. Med. Vet.* **163**, 368-372.

Finot L., Marnet P.G. and Dessauge F. (2011). Reference gene selection for quantitative real-time PCR normalization: application in the caprine mammary gland. *Small Rumin. Res.* **95**, 20-26.

Fleige S. and Pfaffl M.W. (2006). RNA integrity and the effect on the real-time qRT-PCR performance. *Mol. Asp. Med.* **7**, 126-139.

Gachon C., Mingam A. and Charrier B. (2004). Real-time PCR: what relevance to plant studies? *J. Exp. Bot.* **55**, 1445-1454.

Hunter T. and Garrels J.I. (1977). Characterization of themRNAs for alpha-, beta- and gammaactin. *Cell.* **12**, 767-781.

Jarczak J., Kaba J. and Bagnicka E. (2014). The validation of housekeeping genes as a reference in quantitative real time PCR analysis: application in the milk somatic cells and frozen whole blood of goats infected with caprine arthritis encephalitis virus. *Gene.* **549**, 280-285.

Jatav P., Sodhi M., Sharma A., Mann S., Kishore A., Shandilya U.K. and Kumar S. (2016). Identification of internal control genes in milk derived mammary epithelial cells during lactation cycle of Indian zebu cow. *Anim. Sci. J.* **87(3)**, 344-353.

Jedrzejczak M. and Szatkowska I. (2014). Bovine mammary epithelial cell cultures for the study of mammary gland functions. *In vitro. Cell. Dev. Biol. Anim.* **50**, 389-398.

Kadegowda A.K.G., Bionaz M., Thering B.J., Piperova L.S., Erdman R.A. and Loor J.J. (2009). Identification of internal controls for quantitative PCR in mammary tissue of lactating cows receiving lipid supplements. *J. Dairy Sci.* **92(5)**, 2007-2019

Kishore A., Sodhi M., Khate K., Kapila N., Kumari P. and Mukesh M. (2013). Selection of stable reference genes in heat stressed peripheral blood mononuclear cells of tropically adapted Indian cattle and buffaloes. *Mol. Cell. Probes.* **27,** 140-144.

Kizaki K., Shichijo-Kizaki A., Furusawa T., Takahashi T., Hosoe M. and Hashizume K. (2013). Differential neutrophil gene expression in early bovine pregnancy. *Reprod. Biol. Endocrinol.* **11,** 10.

Lee J.W., Bannerman D.D., Paape M.J., Huang M.K. and Zhao X. (2006). Characterization of cytokine expression in milk somatic cells during intramammary infections with *Escherichia coli* or *Staphylococcus aureus* by real-time PCR. *Vet. Res.* **37,** 219-229.

Leutenegger C.M., Alluwaimi A.M., Smith W.L., Perani L. and Cullor J.S. (2000). Quantitation of bovine cytokine mRNA in milk cells of healthy cattle by real-time TaqMan polymerase chain reaction. *Vet. Immunol. Immunopathol.* **77,** 275-287.

Liebe A. (1996). Interrelation between somatic cell counts and concentrations of growth factors, and mastitis in cows kept in different housing systems. Ph D. Thesis. Technical University of Munich, Munich, Germany.

Lisowski P., Pierzchala M., Goscik J., Pareek C.S. and Zwierzchowski L. (2008). Evaluation of reference genes for studies of gene expression in the bovine liver, kidney, pituitary, and thyroid. *J. Appl. Genet.* **49,** 367-372.

Macabelli C.H., Ferreira R.M., Gimenes L.U., de Carvalho N.A.T., Soares J.G., Ayres H., Ferraz M.L., Watanabe Y.F., Sangalli J.R., Smith L.C., Baruselli P.S., Meirelles F.V. and Chiaratti M.R. (2014). Reference gene selection for gene expression analysis of oocytes collected from dairy cattle and buffaloes during winter and summer. *PLoS One.* **9,** e93287.

Mahdipour M., Van Tol H.T., Stout T.A. and Roelen B.A. (2015). Validating reference microRNAs for normalizing qRT-PCR data in bovine oocytes and preimplantation embryos. *BMC Dev. Boil.* **15(1),** 1-11.

Mehta D., Menke A. and Binder E.B. (2010). Gene expression studies in major depression. *Curr. Psychiatry. Rep.* **12,** 135-144.

Modesto P., Peletto S., Pisoni G., Cremonesi P., Castiglioni B., Colussi M. Caramelli M., Bronzo V., Moroni P. and Acutis P.L. (2013). Evaluation of internal reference genes for quantitative expression analysis by real-time reverse

transcription-PCR in somatic cells from goat milk. *J. Dairy Sci.* **96,** 7932-7944.

Pfaffl M.W. (2001). A new mathematical model for relative quantification in real-time RT-PCR. *Nucleic Acids. Res.* **29,** e45.

Pfaffl M.W., Tichopad A., Prgomet C. and Neuvians T.P. (2004). Determination of stable housekeeping genes, differentially regulated target genes and sample integrity:BestKeeper-Excel-based tool using pair-wise correlations. *Biotechnol. Lett.* **26,** 509-515.

Saremi B., Sauerwein H., Danicke S. and Mielenz M. (2012). Technical note: identification of reference genes for gene expression studies in different bovine tissues focusing on different fat depots. *J. Dairy Sci.* **95,** 3131-3138.

Skern R., Frost P. and Nilsen F. (2005). Relative transcript quantification by quantitative PCR: roughly right or precisely wrong? *BMC Mol. Biol.* **6,** 10-18.

Vandesompele J., De Preter K., Pattyn F., Poppe B., Van Roy N., De Paepe A. and Speleman F. (2002). Accurate normalization of real-time quantitative RT-PCR data by geometric averaging of multiple internal control genes. *Genome. Biol.* **3(7),** 1-11

Varshney N., Mohanty A.K., Kumar S., Kaushik J.K., Dang A.K., Mukesh M., Mishra B.P., Kataria R., Kimothi S.P., Mukhopadhyay T.K., Malakar D., Prakash B.S., Grover S. and Batish V.K. (2012). Selection of suitable reference genes for quantitative gene expression studies in milk somatic cells of lactating cows (*Bos indicus*). *J. Dairy Sci.* **95,** 2935-2945.

Verbeke J., Van Poucke M., Peelman L. and De Vliegher S. (2015). Differential expression of *CXCR1* and commonly used reference genes in bovine milk somatic cells following experimental intramammary challenge. *BMC Genet.* **16(1),** 1-9.

Yadav P., Singh D.D., Mukesh M., Kataria R.S., Yadav A., Mohanty A.K. and Mishra B.P. (2012). Identification of suitable housekeeping genes for expression analysis in mammary epithelial cells of buffalo (*Bubalus bubalis*) during lactation cycle. *Livest. Sci.* **147,** 72-76.

18

Transcriptome Sequencing of Guilan Native Cow in Comparison with bosTau4 Reference Genome

M. Moridi[1], S.H. Hosseini Moghaddam[1*] and S.Z. Mirhoseini[1]

[1] Department of Animal Science, Faculty of Agricultural Science, University of Guilan, Rasht, Iran

*Correspondence E-mail: hosseini@guilan.ac.ir

ABSTRACT

RNA-sequencing is a new method of transcriptome characterization of organisms. Based on identity and relatedness, there are large genetic variations among different cattle breeds. The goal of the current study was to sequence the transcriptome of Guilan native cow and compare with available reference genome using RNA-sequencing method. Blood samples were collected from 14 Guilan native cows and then were pooled with same ratios of 3 micrograms per sample. Sequencing of the pooled sample was carried out using Illumina Hiseq 2000 from both end and 100 base pair of reading length. Tophat2 software was used to align the reads with reference genome and identify splice junctions and insertions and deletions. Cufflinks software was used to assemble transcripts and calculate their abundances. Total numbers of sequenced RNA fragments were 28434708 and the overall reading map was 87.4 percent. Total numbers of expressed genes were 24616 genes, which 19994 genes from these were protein coding genes and 3825 genes were non-coding. Adenosine triphosphate synthase 6 (*ATP6*) and ribosomal protein, large, P1 (*RPLP1*) genes showed the highest abundances of all expressed genes. The majority of highly expressed genes were involved in ribosomal structures and translational activities; moreover, they were belonging to housekeeping genes. The current study is a report of leukocytes transcriptome sequencing of Guilan native cow which have been not reported, so far. As Guilan native cow has the biggest population among all native populations in Iran, such studies could help to evaluate the genetic potential of this high precise genetic resource in Western Asia.

KEY WORDS cDNA library, Guilan native cattle, Illumina Hiseq 2000, RNA-sequencing.

INTRODUCTION

Cows could be classified in two major groups including taurine breeds (*Bos taurus*) and indicine breeds (*Bos indicus*) (Zhang *et al.* 2015). The indicine breeds belong to tropical regions where natural selection and adaptation to harsh environmental conditions caused them superior in this environment due to disease resistance and food shortage (Wilson, 2009). Iranian native cows are classified in the indicine breeds and could be grouped in the six groups including Sarabi, Golpaigani, Sistani, Dashtyari, Najdi and

Guilan native cows or Taleshi (Iran's country report on farm animal, 2004). Based on the report of Jihad-Keshavarzi organization of Guilan province in 2015 there are 312107 Guilan native cows in Guilan province that includes 70 percent of all cows in this province. According to the reports of Iranian national animal breeding center and promotion of animal products (INABC-PAP), average of test-day milk yield in Iranian Holstein, crossbred, and native cattle populations are 28.1, 10.3 and 4.68 kg/day, respectively. Because of low milk and meat production in Iranian native cattle populations, the crossbreeding programs have been started around 60 years ago. The basal structure for genetic improvement of Iran cattle populations, such as pedigree registration, recording the traits and artificial insemination has been organized since 50 years ago.

During the last 15 years, while the number of crossbred cattle has increased rapidly (e.g. from 2425000 in 2002 to 4373000 in 2009), the number of native cattle decreased (e.g. from 4337000 in 2002 to 2915000 in 2009). The number of exotic purebred cattle were grown slowly (e.g. from 683000 in 2002 to 961000 in 2009) and keeping exotic purebred cattle had less propensity specially for local breeders, due to management, environmental variation, and feed resource factors (Kamalzadeh *et al.* 2008).

The indigenous cattle breeds play an important economic role in the rural areas by providing milk and meat (Nimbkar *et al.* 2008). However, the production of their milk is too low compared to *Bos taurus* breeds. Due to limitations of forage crop production in Guilan province and genetic resistance of native cattle, still there is an increasing interest in keeping of native cattle and also crossbred of native cattle with taurine (*Bos taurus*) including Holstein and Simmental purebreds. Furthermore, entrance of purebred exotic breed did not succeed in this region due to feed shortages. Therefore, native breeds as an important genetic resource in livestock should be conserved without crossbreeding. A large pool of genetic resource of indigenous breeds justifies the importance of their conservation at the global level (Rewe *et al.* 2015). Approximately, 8 percent of reported livestock breeds have become extinct and an additional 21 percent are considered to be at risk of extinction. Moreover, the situation is presently unknown for 35 percent of breeds, most of which are reared in developing countries (FAO, 2011).

RNA-sequencing (RNA-seq) is a new method for transcriptome characterization of all organisms. This method changed the view of eukaryotic transcriptome complexity and components. RNA-seq provides more accurate and expanded measurement of transcripts and different isoforms of them compared to other methods (i.e. real-time PCR, microarray).

The transcriptome is the complete set of transcripts in a cell, and their quantity, for a specific developmental stage or physiological condition. Understanding the transcriptome is essential for interpreting the functional elements of the genome and revealing the molecular constituents of cells and tissues, and also for understanding development and disease.

RNA-seq is the most powerful method to evaluate genes expression and determine new transcripts, splice junctions and nucleotide variations in RNA sequences (Wang *et al.* 2009). Generally, in RNA-seq method a population of RNA

(total RNA or a portion of total RNA like RNA's with poly-A tail) is transformed into cDNA library with single or both end attached adaptors. Using high-throughput manner each cDNA molecule is sequenced to obtain short sequences. Sequenced reads have length between 30 to 400 base pair. After sequencing, the read results were aligned. The reads were put together to draw genome map includes transcriptome structure and translation levels of genes (Wang *et al.* 2009).

Base on the origin (*Bos taurus* or *Bos indicus*) of different cow breeds there are large genetic variation among different cow breeds (The Bovine HapMap Consortium, 2009). Regardless of large studies on the field of cow transcriptome, the information about different genes expression in various populations and breeds are minor (Huang *et al.* 2012).

Because of the importance of gene expression in shaping the phenotype, understanding of transcriptomic differences of various breeds are important. Although many molecular studies have been performed on Iranian Holstein and native cattle breeds (Ebrahimi *et al.* 2015a; Ebrahimi *et al.* 2015b; Kharrati Koopaei *et al.* 2012; Mohammadabadi *et al.* 2010; Pasandideh *et al.* 2015), but the transcriptome of Guilan native cow has not studied, hence the goal of the current study was sequencing the transcriptome of Guilan native cow and its comparison with available cow reference genome using blood total RNA-seq method.

MATERIALS AND METHODS

The blood samples were obtained from the tail vein of 14 Guilan native cows from station located in Hossein-Kooh of Foman city (Guilan province, Iran). Samples were collected in heparinized venoject tubes and were transferred on ice (4 °C) to the laboratory immediately for RNA extraction. The time interval between collecting the samples and RNA extraction was less than one hour. At least, 4 mL of blood was used for RNA extraction from each sample. First, buffy coat of blood samples was isolated after centrifugation at 2000 × g for 10 min at 4 °C. White blood cells (WBC) were washed with red blood cell (RBC) lysis buffer (1X) twice and centrifuged at 2000 × g for 10 min at 4 °C to obtain white pellet of WBC. Total RNA was extracted using Trizol (Invitrogen) method from WBC (Jiang *et al.* 2013).

The quality and quantity of extracted RNA samples was determined using NanoDrop-2000 (Thermo Scientific) spectrophotometer based on 260 to 280 spectrophotometer ratio and electrophoresis on agarose gel. The same concentration of each sample (3 micrograms/sample) were pooled together to obtain biological average of Guilan native population.

The inherent biological variance between different samples is an important factor to make sound conclusions. Sample pooling will allow us to average out this variance. The quality of pooled sample was assessed using Agilent 2100 Bioanalyzer device and Agilent RNA 6000 Nano kit. Total RNA was digested by DNaseI (NEB) and purified by oligo-dT beads (Dyna beads mRNA purification kit, Invitrogen, USA), then poly (A)-containing mRNA was fragmented into 130 base pair with first strand buffer. First-strand cDNA was synthesized using N6 primer, first strand master mix, and super script II reverse transcription (Invitrogen, USA). Reaction conditions were 25 °C for 10 min, 42 °C for 40 min, and 70 °C for 15 min. The second strand master mix was added to synthesize the second-strand cDNA (16 °C for 1 h). The cDNA was purified using Agencourt® AMPure® XP PCR purification kit (Agencourt, USA). The final library was quantitated in two ways for validation. First, the average of molecule length was determined using the Agilent 2100 Bioanalyzer instrument (Agilent DNA 1000 Reagents), and then the quantification of the library was performed by real-time quantitative PCR (RT-qPCR) using TaqMan Probes. The qualified libraries were amplified on cBot to generate the cluster on the flow cell (TruSeq PE cluster kit V3–cBot–HS, Illumina). The amplified flow cell was sequenced, pair ended, on the HiSeq 2000 device (TruSeq SBS KIT-HS V3, Illumina) with read length of 100 base pair (BGI-Shenzhen, China). The whole dataset is available in NCBI with GSM2101067 GEO accession number.

To assess the quality of sequenced sample we used FastQC software before any analysis. Tophat2 with Bowtie2 (version 2.0.4) was used to align mRNA sequence reads to the reference genome (bosTau4), segment mapping algorithm to discovering splice junctions, and indel search to discover insertions and deletions (Kim *et al.* 2013). The accepted hits outputs of Tophat2 were applied to Cufflinks for assemble transcripts and estimation of their abundances (Trapnell *et al.* 2010). Cufflinks estimates gene expression (fragments per kilo base exon per million mapped fragments, or FPKM). Reference Ensembl *Bos taurus* annotation file was used as guide for assembling transcripts, bias correction using reference genome, and multi-read correction. The minimum for cutoff of abundance was defined 0.1.

RESULTS AND DISCUSSION

In this research, all samples of each population have been pooled to optimize biological differences of samples. Decreased biological variation and increased detection power of differentially expressed genes are the results of samples pooling.

The pooling bias can occur via difference between the calculated value in pooled sample and the average value in each single sample (Rajkumar *et al.* 2015).

Extracted RNA's were shown the average of 350 nanograms per microliter concentration and average of 1.9 for 260/280 ratio. Using Bioanalyzer device, the RIN (RNA Integrity Number) for pooled sample was 8 which showed us a good quality of sample for RNA-seq (Figure 1).

The quality assessment of sequenced sample using FastQC software showed the high sequencing quality in both forward and reverse directions. Summary of mapping reads using reference genome are available in Table 1. Total sequenced fragments and overall read mapping rate were 28434708 and 87.8%, respectively. Numbers of insertions, deletions, splice junctions and aligned pairs with concordant alignments were 144258, 140262, 192100 and 18234548, respectively. Huang *et al.* (2012) reported 21078477, 21358931 and 20940063 total sequenced fragments and 64.4, 67.3 and 78.3 overall read mapping rate for Holstein, Jersey, and Cholistani breeds, respectively. The Holstein and Jersey are *Bos taurus* origin, while Cholistani same as Guilan native cow is classified as *Bos indicus* breeds.

Generally, we observed 24616 expressed genes, from which 19994 genes were protein coding genes and 3825 genes were classified in non coding genes. These non coding genes produce non translated transcripts and have regulatory function in genes expression (Table 2). The results showed us 26740 gene transcripts for Guilan native cow population. The total number of 13409, 13787 and 13666 expressed genes were reported in blood leukocytes transcriptome for Holstein, Jersey, and Cholistani populations, respectively (Huang *et al.* 2012).

The majority of highly expressed genes were housekeeping genes. Housekeeping genes are involved in basic cell maintenance and are expected to maintain constant expression levels in all cells and conditions (Eisenberg and Levanon, 2013). The five more expressed annotated transcripts are represented in Table 3. The majority of these genes categorized in housekeeping genes. The adenosine triphosphate synthase 6 (*ATP6*) and ribosomal protein, large, P1 (*RPLP1*) transcripts were showed highest expression among all the other transcripts. The ATP6 is one subunit of ATP synthase in mitochondria that is responsible for final step of oxidative phosphorylation (Habersetzer *et al.* 2013).

The ATP6 gene transcript showed higher differences with the other genes transcripts. *RPLP1* gene encodes a ribosomal phosphoprotein that is a component of 60S subunit of ribosomes (Martinez-Azorin *et al.* 2008). B2M gene encodes beta-2-microglobuline protein, a component of MHC class I molecules (Hall *et al.* 2015).

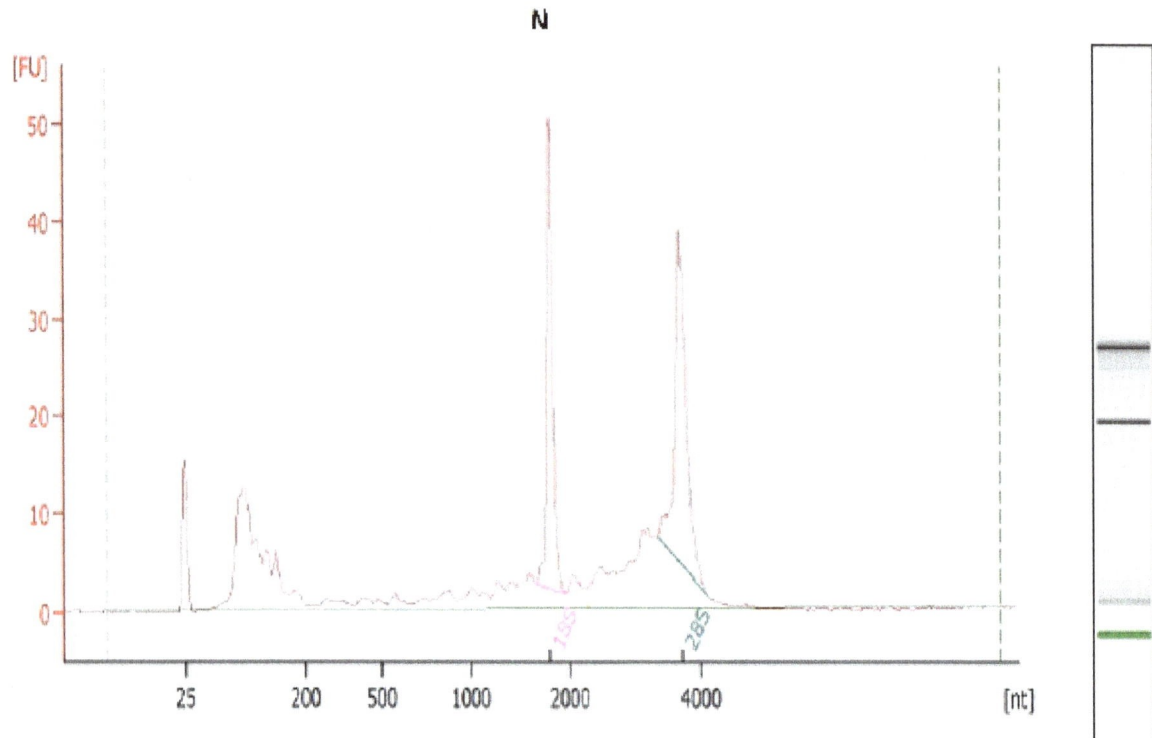

N

[FU]

Overall Result for sample   :   N

RNA Integrity Number (RIN):   8

Fragment table for sample   :   N

Name	Start Size [nt]	End Size [nt]
18S	1,648	1,958
28S	3,346	4,128

Figure 1 Electropherogram of pooled sample from Bioanalyzer device
All 28S, 18S, and 5S bands absorptions and lengths and their gel qualities are obvious
The vertical axe represents the light absorption of nucleotides and the horizontal axe shows the nucleotides length (N= Guilan native cow population)
The right side is the agarose gel electrophoresis of pooled sample

The *B2M* gene was the third highest expressed genes in our case (Table 3). The NADH dehydrogenase 2 (ND2) protein is a subunit of NADH dehydrogenase (ubiquinone), which is located in the mitochondrial inner membrane and is the largest of the five complexes of the electron transport chain (Zickermann *et al*. 2015). MicroRNAs (miRNAs) are small non-coding RNAs of approximately 19 to 25 nucleotides that post-transcriptionally regulate gene expression. MiRNAs bind to the UTR (3'-untranslated region) of their target mRNAs messenger (RNAs) through complementary recognition, which then leads to degradation or repression of protein expression (Yan *et al*. 2014).

Much micro-RNAs have been found to involve in the physiological and pathological processes of angiogenesis (Suarez *et al*. 2008).

Among the top transcripts there was also the micro RNA MiR-126 which is a specific and highly expressed micro-RNA in vessel endothelial cells and has key roles in controlling arteriogenesis (increase in the diameter of existing arterial vessels) and angiogenesis (the physiological process through which new blood vessels form from pre-existing vessels) (Van Solingen *et al*. 2009). In current study MiR-126 was one of the top five expressed transcripts in Guilan native cow population (Table 3).

Table 1 Summary of RNA-seq alignment to the reference genome (bosTau4) for Guilan native cow population

Left reads	Guilan native cattle
Total sequenced Fragments	28434708
Mapped to reference genome	25421000 (89.4% of input)
Sequences with multiple alignments	2170039 (8.5% of input)
Right reads	
Total sequenced Fragments	28434708
Mapped to reference genome	24278459 (85.4% of input)
Sequences with multiple alignments	2064643 (8.5% of input)
Overall read mapping rate	87.40%
Aligned pairs	23140289
Aligned pairs with multiple alignments	1956181 (8.5%)
Aligned pairs with discordant alignments	744972 (3.2%)
Aligned pairs with concordant alignments	18234548 (78.8%)
Number of insertions	140262
Number of deletions	111069
Number of splice junctions	192100

Table 2 Summary of mapped reads of Guilan native cow population

Items	Guilan native cattle
Total number of expressed genes	24616
Number of coding genes	19994
Number of non-coding genes	3825
Number of pseudogenes	797

Table 3 Top five highly expressed transcripts in Guilan native cow population

	Transcript ID	Gene symbol	Value (FPKM*)
1	ENSBTAT00000060539	*ATP6*	36230.2
2	ENSBTAT00000024376	*RPLP1*	8343.5
3	ENSBTAT00000016359	*B2M*	7248.9
4	ENSBTAT00000060548	*ND2*	6235.6
5	ENSBTAT00000036924	*bta-miR-126*	4944.1

* FPKM: fragments per kilobase of exon per million fragments mapped.

Out of first five highly expressed genes, two (*ATP6* and *ND2*) genes belong to electron transport chain in mitochondria and could be the indicative of the importance of this cyclic energy creation biological pathway in cows. The most of the highly expressed genes involved in ribosomal structures and translational activities that could be explained by the housekeeping functions of these genes.

RNA-seq has many advantages over traditional cDNA microarray technologies. RNA-seq is free from probe design or bias from hybridization issues and is more sensitive in detecting genes with very low expression and more accurate in detecting expression of extremely abundant genes (Croucher and Thomson, 2010; Marioni *et al.* 2011).

With manufacturers predicting increased read lengths, reduced costs and faster sequencing relative to existing platforms, the future of RNA-Seq technology appears to be promising and routinely affordable. It is expected that once the issues with the widespread use of RNA-Seq are overcome (e.g. higher cost, high data-storage requirements, and the absence of a standard for analysis) this technique will become the predominant tool for transcriptome analysis.

Marioni *et al.* (2011) showed that the correlation between RNA-Seq and qRT-PCR could reach 0.93, offering that the RNA-Seq technique is accurate and reproducible. This technique may also consider for discovering variations reasons in economically important traits and an accurate tool for designing of a cattle breeding program.

The negative consequences of genetic erosion and inbreeding depression may be manifested by loss of viability, fertility and disease resistance, and the frequent occurrence of recessive genetic diseases (FAO, 2011). Thus, more attention should be paid to animal genetic resources. More accurate genetic information can be obtained to better understanding of existing animal genetic resources by applying for advances in molecular biotechnology (Yang *et al.* 2013).

CONCLUSION

This study investigated the complexity of the blood transcriptome using RNA-seq technique and it was the first report of blood leukocytes transcriptome sequencing of

Guilan native cow population. Guilan native cow has the biggest population among all native cattle populations in Iran and there is a lot of interest to us to explore the genetic potential of this population. Such studies could help to detect and introduce highly expressed genes and unique transcripts in Guilan native cow population and flaunt the genetic potential of this population with other indicine breeds with high resistance to harsh environments, diseases, and feed shortages. Furthermore, the results of this study can be used as a valuable resource for further investigations on cattle aimed at finding genes expression patterns and RNA sequence variations.

ACKNOWLEDGEMENT

This work was supported by Iran national science foundation under grant No. 93012544.

REFERENCES

Croucher N.J. and Thomson N.R. (2010). Studying bacterial transcriptomes using RNA-seq. *Curr. Opin. Microbiol.* **13,** 619-624.

Ebrahimi Z., Mohammadabadi M.R., Esmailizadeh A.K., Khezri A. and Najmi Noori A. (2015a). Association of *PIT1* gene with milk fat percentage in Holstein cattle. *Iranian J. Appl. Anim. Sci.* **5,** 575-582.

Ebrahimi Z., Mohammadabadi M.R., Esmailizadeh A. and Khezri A. (2015b). Association of *PIT1* gene and milk protein percentage in Holstein cattle. *J. Livest. Sci. Technol.* **3,** 41-49.

Eisenberg E. and Levanon E.Y. (2013). Human housekeeping genes, revisited. *Trands Genet.* **29,** 569-574.

FAO. (2011). Molecular genetic characterization of animal genetic resources. Food and Agriculture Organization of the United Nations (FAO), Rome, Italy.

Habersetzer J., Ziani W., Larrieu I., Stines-Chaumeil C., Giraud M.F., Brèthes D., Dautant A. and Paumard P. (2013). ATP synthase oligomerization: from the enzyme models to the mitochondrial morphology. *Int. J. Biochem. Cell. Biol.* **45,** 99-105.

Hall Z., Schmidt C. and Politis A. (2015). Uncovering the early assembly mechanism for amyloidogenic β2-microglobulin using cross-linking and native mass spectrometry. *J. Biol. Chem.* **291,** 4626-4637.

Huang W., Nadeem A., Zhang B., Babar M., Soller M. and Khatib H. (2012). Characterization and comparison of the leukocyte transcriptomes of three cattle breeds. *PLoS One.* **7,** 30244-30251.

Iran's Country Report on Farm Animal Genetic Resources. (2004). Draft Iran's Country Report on Farm Animal Genetic Resources. Animal Science Research Institute of Iran, Tehran, Iran.

Jiang Z., Uboh C.E., Chen J. and Soma L.R. (2013). Isolation of RNA from equine peripheral blood cells: comparison of methods. *SpringerPlus.* **2,** 1-6.

Kamalzadeh A., Rajabbaigy M. and Kiasat A. (2008). Livestock production systems and trends in livestock industry in Iran. *J. Agric. Soc. Sci.* **4,** 183-188.

Kharrati Koopaei H., Mohammad Abadi M.R., Ansari Mahyari S., Tarang A.R., Potki P. and Esmailizadeh A.K. (2012). Effect of *DGAT1* variants on milk composition traits in Iranian Holstein cattle population. *Anim. Sci. Pap. Rep.* **30,** 231-240.

Kim D., Pertea G., Trapnell C., Harold P., Kelley R. and Salzberg S.L. (2013). TopHat2: accurate alignment of transcriptomes in the presence of insertions, deletions and gene fusions. *Genome. Biol.* **14,** 36-42.

Marioni J.C., Mason C.E., Mane S.M., Stephens M. and Gilad Y. (2011). RNA-seq: an assessment of technical reproducibility and comparison with gene expression arrays. *Genome. Res.* **18,** 1509-1517.

Martinez-Azorin F., Remacha M. and Ballesta J.P. (2008). Functional characterization of ribosomal P1/P2 proteins in human cells. *Biochem. J.* **413,** 527-534.

Mohammadabadi M.R., Torabi A., Tahmourespoor M., Baghizadeh A., Esmailizadeh Koshkoie A. and Mohammadi A. (2010). Analysis of bovine growth hormone gene polymorphism of local and Holstein cattle breeds in Kerman province of Iran using polymerase chain reaction restriction fragment length polymorphism (PCR-RFLP). *African J. Biotechnol.* **9,** 6848-6852.

Nimbkar C., Gibson J., Okeyo M., Boettcher P. and Soelkner J. (2008). Sustainable use and genetic improvement. *Anim. Genet. Resour.* **42,** 49-70.

Pasandideh M., Mohammadabadi M.R., Esmailizadeh A.K. and Tarang A. (2015). Association of bovine *PPARGC1A* and *OPN* genes with milk production and composition in Holstein cattle. *Czech J. Anim. Sci.* **60,** 97-104.

Rajkumar A.P., Qvist P., Lazarus R., Lescai F., Ju J., Nyegaard M., Mors O., Børglum A.D., Li Q. and Christensen J.H. (2015). Experimental validation of methods for differential gene expression analysis and sample pooling in RNA-seq. *BMC Genom.* **16,** 548-552.

Rewe T.O., Peixoto M.G.C.D., Cardoso V.L., Vercesi Filho A.E., El Faro L. and Strandberg E. (2015). Gir for the Giriama: the case for Zebu dairying in the tropics-a review. *Livest. Res. Rural Dev.* Available at: http://www.lrrd.org/lrrd27/8/rewe27150.html.

Suarez Y., Fernandez-Hernando C., Yu J., Gerber S.A., Harrison K.D., Pober J.S., Iruela-Arispe M.L., Merkenschlager M. and Sessa W.C. (2008). Dicer-dependent endothelial microRNAs are necessary for postnatal angiogenesis. *Proc. Natl. Acad. Sci. USA.* **105,** 14082-14087.

The Bovine HapMap Consortium. (2009). Genome-wide survey of SNP variation uncovers the genetic structure of cattle breeds. *Science.* **324,** 528-532.

Trapnell C., Williams B.A., Pertea G., Mortazavi A., Kwan G., Van Baren M.J., Salzberg S.L., Wold B.J. and Pachter L. (2010). Transcript assembly and quantification by RNA-Seq reveals unannotated transcripts and isoform switching during cell differentiation. *Nat. Biotechnol.* **28,** 511-515.

Van Solingen C., Seghers L., Bijkerk R., Duijs J.M., Roeten M.K., van Oeveren-Rietdijk A.M., Baelde H.J., Monge M., Vos J.B., de Boer H.C., Quax P.H., Rabelink T.J. and van Zonneveld A.J. (2009). Antagomir-mediated silencing of endothelial cell specific microRNA-126 impairs ischemia-induced angiogenesis. *J. Cell. Mol. Med.* **13,** 1577-1585.

Wang Z., Gerstein M. and Snyder M. (2009). RNA-seq: a revolutionary tool for transcriptomics. *Nat. Rev. Genet.* **10,** 57-63.

Wilson R.T. (2009). Fit for purpose-the right animal in the right place. *Trop. Anim. Health Prod.* **41,** 1081-1090.

Yan S., Yim L.Y., Lu L., Lau C.S. and Chan V.S.F. (2014). MicroRNA regulation in systemic lupus erythematosus pathogenesis. *Immune. Netw.* **14,** 138-148.

Yang W., Kang X., Yang O., Lin Y. and Fang M. (2013). Review on the development of genotyping methods for assessing farm animal diversity. *J. Anim. Sci. Biotechnol.* **4,** 2-9.

Zhang L., Jia S., Plath M., Huang Y., Li C., Lei C., Zhao X. and Chen H. (2015). Impact of parental *Bos taurus* and *Bos indicus* origins on copy number variation in traditional Chinese cattle breeds. *Genome. Biol. Evol.* **7,** 2352-2361.

Zickermann V., Wirth C., Nasiri H., Siegmund K., Schwalbe H., Hunte C. and Brandt U. (2015). Mechanistic insight from the crystal structure of mitochondrial complex I. *Science.* **347,** 44-49.

19

A Research on Association between *SCD1* and *OLR1* Genes and Milk Production Traits in Iranian Holstein Dairy Cattle

M. Hosseinpour Mashhadi[1*]

[1] Department of Animal Science, Mashhad Branch, Islamic Azad University, Mashhad, Iran

*Correspondence E-mail: hosseinpour@mshdiau.ac.ir

ABSTRACT

The present study was carried out to investigate the association of C/T single nucleotide polymorphism (SNP) in exon 5 of stearoyl-CoA desaturase 1 (*SCD1*) gene and A/C SNP in the 3' untranslated region of oxidized low density lipoprotein receptor 1 (*OLR1*) gene with milk production traits in Iranian Holstein dairy Cattle. The blood samples of 153 (for *OLR1*) and 308 (for *SCD1*) dairy cattle from three different farms were used for genotyping. A 146 bp fragment of 3'UTR of *OLR1* gene and a 400 bp fragment of exon 5 of *SCD1* gene were amplified by standard PCR. Single nucleotide polymorphism of *OLR1* and *SCD1* gene was determined by polymerase chain reaction single-restriction fragment length polymorphism (PCR-RFLP) technique. The association between genotypes of *OLR1* and *SCD1* genes with milk production traits was studied by general linear models (GLM) procedure of SAS package and Duncan test was used for comparing means of traits. The frequency of AA, AV and VV genotype of *SCD1* gene were 0.60, 0.32 and 0.08 respectively. The frequency of alleles A and V were 0.76 and 0.24. The genotype frequencies of AA, AC and CC in *OLR1* gene were 0.22, 0.50 and 0.28 respectively. The frequency of allele A and C was 0.47 and 0.53. Thus, this population was in Hardy-Weinberg equilibrium for *OLR1* but not for *SCD1*. The means of fat percentage for *SCD1* genotypes were 3.43% and 3.33 for VV and AA respectively (P<0.05). The means of *OLR1* genotypes were 8273 kg (CC), 8344 kg (AC) and 7178 kg (AA) for milk yield; 276.3 kg (CC), 277.6 kg (AC) and 239.7 kg (AA) for fat yield and for protein yield were 286.7 kg (CC), 290.5 kg (AC) and 253 kg (AA) (P<0.05). The results revealed these two SNP are appropriate for marker assisted selection.

KEY WORDS gene, Holstein, *OLR1*, *SCD1*.

INTRODUCTION

Quantitative traits are controlled by large numbers of genes and also influenced by environmental factors. Recent researches with farm animals have quantified the effect of candidate genes on economically important traits. Identification of genes with large effects on milk production traits would be useful for genetic improvement programs in dairy cattle (Wang *et al.* 2016; Cecchinato *et al.* 2015; Hosseinpour *et al.* 2013). Stearoyl-CoA desaturase (SCD) is an enzyme that plays an important role in the biosynthesis of fatty acids. This enzyme belongs to a large family of enzymes that are involved in the synthesis of saturated fatty acids and they are found in both animals and plants. Stearoyl-CoA desaturase (SCD) is the enzyme responsible for conversion of saturated fatty acids into 9-monounsaturated fatty acids in mammalian adiposities (Campbell *et al.* 2001; Taniguchi *et al.* 2004). Two SCD isoforms are known in cattle, *SCD5* is expressed in brain and is located on chromosome 6 and *SCD1* is located on chromosome 26 and expressed in adipose and mammary tissue (Schennink *et al.* 2008). The cytosine to thymine substitution causes alanine change to valine amino acids on protein (A293V). Several researches have shown significant effects of genotypes SCD1A293V in exon 5 on the composition of fatty acids in milk and milk production traits (Moioli *et al.* 2007; Kgwatalala *et al.* 2009; Clark *et al.* 2010).

The major protein oxidized low density lipoprotein receptor1 (*OLR1*), was initially identified in bovine aortic endothelial cells, this protein binds, internalizes, and degrades oxidized low-density lipoprotein (Sawamura *et al.* 1997). The oxidized form of the low density lipoprotein (oxLDL) is involved in endothelial cell injury, dysfunction, and activation, all of which are implicated in the development of atherosclerosis (Mehta and Li, 1998).The oxLDL and its lipid constituents have numerous damaging effects on secretory activities of the endothelium, including induction of apoptosis (Imanishi *et al.* 2002). The *OLR1* gene encodes a vascular endothelial cell-surface receptor that binds and degrades the oxidized forms of low-density lipoproteins (oxLDL) (Mehta and Li, 2002). The genomic sequence of bovine *OLR1*, released by Baylor College of Medicine, contains five exons and located on chromosome 5. The length of this gene is 11373 base pairs (GenBank accession no. NW_215807). The bovine *OLR1* gene encodes 270 AA that has a 72% identity to the human protein (Sawamura *et al.* 1997).

Many studies have been conducted on QTL in bovine chromosome 5 near *OLR1* affecting milk production traits between 1999 and 2004 (Heyen *et al.* 1999; Olsen *et al.* 2002; Ashwell *et al.* 2004; De Koning *et al.* 2001; Viitala *et al.* 2003; Rodriguez-Zas *et al.* 2002). Direct cDNA and genomic sequencing of *OLR1* revealed 2 single nucleotide polymorphisms (SNP) in exon 4, 5 SNP in intron 4 and 1 in the 3' untranslated region (3'UTR) (Khatib *et al.* 2006). Some researchers reported that allele C of SNP in the 3'UTR had significant effects on fat yield and fat percentage. Khatib *et al.* (2006) reported significant effects of A/C SNP in the 3'-untranslated region of OLR1 on milk fat yield and fat percentage in a granddaughter-design Holstein bull population. Association between *OLR1* haplotypes and milk production traits was further confirmed in a daughter-design study of Holstein cows and in an Italian Brown Swiss population (Khatib *et al.* 2007). Schennink *et al.* (2009) reported a significant association between *OLR1* and milk fat percentage in Dutch Holstein-Friesian cattle.

Because of the role of *OLR1* in lipid metabolism and degradation of Ox-LDL, the *OLR1* gene has been regarded as a candidate gene affecting milk production traits in dairy cattle (Khatib *et al.* 2006). Thus, the objective of this research was to study the association between genes *OLR1* and *SCD1* and milk production traits in Iranian Holstein dairy cattle.

MATERIALS AND METHODS

Samples and phenotypic data

Blood samples of 153 (for *OLR1*) and 308 (for *SCD1*) of Iranian Holstein dairy cattle were randomly selected for genotyping from three different farms in the Khorasan Razavi province of Iran. Records of 136 (*OLR1*) and 274 (*SCD1*) cows were used for the association study. The Studied traits were 305-day milk, fat and protein yield and fat and protein percentage.

Genotyping of SNP

Approximately 5 mL of blood were collected from the jugular vein of each animal in EDTA tubes. The aliquots of whole blood were stored at -20 °C. The genomic DNA from blood samples was extracted using the GuSCN-Silica Gel method and standard protocol with commercial kit of DIAtom DNA Prep (Biokom). The quality of DNA was examined by agarose gel electrophoresis. Genotyping was carried out using the PCR-RFLP technique. The 146 bp fragment of *OLR1* gene from 3'UTR (GenBank accession no. NC_007303.4; 107080852; 107092157 described in Khatib *et al.* 2006) was amplified by standard PCR by Biometra thermo cycler (Germany). The sequences of the primers were (F 5′-TCCCTAACTTGTTCCAAGTCCT-3′) and (R 5′-CTCTACAATGCCTAGAAGAAAGC-3'), respectively (Komisarek and Dorynek, 2009). The total volume of reaction was 25 μL that contained one unit (0.2 μL) of *Taq* polymerase, 200 μM (0.5 μL) of dNTP, 2 mM MgCl$_2$, 10 pM (3 μL) primer mix and 2.5 μL standard buffer in 13.8 μL dH$_2$O. Fifty nano grams (5 μL) DNA were added to the reaction mix. The thermal cycling conditions for the PCR were as follows: initial denaturation at 94 °C for 5 min, cyclic denaturation at 94 °C for 30 s, cyclic annealing of primers at 62 °C for 30 s, cyclic elongation at 72 °C for 45 s (for 30 cycles) and final elongation at 72 °C for 5 min (Komisarek and Dorynek, 2009). The PCR products were separated by electrophoresis on 2% agarose gel and visualized on gel documentation system (UVP, USA). The PCR product was digested with *PstI* restriction enzyme (Fermentas). The 5 μL of PCR product was mixed with 2 μL 10X buffer, 5 μL dH$_2$O and 2 units of *PstI* enzyme and digested over 5 hours at 37 °C. Polyacrylamide gel electrophoresis methods were used to identify genotypes. The 400bp fragment of *SCD1* gene that included exon 5 was amplified with standard PCR based on GenBank accession no. AY241932. The primers were as follow: F (5′-CCC ATT CGC TCT TGT TCT GT-3′) and R (5′-CCC ATT CGC TCT TGT TCT GT-3′).

The total volume of PCR reaction was 25 μL similar to *OLR1* fragment. The thermal program of PCR reaction was initial denaturation at 94 °C for 3 min, cyclic denaturation at 94 °C for 45 s, cyclic annealing of primers at 54 °C for 30 s , cyclic elongation at 72 °C for 90 s (for 34 cycles) and final elongation at 72 °C for 10 min (Kgwatalala *et al.* 2009).

The PCR products were separated by electrophoresis on 2% agarose gel and visualized on UVP gel documentation system. The PCR product was digested with restriction enzyme *NcoI* (Fermentas). The 3 μL of PCR product was mixed with 2 μL 10X buffer, 4.5 μL dH$_2$O and 2 units of *NcoI* and digested over 5 hours at 37 °C . The digested fragment was loaded in 2% agarose gel and visualized on UVP gel documentation system.

Statistical analysis

The Hardy-Weinberg equilibrium for allele and genotype frequencies was analyzed with Chi-square test using POPGENE software (Yeh *et al.* 1999). The association between genotypes and traits was assessed with the (GLM) procedure of SAS (2004) according to the following general linear model:

$$Y_{ijnl} = \mu + G_i + HYS_j + L_k + S_l + e_{ijkl}$$

Where:

Y_{ijkl}: value for each milk-related trait.

μ: overall mean.

G_i: fixed effect of the ith genotype (3 genotypes for each gene: AA, AV and VV for SCD1 or AA, AC and CC for *OLR1*).

HYS_j: fixed effect of herd (1, 2, 3), year and season of parturition.

L_k: k^{th} lactation (1 and 2).

S_l: random effect of sire (1,...,128).

e_{ijkl}: residual effects.

Due to use the records of traits, the fixed effects and the sire effect (as random genetic effect) were considered in the model. Genotype means were compared with the Duncan test (P<0.05).

RESULTS AND DISCUSSION

Gene and genotypic frequencies

Genotype and allele frequencies are shown in Table 1. The *SCD1* genotypic frequencies for AA, AV and VV were 0.60, 0.32 and 0.08, respectively. The frequencies of alleles A and V of *SCD1* were 0.76 and 0.24. The population was not in Hardy Weinberg equilibrium for gene *SCD1*. Be-

cause of population was under selection and probably other dispersive factor such as population size was affected.

The higher frequency of the A allele is in agreement with results reported by other studies. Schennink *et al.* (2008) reported the frequency of allele A and V as 0.73 and 0.27. Kgwatalala *et al.* (2009) genotyped 525 Canadian Jersey cows for the *SCD1* gene and reported genotypic frequencies as 0.686, 0.244 and 0.07 for the AA, AV and VV genotypes, respectively and frequencies of alleles A and V of 0.808 and 0.192. Genotype frequencies were in Hardy-Weinberg equilibrium. Cows of 3 breeds of northern Italy, Jersey, Valdostana and Piedmontese were genotyped at exon 5 of the *SCD* gene; the frequency of A allele was 0.94 for Jersey, 0.65 for Valdostana and 0.42 for Piedmontese (Moioli *et al.* 2007). Clark *et al.* (2010) determined the polymorphism of 143 and 215 cows in two studies. The distribution of genotypes among 143 dairy cows was 72 AA, 60 AV, and 11 VV animals. Therefore, allele frequencies were 0.71 (A) and 0.29 (V). In study 2, the distribution of genotypes for 215 dairy cows was 111 (AA), 85 (AV) and 19 (VV) animals. Allele frequencies were 0.71 and 0.29 for the A and V alleles. Frequencies of *SCD* genotypes in the sample of Italian Friesian cows were 0.27, 0.6 and 0.13 for AA, AV and VV genotypes, respectively (Mele *et al.* 2007). The higher frequency of the A allele (0.57) compared to allele V (0.43) agreed with the result of this study.

The genotypic frequencies of gene *OLR1* (AA, AC and CC) were 0.22, 0.5, and 0.28 respectively. Frequencies of A and C alleles were 0.47 and 0.53. The Chi-square test revealed that the population was in Hardy-Weinberg equilibrium, the two alleles frequencies were near to 0.5 so the population was in equilibrium and probably the selection and other factors were not effective. Khatib *et al.* (2006) and Komisarek and Dorynek (2009) reported the same frequency of 0.46 and 0.43 for allele A and 0.54 and 0.57 for allele C in the US, Polish and Holstein cattle populations respectively. Soltani-Ghombavan *et al.* (2013) estimated the frequency of allele C and A to be 0.483 and 0.517 in Holstein dairy cattle from 5 farms in the Esfahan province of Iran. Schennink *et al.* (2009) reported frequencies of 0.29 and 0.71 for alleles A and C in a Dutch Holstein population.

Association analysis

In general, the used regression model was significant for all studied traits with *OLR1* or *SCD1* genotypes included in model. The coefficient of determination (R^2) was higher than 0.8 and 0.9 for most of traits that showed the fixed and random effects in general linear model justify the variation in traits. Trait means, standard errors, and coefficients of variation are also shown in Table 2.

Table 1 Genotype and allele frequencies

Gene	Genotype			Allele		Index		HWE
						Shanon	Nei	χ^2
SCD1	AA	AV	VV	A	V			
	0.60	0.32	0.08	0.76	0.24	0.55	0.36	20.37**
OLR1	AA	AC	CC	A	C	-	-	-
	0.22	0.50	0.28	0.47	0.53	0.69	0.50	0.17^{ns}

SCD1: stearoyl-CoA desaturase 1 and OLR1: oxidized low density lipoprotein receptor 1.
HWE: Hardy-Weinberg equilibrium.
** (P<0.01).
NS: non significant.

Table 2 Analysis of general linear model for studied traits including SCD1 or OLR1 genotypes

Model	Parameter	MY (kg)	FY (kg)	PY (kg)	FP %	PP %
With SCD1 genotype	Pr > F	0.0002	0.0001	0.0001	0.0019	0.0001
	R^2	0.77	0.82	0.80	0.74	0.87
	(Mean±SE)	8115±1364	272±42.8	283±33.0	3.36±0.231	3.5±0.254
	CV%	16.8	15.7	18.8	6.9	7.3
With OLR1 genotype	Pr > F	0.0175	0.0008	0.0018	0.0219	0.0001
	R^2	0.88	0.92	0.91	0.88	0.94
	(Mean±SE)	8077±1302	269±39.3	281±48.6	3.33±0.208	3.47±0.243
	CV %	16	14.6	17.2	6.25	7.02

SCD1: stearoyl-CoA desaturase 1 and OLR1: oxidized low density lipoprotein receptor 1.
MY: milk yield; FY: fat yield; PY: protein yield; FP: fat percentage and PP: protein percentage.
SE: standard error.
CV: coefficient of variation.

The coefficient of variation range for 305-d milk, fat and protein yield were between 14.7 (for fat yield) to 18.8 (for protein yield), this value for fat and protein percentage were lower and the range was between 6.25 (for fat percentage) and 7.3 (for protein percentage). The results of comparing traits means with different SCD1 and OLR1 genotype by Duncan multiple rang test are shown in Table 3.

SCD1

The SCD1 genotypes showed significant effects on fat percentage but no effect on the other traits. The means of fat percentage for VV and AA genotype were 3.43% and 3.33% (P<0.05), this result was similar to Komisark and Dorynek (2009), that showed a positive effect of VV genotype on fat percentage. We found 305-d milk and protein yield means of cows with genotype AA to be higher than those of cows with the VV genotype. Kgwatalala et al. (2009) reported that allele A was positively associated with increased 305-d milk and protein yields. Effects of SCD genotype were not observed for milk yield or composition (Clark et al. 2010). The SCD1 genotype did not significantly affect fat or protein percentage and fat, protein, or milk yield (Schennink et al. 2008).

OLR1

The comparison of traits mean for OLR1 genotypes are shown in Table 3. The milk yield mean of cows with genotype AA (7178 kg) was significantly higher (P<0.05) than those of cows with genotypes AC (8344 kg) and CC (8273 kg).

The means of fat and protein yield for cows with the AC and CC genotypes were higher than the corresponding values for cows with the AA genotype (P<0.05). The means for fat and protein percentage in AA cows were somewhat higher than those for AC and CC cows, but differences were not significant (P>0.05). Although other SNP were identified in OLR1, only the 3'-UTR SNP was found to be associated with milk traits in the Holstein population. Khatib et al. (2006) reported the positive effect of allele C on fat and protein percentage and fat yield (P<0.05). Soltani-Ghombavani et al. (2013) revealed a significant effect of OLR1 gene on fat and protein percentage and breeding value of fat yield in Iranian Holstein cattle, they found the value of milk yield for CC and AC genotype more than AA genotype but there was no significant association. Interestingly, quantitative real-time PCR analysis revealed that the expression level of OLR1 was higher in individuals bearing the CC genotype compared with the AA genotype of the 3'-UTR SNP, suggesting that C is the nucleotide causing increased expression of OLR1 or is in strong linkage disequilibrium with the causative SNP (Khatib et al. 2006).

Komisarek and Dorynek (2009) reported a significant association between C allele and fat yield and fat percentage. The significant effect of allele C on fat yield in this study was similar to those reported by Khatib et al. (2006); Komisarek and Dorynek (2009) and Soltani-Ghombavan et al. (2013). The association of CC genotype with milk yield shown in this study disagreed with the findings of Khatib et al. (2006) and Komisarek and Dorynek (2009).

Table 3 Compare genotypes mean for milk production traits with different genotypes of *SCD1* or *OLR1* gene

Gene	Genotype	(MY (kg)±SE)	(FY (kg))±SE	(PY (kg)±SE)	(FP %±SE)	(PP %±SE)
SCD1	AA (no 172)	8130±134	271±4.8	285.8±5.8	3.33[b]±0.02	3.49±0.03
	AV (no 72)	8113±202	276±7.2	285.6±8.7	3.40[ab]±0.03	3.5±0.05
	VV (no 30)	8036±285	274 ±10.5	283.7±10.6	3.43[a]±0.06	3.53±0.04
P-value	-	0.31	0.41	0.50	0.048	0.68
OLR1	AA (no 29)	7178[b]±267	239[b]±9.4	253[b]±12	3.35±0.052	3.52±0.09
	AC (no 73)	8344[a]±209	277[a]±7.5	290[a]±9.2	3.33±0.032	3.46±0.06
	CC (no 34)	8273[a]±291	276[a]±8.3	286[a]±11.7	3.34±0.045	3.46±0.06
P-value	-	0.049	0.038	0.039	0.41	0.71

SCD1: stearoyl-CoA desaturase 1 and *OLR1*: oxidized low density lipoprotein receptor 1.
MY: milk yield; FY: fat yield; PY: protein yield; FP: fat percentage and PP: protein percentage.
The means within the same column with at least one common letter, do not have significant difference (P>0.05).
SE: standard error.

Conversely, Schennink *et al.* (2009) also found no association between *OLR1* gene and fat percentage in Dutch Holstein Friesian cows agreeing on results here. Similarly, no association was observed between polymorphism in *OLR1* and either milk production traits or reproduction traits in a Czech Fleckvieh population (Rychtarova *et al.* 2014).

CONCLUSION

The frequency of allele A of *SCD1* was 0.76 in the sample of Iranian Holstein dairy cattle studied here, and the association of genotype with fat percentage was significant. The population was in linkage disequilibrium for *SCD1*, perhaps due to selection for fat percentage in this population. Although the frequency of the C allele of gene *OLR1* was somewhat higher than that of the A allele, the population was in linkage equilibrium. However, the C allele had significant effects on milk, fat and protein yields, it could be concluded the favorable allele for *SCD1* and *OLR1* gene are A and C allele, thus these alleles could be used in a marker-assisted selection program to increase favorable alleles frequencies in the population and also these SNPs are proper for having on commercial SNP panels.

REFERENCES

Ashwell M.S., Heyen D.W., Sonstegard T.S., Van Tassell C.P., Da Y., VanRaden P.M., Ron M., Weller J.I. and Lewin H.A. (2004). Detection of quantitative trait loci affecting milk production, health, and reproductive traits in Holstein cattle. *J. Dairy Sci.* **87,** 468-475.

Campbell E.M.G., Gallagher D.S., Davis S.K., Taylor J.F. and Smith S.B. (2001). Mapping of the bovine stearoyl coenzyme A desaturase (SCD) gene to BTA26. *J. Anim. Sci.* **79,** 1954-1955.

Cecchinato A., Chessa S., Ribeca C., Cipolat-Gotet C., Bobbo T., Casellas J. and Bittante G. (2015). Genetic variation and effects of candidate-gene polymorphisms on coagulation properties, curd firmness modeling and acidity in milk from Brown Swiss cows. *Animal.* **9(7),** 1104-1112.

Clark L.A., Thomson J.M., Moore S.S. and Oba M. (2010). The effect of Ala293Val single nucleotide polymorphism in the stearoyl-CoA desaturase gene on conjugated linoleic acid concentration in milk fat of dairy cows. *Canadian J. Anim. Sci.* **90,** 575-584.

De Koning D.J., Schulmant N.F., Elo K., Moisio S., Kinos R., Vilkki J. and Maki-Tanila A. (2001). Mapping of multiple quantitative trait loci by simple regression in half-sib designs. *J. Anim. Sci.* **79,** 616-622.

Heyen D.W., Weller J.I., Ron M., Band M., Beever J.E., Feldmesser E., Da Y., Wiggans G.R., VanRaden P.M. and Lewin H.A. (1999). A genome scans for QTL influencing milk production and health traits in dairy cattle. *Physiol. Genomics.* **1,** 165-175.

Hosseinpour Mashhadi H., Nassiri M.R., Emam Jome Kashan N. and Vaez Torshizi R. (2013). Association between DGAT1 genotype and breeding value of milk production traits in Iranian Holstein bulls. *Iranian J. Appl. Anim. Sci.* **3,** 811-815.

Imanishi T., Hano T., Sawamura T., Takarada S. and Nishio I. (2002). Oxidized low density lipoprotein potentiation of Fas-induced apoptosis through lectin-like oxidized-low density lipoprotein receptor-1 in human umbilical vascular endothelial cells. *Circ. J.* **66,** 1060-1064.

Kgwatalala P.M., Ibeagha-Awemu E.M., Mustafa A.F. and Zhao X. (2009). Influence of stearoyl-coenzyme a desaturase 1genotype and stage of lactation on fatty acid composition of Canadian Jersey cows. *J. Dairy Sci.* **92,** 1220-1228.

Khatib H., Leonard S.D., Schutzkus V., Luo W. and Chang Y.M. (2006). Association of the OLR1 gene with milk composition in Holstein dairy cattle. *J. Dairy Sci.* **89,** 1753-1760.

Khatib H., Rosa G.J., Weigel K., Schiavini F., Santus E. and Bagnato A. (2007). Additional support for an association between OLR1 and milk fat traits in cattle. *Anim. Genet.* **38,** 308-310.

Komisarek J. and Dorynek Z. (2009). Effect of *ABCG2, PPARGC1A, OLR1* and *SCD1* gene polymorphism on estimated breeding values for functional and production traits in Polish Holstein-Friesian bulls. *J. Appl. Genet.* **50,** 125-132.

Mehta J.L. and Li D. (2002). Identification, regulation and function of a novel lectin–like oxidized low–density lipoprotein receptor. *J. Am. Coll. Cardiol.* **39,** 1429-1435.

Mehta J.L. and Li D. (1998). Identification and autoregulation of receptor for OX-LDL in cultured human coronary artery endothelial cells. *Biochem. Biophys. Res. Commun.* **248,** 511-514.

Mele M., Conte G., Castiglioni B., Chessa S., Macciotta N.P.,

Serra A., Buccioni A., Pagnacco G. and Secchiari P. (2007). Stearoyl-CoA desaturase gene polymorphism and milk fatty acid composition in Italian Holsteins. *J. Dairy Sci.* **90,** 4458-4465.

Moioli B., Contarini G., Avalli A., Catillo G., Orru L., De Matteis G., Masoero G. and Napolitano F. (2007). Short communication: effect of stearoyl-coenzyme a desaturase polymorphism on fatty acid composition of milk. *J. Dairy Sci.* **90,** 3553-3558.

Olsen H.G., Gomez-Raya L., Vage D.I., Olsaker I., Klungland H., Svendsen M., Adnoy T., Sabry A., Klemetsdal G., Schulman N., Kramer W., Thaller G., Ronningen K. and Lien S. (2002). A genome scan for quantitative trait loci affecting milk production in Norwegian dairy cattle. *J. Dairy Sci.* **85,** 3124-3130.

Rodriguez-Zas S.L., Southey B.R., Heyen D.W. and Lewin H.A. (2002). Interval and composite interval mapping of somatic cell score, yield, and components of milk in dairy cattle. *J. Dairy Sci.* **85,** 3081-3091.

Rychtarova J., Sztankoova Z., Kyselova J., Zink V., Stipkova M., Vacek M. and Stolc L. (2014). Effect of *DGAT1*, *BTN1A1*, *OLR1* and *STAT1* genes on milk production and reproduction traits in the Czech Fleckvieh breed. *Czech J. Anim. Sci.* **59 (2),** 45-53.

SAS Institute. (2004). SAS®/STAT Software, Release 9.1. SAS Institute, Inc., Cary, NC. USA.

Sawamura T., Kume N., Aoyama T., Moriwaki H., Hoshikawa H., Aiba Y., Tanaka T., Miwa S., Katsura Y., Kita T. and Masaki T. (1997). An endothelial receptor for oxidized low-density lipoprotein. *Nature.* **386,** 73-77.

Schennink A., Heck J.M., Bovenhuis H., Visker M.H., Van Valenberg H.J. and Van Arendonk J.A. (2008). Milk fatty acid unsaturation: genetic parameters and effects of stearoyl-CoA desaturase (SCD1) and acyl-CoA: diacylglycerol acyltransferase 1 (*DGAT1*). *J. Dairy Sci.* **91,** 2135-2143.

Schennink A., Bovenhuis H., Le´on-Kloosterziel K.M., van Arendonk J.A.M. and Visker M.H.P.W. (2009). Effect of polymorphisms in the *FASN, OLR1, PPARGC1A, PRL* and *STAT5A* genes on bovine milk-fat composition. *Anim. Genet.* **40,** 909-916.

Soltani-Ghombavan M., Ansari-Mahyari S., Ghorbani G.R. and Edriss M.A. (2013). Association of a polymorphism in 3' untranslated region of *OLR1* gene with milk fat and protein in dairy cows. *Arch. Tierz.* **56,** 32-40.

Taniguchi M., Utsumi T., Oyama K., Mannen H., Kobayashi M., Tanabe Y., Ogino A. and Tsuji S. (2004). Genotype of stearoyl-CoA desaturase is associated with fatty acid composition in Japanese Black cattle. *Mamm. Genome.* **14,** 142-148.

Viitala S.M., Schulman N.F., de Koning D.J., Elo K., Kinos R., Virta A., Virta J., Maki-Tanila A. and Vilkki J.H. (2003). Quantitative trait loci affecting milk production traits in Finnish Ayrshire dairy cattle. *J. Dairy Sci.* **86,** 1828-1836.

Wang Q., Hulzebosch A. and Bovenhuis H. (2016). Genetic and environmental variation in bovine milk infrared spectra. *J. Dairy Sci.* **99(8),** 6793-6803.

Yeh F.C., Yang R.C. and Boyle T. (1999). POPGENE. Microsoft Windows-Based Freeware for Population Genetic Analysis. Release 1.31. Univ Alberta, Edmonton, Canada.

20

Ruminal Methane Emission, Microbial Population and Fermentation Characteristics in Sheep as Affected by Malva sylvestris Leaf Extract: in vitro Study

S. Khamoshi[1], F. Kafilzadeh[1], H. Jahani-Azizabadi[2*] and V. Naseri[1]

[1] Department of Animal Science, Faculty of Agriculture, Razi University, Kermanshah, Iran
[2] Department of Animal Science, Faculty of Agriculture, University of Kurdistan, Sanandaj, Iran

*Correspondence E-mail: ho.jahani@uok.ac.ir

ABSTRACT

The objective of this study was to investigate *in vitro* effect of *Malva sylvestris* leaf extract (at 0, 25, 50 and 100 µL/30 mL of medium) on sheep ruminal cellulolytic and total viable bacteria growth, protozoa populations, methane production, neutral detergent fiber degradability (NDFD) and fermentation efficiency of oat hay. The addition of *Malva sylvestris* leaf extract at 25, 50 and 100 µL led to a linear increase (P<0.01) *in vitro* truly degraded dry matter (INTDDM), NDFD and partitioning factor (PF) of oat hay and decrease (P<0.01) methane emission after 24 hours of incubation. The addition of *Malva sylvestris* leaf extract resulted in a decrease (P<0.01) in potential of gas production at 25 and 50 µL and increase in lag time (0.95, 1.01 and 1.13 h relative to 0.61 h), constant rate of gas production (*b*) and gas produced at half-life (*c*) at 25, 50 and 100 µL. The addition of this extract decreased a number of total protozoa and *Entodinium, Isotrichae, Diplodinium* and *Ophryoscolex* species (P<0.01). The number of total and cellulolytic bacteria was not influenced by the addition of *Malva sylvestris* leaf extract. The result of this study demonstrated that *Malva sylvestris* extract had some potential for improving the rumen fermentation.

KEY WORDS cellulolytic bacteria, kinetics, partitioning factor, protozoa, rumen.

INTRODUCTION

With increase in the probability of transmissible antibiotic resistance to human, the use of growth-promoter antibiotics in animal production systems has been banned in most developed countries (European Union, 2003). The removal of antibiotics has led the scientists' interest on investigation for the alternatives (Busquet *et al.* 2005; Iason, 2005; Durmic *et al.* 2014). Medicinal plants and their extracts as natural alternative to antibiotics showed the potential of manipulating the rumen fermentation to improve the utilization of nutrients (Busquet *et al.* 2005; Kim *et al.* 2013; Kim *et al.* 2015). The bioactive constituents of these compounds are secondary metabolites that are produced by plants to protect themselves against plant diseases and insects (Wallace, 2004). *Malva sylvestris* usually known as common Mallow is native to Asia, Europe, and north Africa and have been used as a medicinal plant since a long time ago. In the Mediterranean region, *Malva sylvestris* has a long history of use as food and potent drug in traditional and ethno veterinary medicines because of its anti-inflammatory, antioxidant, anti-complementary, anticancer and skin tissue integrity characteristics (Gasparetto *et al.* 2011). The whole aerial parts of *Malva sylvestris* or its leaves have been administrated to ruminant to treat colic, blocked rumen, mastitis, young calves' diarrhea, respiration problem and reproductive disorder (Gonzalez *et al.* 2010; Idolo *et al.* 2010).

Enteric methane arising due to fermentation of feeds in the rumen not only is a great source of energy loss but contributes substantially to the greenhouse gas emissions. In the recent years with increase horse farm in Kermanshah province the request for oat grain increased. Therefore, increase in oat cultivation resulted to enhance in oat hay and straw production that mostly have been used in the sheep nutrition. Thus, like an evaluation of chemical composition and nutritive values of feeds, methane production potential of each feed should be determined (Patra *et al.* 2015). The effect of *Malva sylvestris* on ruminant colic and improve the blocked rumen, may be because of its effect on rumen microbial activity. Therefore, the objective of the present experiment was to investigate the effect of *Malva sylvestris* leaf extract on rumen microbial populations, methane production and microbial fermentation of oat hay in sheep rumen liquor.

MATERIALS AND METHODS

Malva sylvestris plant was collected from Kermanshah province, Iran. For juice extraction, the fresh leaves of *Malva sylvestris* (pre-flowering) were finely grounded and blended in a commercial blender, squeezed through 2 layers of cheesecloth (Davys *et al.* 1969). The extract centrifuged at 454 × g for 15 minutes and the supernatant collected. The supernatant was saved and kept frozen at -80 °C until use. The effects of four doses [0.0 (as control), 25, 50 and 100 μL] of *Malva sylvestris* leaf extract were examined *in vitro* using mixed ruminal microbiota from sheep rumen liquor. The substrate used for batch cultures was oaten hay (crude protein (CP)=15.5, neutral detergent fiber (NDF)=54.2 and acid detergent fiber (ADF)=29.4% of dry matter (DM)) wich was grounded to pass from 1 mm screen. Rumen content was obtained from three fistulated sheep (39±4.5 kg body weight) before morning feeding. Animals were fed 0.6 kg alfalfa hay and 0.6 kg concentrate (16.5% CP, 25% NDF and 45% nonfiber carbohydrates (NFC)). For eliminating of large feed particles, immediately rumen content was filtered through two layers of cheesecloth and transferred to the laboratory in a prewarmed thermos (38 °C). In the laboratory, under anaerobic conditions, 30 mL of buffered rumen fluid [ratio of rumen fluid to buffer was 1:2, buffer prepared as proposed by McDougall (1948)], using pipettor pump was added into 125 mL bottles containing 0.2 g of oaten hay (12 replicates for each treatment in two runs). Then, bottles were sealed by a rubber stopper and aluminum cap and placed in shaking water bath for 96 h at 38.6 °C. Accumulation of gas produced in the bottles, head space was measured by a pressure transducer at 2, 4, 8, 12, 18, 24, 36, 48, 72 and 96 h after the incubation and the gas released (Theodorou *et al.* 1995).

After 24 h of incubation 8 tubes from each treatment were withdrawn to determine *in vitro* truly degraded DM (INTDDM), partitioning factor (PF) and NDF degradability (NDFD) and enumeration of total viable and cellulolytic bacteria and rumen protozoa. Then each bottle content was filtered (42 μm pore size) and solid residues were used for determining IVDMD, PF and NDFD. A 10 mL sample of each filtrate bottle was taken and transfer into the separate flask (50 mL, 4 replicate/treatment). The flasks were closed and allowed to stand for 1 h at 38.6 °C. Rumen fluid was collected by suction from the middle of each flask. The ruminal fluid contained bacteria were serially diluted (10-fold increments) in the liquid version of mediums in the Hungate tubes (3 replicate). For enumeration of total viable and cellulolytic bacteria, the anaerobic technique proposed by Bryant (1972) was used for preparation medium. For cellulolytic bacteria ball-milled cellulose used as a single source of energy. The tubes were incubated at 38.6 °C for 72 h and 14 d (for cellulolytic bacteria) and in the end of incubation, growth was scored (+ or -) by the increase in optical density (650 nm). The cellulolytic and total viable bacteria population size was estimated using most probable number (MPN) procedure from replicate (3 tubes/dilution) dilutions. For protozoa enumeration, 5 mL of filtered rumen fluid was preserved using 5 mL of 50% Formalin (18.5% concentration of formaldehyde) as described by Dehority *et al.* (1984). Two drops of Brilliant green dye was added to 1 mL of rumen fluid (n=2) and stored overnight at the laboratory temperature. Then, 9 mL of 30% glycerol solution was added. Protozoa were enumerated microscopically in a Sedgwick-Rafter counting chamber. Protozoa species were identified from photographs and descriptions given by Ogimoto and Imai (1981).

Calculation and statistical analysis

Gas pressure was converted into volume using an experimentally calibrated curve. To estimate kinetic parameters of gas production, gas production data were fitted using France *et al.* (1993) equation as:

$$Y = A \times ((1-\exp)-(b(t-L))-(c(t^{1/2}-L^{1/2})))$$

Where:

A: volume of gas produced from quickly and slowly degradable fraction.

b and c: constants of the fractional rate (%/h).

t: incubation time (h).

L: lag time (h).

Y: volume of gas produced at time t.

The half-life ($t^{1/2}$, h) of the fermentable fraction of each substrate was calculated as the time taken for gas accumulation to reach 50% of its asymptotic value. Methane production was measured according to the method proposed by Fievez *et al.* (2005).

True substrate degradability was measured following the procedure of Blummel *et al.* (1997). Briefly, the contents of the bottles were transferred into the Berzelius beaker by repeated washings with 50 mL of neutral detergent solution (double strength), refluxed for 1 h, filtered through silica crucibles (Grade 1) and then for determining organic matter (OM) concentration, residues were burnt in a furnace at 600 °C for 3 h.

Then INTDDM, INTDOM and NDFD were calculated as the ratio of residues DM or OM after 24 h of incubation/incubated DM or OM. PF was calculated as the ratio of milligram organic matter truly degraded/mL gas produced after 24 h of incubation (Blummel *et al.* 1997).

The population size of total viable and cellulolytic bacteria were calculated using most-probable-number tables (Alexander, 1982) with values derived from the number of tubes that showed positive growth.

Data were analyzed using the SAS (SAS, 2002) and significance between individual means was identified using Duncan multiple range test. For enumeration total and cellulolytic bacteria and protozoa, least significant difference (LSD) was used to compare the means.

RESULTS AND DISCUSSION

In vitro rumen fermentation

Effect of addition of *Malva sylvestris* leaf extract on INTDDM, NDFD and PF of oat hay are summarized in Table 1. Relative to the control, the addition of *Malva sylvestris* leaf extract at 25, 50 and 100 µL/30 mL of medium resulted in a linear increase (P<0.01) in INTDDM and NDFD. Relative to the control, the PF was increased (P<0.05) with addition *Malva sylvestris* leaf extract at 25, 50 and 100 µL (2.74 *vs.* 2.81, 2.82 and 2.94).

As it is shown in Table 1, the addition of *Malva sylvestris* leaf extract at 25, 50 and 100 µL, after 24 h of incubation decreased (P<0.01) the methane production (except at 50 µL doses) relative to the control. Gas production not affected with *Malva sylvestris* leaf extract supplementation, except at 50 µL (41.22 *vs.* 40.16 in control, P-value=0.001). According to results shown in Table 2, the addition of *Malva sylvestris* leaf extract resulted in a decrease in half-life (*h*) of gas production at 25 and 50 µL and increase at 100 µL.

In addition, supplementation of *Malva sylvestris* at 25, 50 and 100 µL increased the lag time (*h*) and constant rate of gas production at half-life (*c*) (P<0.01). Half-life was decreased as a result of increased constant rate of gas production (*c*) during the half of incubation time (P<0.01).

Total viable and cellulolytic bacteria and protozoa population counting

Effect of *Malva sylvestris* leaf extract on genus diversity and total population of protozoa, total viable and cellulolytic bacteria population are shown in Tables 3 and 4. Results of the present study showed that the addition of *Malva sylvestris* leaf extract resulted in a decrease (P<0.01) in total protozoa (highest at 100 µL and lowest at 25 µL) and *Entodinium*, *Isotrichae*, *Diplodinium* and *Ophryoscolex* genus.

Relative to the control, the addition of *Malva sylvestris* leaf extract at 25 and 50 µL increased and at 100 µL decreased total viable and cellulolytic bacteria (P>0.05).

Partitioning factor (PF) is reported to be valuable in the accuracy of voluntary dry matter intake (DMI) prediction and microbial biomass synthesis efficiency of temperate-tropical crop residues and Mediterranean hays and forages with high PF result in high dry matter intake (DMI) (Blummel *et al.* 1997). The PF is known as the index of microbial biomass synthesis efficiency (Blummel *et al.* 1997). Therefore, results of the present study showed that *Malva sylvestris* leaf extract had a potential to increase the efficiency of ruminal microbial protein synthesis. The increase in PF by supplementation of *Malva sylvestris* leaf extract demonstrated that proportionally more of the truly degraded substrate was incorporated in microbial mass which can increase the efficiency of microbial protein synthesis. Blummel *et al.* (2005) reported 48.8% and 2.81 mg/mL for INTDDM and PF (at 24 h incubation) in oat hay, respectively.

The inhibitory effect of *Malva sylvestris* leaf extract on methane production confirmed their anti-methanogens potent on ruminal microbes involved in methanogenesis. In ruminants, methane emission represents 8 to 12% loss of intake energy (Johnson and Johnson, 1995). On the other hands, methane is a greenhouse gas that has a global warming potential 21 times that of CO_2 (Crutzen *et al.* 1995).

Therefore, supplementation of additives that decrease ruminal methane emission can increase the efficiency of energy utilization in ruminant and decrease environment contamination (Kim *et al.* 2013; Kim *et al.* 2015). In addition, utilization of low methane producing feeds that are available at livestock farms could be strategically considered to feed ruminants decreasing environmental impacts (Patra *et al.* 2015). Increase in gas production with the addition of 50 µL of *Malva sylvestris* leaf extract and decrease in gas production at 100 µL may be due to raise in the concentration of antimicrobial secondary compounds of *Malva sylvestris* that added to the medium at this level. Our findings agree with previous *in vitro* batch culture reports where high doses of essential oils or extract have been tested (Jahani-Azizabadi *et al.* 2011; Busquet *et al.* 2005).

Table 1 Effect of *Malva sylvestris* leaf extract on *in vitro* fermentation characteristics of oat hay

Items	*Malva sylvestris* leaf extract (µL)				SEM	P-value
	0.0	25	50	100		
Gas production (mL/24 h/0.2 g)	40.16[b]	40.16[b]	41.22[a]	40.03[b]	0.15	0.001
Methane production (mL)	11.82[b]	11.02[c]	12.32[a]	10.75[d]	0.19	0.001
Methane reduction potential (%)	-	6.34[b]	-12.29[c]	13.10[a]	2.82	0.001
IVTDMD (g/kg DM)	527.10[d]	534.60[c]	542.20[b]	573.30[a]	5.33	0.001
NDFD (g/kg DM)	615.30[b]	623.20[b]	625.30[b]	674.30[a]	8.66	0.03
PF (mg/mL)	2.74[c]	2.81[b]	2.82[b]	2.92[a]	0.02	0.001

IVTDMD: *in vitro* truly degraded dry matter; NDFD: nuteral detergent fiber degradability; PF: partitioning factor and DM: dry matter.
The means within the same column with at least one common letter, do not have significant difference (P>0.05).
SEM: standard error of the means.

Table 2 Effect of *Malva sylvestris* leaf extract on *in vitro* gas production parameters

Parameters	*Malva sylvestris* leaf extract (µL)				SEM	P-value
	0.0	25	50	100		
A (mL/0.2 g DM)	60.57[b]	60.04[c]	56.43[d]	61.12[a]	0.03	0.001
b (mL/h)	0.06[a]	0.05[b]	0.05[b]	0.06[a]	0.002	0.001
c (mL/h$^{1/2}$)	0.03[b]	0.05[a]	0.04[a]	0.04[a]	0.004	0.002
L (h)	0.61[c]	0.95[b]	1.01[b]	1.13[a]	0.06	0.002
Half-life (h)	11.13[b]	11.05[c]	10.50	11.30[a]	0.01	0.001
Fermentation rate (T$^{1/2}$/h)	0.011[b]	0.013[b]	0.016[a]	0.007[c]	0.001	0.001

A: potential gas production; b,c: constant rate; L: lag time and DM: dry matter.
The means within the same column with at least one common letter, do not have significant difference (P>0.05).
SEM: standard error of the means.

Table 3 Effect of *Malva sylvestris* leaf extract supplementation on *in vitro* protozoa population ($\times 10^5$/mL)

Protozoa population	*Malva sylvestris* leaf extract (µL)				SEM	P-value
	0.0	25	50	100		
Total	2.89[a]	2.62[b]	2.14[c]	1.17[d]	0.20	0.001
Entodinium	2.34[a]	2.23[a]	1.88[b]	0.94[c]	0.17	0.001
Isotrichae	0.20[a]	0.10[b]	0.05[c]	0.12[b]	0.02	0.001
Diplodinium	0.05[a]	0.05[a]	0.03[b]	ND	0.006	0.001
Ophryoscolex	0.17[a]	0.11[b]	0.05[c]	0.03[d]	0.02	0.001

ND: not detected.
The means within the same row with at least one common letter, do not have significant difference (P>0.05).
SEM: standard error of the means.

Table 4 Effect of *Malva sylvestris* leaf extract on *in vitro* total and cellulolytic bacteria (log10/mL of medium)

Items	*Malva sylvestris* leaf extract (µL)				SEM	P-value
	0.0	25	50	100		
Total viable bacteria	8.1	8.56	8.58	7.70	0.35	0.08
Cellulolytic	7.1	7.21	7.49	6.95	0.15	NS

The means within the same column with at least one common letter, do not have significant difference (P>0.05).
NS: non significant.
SEM: standard error of the means.

Linearly increasing in lag time withe increase in the concentration of *Malva sylvestris* leaf extract could be due to the increase in time required for bacterial colonization on feed particles. Therefore, results of the present study demonstrated that high doses of *Malva sylvestris* leaf extract in high-forage diets can decrease in dry matter intake due to increase in colonization time and rumen filling.

Results of the present study demonstrated that *Entodinium* species accounted for around 90% of total counted protozoa. This finding was in agreement with values observed by other researchers (Santra *et al.* 1998; Hindrichsen *et al.* 2002).

In addition, results of the present study suggested that *Entodinium*, *Diplodinum* and *Ophryoscolex* species were more sensitive to the addition of *Malva sylvestris* leaf extract. Relative to the control, abundance of *Entodinium*, *Diplodinum* and *Ophryoscolex* species were decreased by 59.8%, near to 100% and 82.3%, respectively. Therefore, in the present study decrease in methane production with addition of *Malva sylvestris* leaf extract was accompanied by a significant decrease in protozoa count. Around 10-20% of ruminal methanogenic bacteria are attached to the surface of protozoa, especially *Entodinium* species (Stumm *et al.* 1982).

The methanogens take up hydrogen, which appears too toxic for *Entodinium* species activity. Previous studies have demonstrated that rumen protozoa defaunation reduced methane production ranging from 13 to 35% (Morgavi *et al.* 2008; Morgavi *et al.* 2012) and 9 to 25% *in vitro* (Newbold *et al.* 1995).

Neutral detergent fiber degradability (NDFD) is an important index of forage quality. Increased NDFD may result to greater voluntary feed intake by reducing physical fill in the rumen (Dado and Allen, 1995). *In vitro* or *in situ* one unit increase in forage NDFD was associated with 0.17 kg increase in dry matter intake (Oba and Allen, 1999).On the other hand, results of the present study showed that addition of *Malva sylvestris* leaf extract resulted in a significant linear (P<0.01) increase in NDF disappearance. No significant changes in rumen cellulolytic bacteria and increase in NDF disappearance of oaten hay suggest that probably, the addition of *Malva sylvestris* leaf extract could result in an increase in the relative abundance of obligate cellulolytic bacteria with higher cellulolytic activity and persistence to main secondary compounds of *Malva sylvestris* leaf extract. However, unfortunately, there is not any information about the effect of *Malva sylvestris* leaf extract on rumen microbial fermentation.

CONCLUSION

In conclusions, *Malva sylvestris* leaf extract used in the present study resulted in the valuable effect on rumen microbial fermentation. For determine the mechanisms that *Malva sylvestris* leaf extract affects ruminal degradation of NDF and methane emission, more studies on the main rumen cellulolytic and methanogens bacteria are needed. As regards that this is first report about effect of *Malva sylvestris* extract on rumen microbial fermentation, future research is required to determine optimal doses and the effect of *Malva sylvestris* organic extract and essential oil on *in vitro* and *in vivo* rumen microbial fermentation patterns and animal performance.

ACKNOWLEDGEMENT

The authors thank the University of Razi (Kermanshah, Iran) for the financial support.

REFERENCES

Alexander M. (1982). Most probable number method for microbial populations. Pp. 815-820 in Methods of Soil Analysis, Part 2. A.L. Page, R.H. Miller and D.R. Keeney, Eds. American Society of Agronomy, Madison, USA.

Blummel M., Cone J.W., Van Gelder A.H., Nshalai I., Umunna N.N., Makkar H.P.S. and Becker K. (2005). Prediction of forage intake using *in vitro* gas production methods comparison of multiphase fermentation kinetics measured in an automated gas test and combined gas volume and substrate degradability measurements in a manual syringe system. *Anim. Feed Sci. Technol.* **5**, 123-124.

Blummel M., Makkar H.P.S. and Becker K. (1997). *In vitro* gas production: a technique revisited. *J. Anim. Physiol. Anim. Nutri.* **77**, 24-34.

Bryant M.P. (1972). Commentary on the Hungate technique for culture of anaerobic bacteria. *Am. J. Clin. Nutr.* **25(12)**, 1324-1328.

Busquet M., Calsamiglia S., Ferret A. and Kamel C. (2005). Screening for effects of plant extracts and active compounds of plants on dairy cattle rumen microbial fermentation in a continuous culture system. *Anim. Feed Sci. Technol.* **124**, 597-613.

Crutzen P.J. (1995). The role of methane in atmospheric chemistry and climate. Pp. 291-315 in Ruminant Physiology: Digestion, Metabolism, Growth and Reproduction. W.V. Engelhardt, S. Leonhard-Marek, G. Breves and D. Giesecke, Eds. Ferdinand Enke Verlag Stuttgart, Germany.

Dado R.G. and Allen M.S. (1995). Intake limitation, feeding behaviour and rumen function of cows challenged with rumen fill from dietary fiber or inert bulk. *J. Dairy Sci.* **78**, 118-113.

Davys M.N.G., Pirie N.W. and Street G. (1969). A laboratory-scale press for extracting juice from leaf pulp. *Biotech. Bioeng.* **11**, 529-538.

Dehority B.A. (1984). Evaluation of subsampling and fixation procedures used for counting rumen protozoa. *Appl. Environ. Microbiol.* **48**, 182-185.

Durmic Z., Moate P.L., Eckard R., Revell D.K., Williams R. and Vercoe P. (2014). *In vitro* screening of selected feed additives, plant essential oils and plant extracts for rumen methane mitigation. *J. Sci. Food Agric.* **94**, 1191-1196.

European Union. (2003). Regulation (EC) No 1831/2003 of the European Parliament and of the Council of 22 September 2003 on Additives for Use in Animal Nutrition. Council of the European Union , European Parliament.

Fievez V., Babayemi O.J. and Demeyer D. (2005). Estimation of direct and indirect gas production in syringes: a tool to estimate short chain fatty acid production that requires minimal laboratory facilities. *Anim. Feed Sci. Technol.* **123**, 197-210.

France J., Dhanoa M.S., Theodorou M.K., Lister S.J., Davies D.R. and Isaac D.A. (1993). Model to interpret gas accumulation profiles associated with *in vitro* degradation of ruminant feeds. *J. Theor. Biol.* **163**, 99-111.

Gasparetto J.C., Martins A.F., Hayashi S.S., Otuky M.F. and Pontarolo R. (2011). Ethnobotanical and scientific aspects of *Malva sylvestris*: a millennial herbal medicine. *J. Pharm. Pharmacol.* **64**, 172-182.

Gonzalez J.A., Barriuso M.G. and Amich F. (2010). Ethnobotanical study of medicinal plants traditionally used in the Arribesdel Duero, western Spain. *J. Ethnopharmacol.* **131**, 343-355.

Hindrichsen I.K., Osuji P.O., Odenyo A.A., Madsen J. and Hvelplund T. (2002). Effects of supplementation of a basal diet of maize stover with different amounts of

*Leucaenadiversifolia*on intake, digestibility and nitrogen metabolism and rumen parameters in sheep. *Anim. Feed Sci. Technol.* **98**, 131-142.

Iason G. (2005). The role of plant secondary metabolites in mammalian herbivory: ecological perspectives. *Proc. Nutr. Soci.* **64**, 123-131.

Idolo M., Motti R. and Mazzoleni S. (2010). Ethnobotanical and phytomedicinal knowledge in a long history protected area, the Abruzzo, Lazio and Molise National Park (Italian Apennines). *J. Ethnopharmacol.* **127**, 379-395.

Jahani-Azizabadi H., Mesgaran M.D.,Vakili A.R., Rezayazdi K. and Hashemi M. (2011). Effect of various medicinal plant essential oils obtained from semi-aridclimate on rumen fermentation characteristics of a high forage diet using *in vitro* batch culture. *African J. Microbiol. Res.* **5**, 4812-4819.

Johnson K.A. and Johnson D.E. (1995). Methane emission from cattle. *J. Anim. Sci.* **73**, 2483-2492.

Kim E.T., Guan Le L., Lee S.J., Lee S.M., Lee S.S., Lee I.D., Lee S.K. and Lee S.S. (2015). Effects of flavonoid-rich plant extracts on *in vitro* ruminal methanogenesis, microbial populations and fermentation characteristics. *Asian-Australas J. Anim. Sci.* **28**, 530-537.

Kim E.T., Min K.S., Kim C.H., Moon Y.H., Kim S.C. and Lee S.S. (2013). The effect of plant extracts on *in vitro* ruminal fermentation, methanogenesis and methane-related microbes in the rumen. *Asian-Australas J. Anim. Sci.* **26**, 517-522.

McDougall E.I. (1948). Studies on ruminant saliva. 1. The composition and output of sheeps saliva. *Biochem. J.* **43**, 99-109.

Morgavi D.P., Forano E., Martin C. and Newbold C.J. (2008). Microbial ecosystem and methanogenesis in ruminants.

Animal. **4**, 1024-1036.

Morgavi D.P., Martin C., Jouany J.P. and Ranilla M.J. (2012). Rumen protozoa and methanogenesis: not a simple cause-effect relationship. *Br. J. Nutr.* **107**, 388-397.

Newbold C.J., Lassalas B. and Jouany J.P. (1995). The importance of methanogens associated with ciliate protozoa in ruminal methane production *in vitro*. *Lett. Appl. Microbiol.* **21**, 230-234.

Oba M. and Allen M.S. (1999). Evaluation of the importance of NDF digestibility: effects on dry matter intake and milk yield of dairy cows. *J. Dairy Sci.* **82**, 589-596.

Ogimoto K. and Imai S. (1981). Atlas of Rumen Microbiology. Japanese Science Society Press, Tokyo, Japan.

Patra P.K. and Sahoo A.K. (2015). Evaluation of feeds from tropical origin for *in vitro* methane production potential and rumen fermentation *in vitro*. *Spanian J. Agric. Res.* **13**, 1-11.

Santra A., Karim S.A., Mishra A.S., Chaturvedi O.H. and Prasad R. (1998). Rumenciliate protozoa and fibre digestion in sheep and goats. *Smal Rumin. Res.* **30**, 13-18.

SAS. (2002). SAS®/STAT Software, Release 9. SAS Institute, Inc., Cary, NC. USA.

Stumm C.K., Gijzen H.J. and Vogels G.D. (1982). Association of methanogenic bacteria with ovine rumen ciliates. *Br. J. Nutr.* **47**, 95-99.

Theodorou M.K., Williams B.A., Dhanoa M.S., McAllan A.B. and France J. (1995). A simple gas production method using a pressure transducer to determine the fermentation kinetics of ruminant feeds. *Anim. Feed Sci. Technol.* **48**, 185-197.

Wallace R.J. (2004). Anti-microbial properties of plant secondary metabolites. *Proc. Nutr. Soc.* **63**, 621-629.

Effects of Barley Grain Particle Size on Ruminal Fermentation and Carcass Characteristics of Male Lambs Fed High Urea Diet

S.R. Ebrahimi-Mahmoudabad[1] and M. Taghinejad-Roudbaneh[2]

[1] Department of Animal Science, Shahr-e-Qods Branch, Islamic Azad University, Tehran, Iran
[2] Department of Animal Science, Tabriz Branch, Islamic Azad University, Tabriz, Iran

*Correspondence E-mail: sayyedroohollah.ebrahimi@yahoo.com

ABSTRACT

Two experiments were conducted to evaluate effects of barley grains particle size on ruminal pH and ammonia concentration of rams (experiment 1) and carcass characteristics (experiment 2) of male lambs fed high urea diet. Treatments in two experiments were (1: basal diet + whole barley grains, 2: basal diet + ground barley grains with a 5 mm screen, 3: basal diet + ground barley grains with a 3 mm screen and 4: basal diet + ground barley grains with a 1 mm screen). Basal diet (on a dry matter (DM) basis) consisted of 365 g/kg corn silage, 10 g/kg limestone, 10 g/kg urea, 5 g/kg salt, 10 g/kg a vitamin-mineral premix and 600 g/kg of barley. In experiment 1, three 2-year old rams were fistulated for measuring ruminal pH and ruminal ammonia concentration. Ruminl pH was decreased by feeding ground barley grains through a 1 mm screen compared to feeding the whole barley grains (P<0.05). However, Ruminal ammonia concentrations were similar for all groups. In experiment 2, twenty four male lambs were used in a completely randomized design. Lambs were fed with the above mentioned diets for 90 days. Average daily gain, feed conversion ratio, average DM intake was significantly affected by treatments (P<0.05). Final body weight, cold carcass weight, dressing percentage, back fat thicknesses, carcass cuts (leg, shoulder, back and neck weights), internal organs (kidney, lungs, heart and gastrointestinal tract) weights were not significantly different between diets (P>0.05). Lambs fed diet containing ground barley with 3 mm of screen had (P<0.05) higher longissimus muscle area compared to lambs fed whole barley grain. Consumption of whole barley grains increased DM intake and pelvic and abdominal fats. As a conclusion, the consumption of ground barley grain with a 3 mm or a 5 mm screen is suggested for feeding lambs fed high urea diet.

KEY WORDS carcass, performance, processing, ruminal ammonia.

INTRODUCTION

Barley is an important feed grain for fattening of lambs in Iran as it is a readily available source of dietary energy. However, the endosperm of the barley kernel is surrounded by the pericarp overlain by a fibrous hull, which is extremely resistant to microbial degradation in the rumen. Processing makes the starch more accessible to microbes, and increases the rate and extent of starch degradation in the rumen. Although processing is essential to maximize the utilization of barley grain by sheep, extensive grain processing increases ruminal starch degradation, which often decreases feed intake in ruminants (McAllister *et al.* 1994; Allen, 2000). The price of sources of vegetable protein as oil seed meals in some countries is very high and non-protein nitrogen (NPN) is used to increase the dietary crude protein (CP). Urea can be used as a nitrogen source in ration to meet 25% of the total CP requirement of lambs (Stanto and Whittie, 2006). Moreover, Canpolat and Karabulut (2010) reported that supplementation of diet with urea had a positive effect on feedlot performance of lambs.

It has been reported that a synchronous supply of energy and nitrogen to the rumen enhance microbial CP synthesis (Sinclair *et al.* 1993; Trevaskis *et al.* 2001). Microbial protein can supply from 70% to 100% of amino acids requirements of ruminants (AFRC, 1992). Processing such as grinding generally increases ruminal degradable organic matter of grains for rumen bacteria for microbial crude protein synthesis.

Although the effect of processing of barley grain on feedlot performance of lambs has been evaluated in some studies (Economides *et al.* 1990; Petit, 2000; Voia *et al.* 2009), there is little information concerning the effect of barley grains particle size on performance of male lambs fed high urea diets. Therefore, the aim of this study was to evaluate the effects of feeding whole and ground barley grains on ruminal fermentation, growth and carcass characteristics of lambs fed high urea diet.

MATERIALS AND METHODS

Experiment 1

Three fistulated Mehraban rams aged 2 years and weighing 48 kg were used in change over design to determine rumen pH and ammonia concentration. Treatments were 1: basal diet + whole barley grains, 2: basal diet + ground barley grains through a 5 mm screen, 3: basal diet + ground barley grains through a 3 mm screen and 4: basal diet + ground barley grains through a 1 mm screen. The rams were fed with a basal diet on DM basis containing 365 g/kg corn silage, 600 g/kg barley, 10 g/kg limestone, 10 g/kg urea, 5 g/kg salt and 10 g/kg a vitamin-mineral premix. A 14-day period of adaptation to the diet was employed. The diet was formulated according to National Research Council (NRC, 1985) specifications to contain 124 g CP/kg of DM and fed twice daily at 07:00 and 15:00 h. Water was available *ad libitum*. Samples of rumen fluid were collected at 0 (before morning feeding) and 0.5, 2, 4 and 5 h after feeding. Samples of rumen fluid were strained through two layers of cheesecloth, and pH measured immediately with a pH meter (Metrohm, 744 Swiss). For determination of ammonia nitrogen (NH_3-N) in rumen fluid, 50 mL subsamples of strained rumen fluid were preserved by addition of 1 mL sulfuric acid 97% and stored at $-20°C$. Just before analysis, samples were thawed and analyzed for ammonia (Crooke and Simpson, 1971).

Experiment 2

Twenty four Mehraban male lambs (7 to 8 months old) with an initial live weight of 39.5 ± 4.17 kg were used to determine the effects of barley grains particle size on feedlot performance and carcass characteristics of lambs.

The lambs were allotted randomly to 24 pens (140×120 cm). They were maintained at ambient temperature and natural day length and water was available *ad libitum*. The lambs were adapted to the diets for 2 weeks (to limit the risk of digestive upsets) followed by growth trial of 90 days. Lambs were fed 4 diets as follows 1: basal diet + whole barley grain, 2: basal diet + ground barley grains through a 5 mm screen, 3: basal diet + ground barley grains through a 3 mm screen, and 4: basal diet + ground barley grains through a 1 mm screen). Basal diet on DM basis containing 365 g/kg corn silage, 10 g/kg limestone, 10 g/kg urea, 5 g/kg salt and 10 g/kg a vitamin-mineral premix. Chemical composition of the diets (on DM basis) was similar and contained 2.63 MJ ME per kg, 12.4% CP, 0.70% calcium and 0.28% phosphorus. The ME concentration of diet was calculated based on the digestible energy and acid detergent fiber content (Khalil *et al.* 1986). Vitamins and minerals were added to experimental diets to meet the requirements (NRC, 1985). Quarter of the total CP requirement of lambs was supplied by adding urea to the diet (Stanto and Whittie, 2006). The diets were fed as a totally mixed rations twice daily at 07:00 and 15:00 h in equal amounts to ensure 5% orts.

Amounts of feed offered and refused were recorded daily. At the beginning of experiment 2, lambs were weighed and lamb weights were recorded monthly before the 07:00 h feeding to monitor body weight change throughout the experiment. At the end of the experiment, the feed and water were removed for 18 h and then the lambs were weighed and slaughtered according to local practices (Zamiri and Izadifard, 1997). Cold carcass weight was determined after chilling hot carcass at 4 °C for 24 h. The cold carcass weight and live weight were used for determination of the dressing percentage. The abdominal and pelvic fats were removed and weighed. Fat depth over carcass was measured at the cross section of the 12th and 13th thoracic ribs at 4 points by caliper and the values were averaged as a measure of subcutaneous fat depth and then, longissimus muscle area were measured. The carcass was then split into the right and left sides, each side was cut into leg, shoulder, neck, brisket, flap and back joints (Farid, 1989) and each cut was weighed separately. Internal organs (kidney, liver, lungs, heart and gastrointestinal tract) were separated and weighed.

Rumen pH and ruminal ammonia concentration data were analyzed according to Proc Mixed as Repeated Measures of SAS (1996). Data for feedlot performance and carcass characteristics were analyzed as a completely randomized design according to the GLM procedure of SAS (1996) with the statistical model of:

$$Y_{ij} = \mu + T_i + e_{ij}$$

Where:

Y_{ij}: dependent variable.

μ: overall mean.

T_i: screen size effect.

e_{ij}: residual error.

Least significant difference was used to compare means. Treatment differences were considered significant when (P<0.05).

RESULTS AND DISCUSSION

Experiment 1

Effects of barley particle size on rumen pH and ruminal ammonia concentration are presented in Table 1. Rumen pH was decreased by feeding ground barley grains through a 1 mm screen compared to feeding the whole barley grains. Our results are consistent with Tagawa et al. (2016), who reported that decreasing barley grains particle size in in vitro assay decreased fermentation pH. Grinding the cereals may increase availability and surface area of nutrients to rumen microbes compared with whole grains (Callison et al. 2001). Consequently, microbes could digest the nutrients more easily in ground grains and therefore rumen pH has the potential to decrease in ground vs. whole grains. Average ruminal NH_3-N concentration in rams fed whole barley grain was 14.08 mg/100 mL. This is relatively higher than rumen NH_3-N concentrations of 5 mg/100 mL reported by Satter and Slyter (1974) as a minimal concentration for optimum microbial protein synthesis. The relatively higher level of ruminal NH_3-N was probably due to the high proportion of non-protein nitrogen (urea) in the diet (10 g/kg), which is highly soluble in the rumen (Van Soest, 1994). The ruminal ammonia nitrogen concentrations were not affected by treatments.

This result is supported by Ebrahimi-Mahmoudabad and Taghinejad-Roudbaneh (2015), who reported that decreasing barley grain particle size had not significant effect on maximum potential degradability of CP. The concentrations of ammonia nitrogen in rumen fluid depend on the source and amount of the degradable protein in rumen (Reynal et al. 2007). Vanhatalo et al. (2003) found a numerical increase in rumen NH_3-N concentration with abomasal infusion of casein in comparison with water infusion (control group).

Experiment 2

Effects of barley grains particle size on performance of lambs are presented in Table 2. Treatments had not significant effects on final body weight of lambs. Moreover, Salo et al. (2016) reported that final body weight of lambs was not affected by barley grain processing.

However, average DM intake, average daily gain (ADG) and feed conversion ratio (FCR) (kg DM to kg live weight gain) of lambs were affected by treatments (P<0.05). Our results are in agreement with other studies (Economides et al. 1990; Petit, 2000).

There were differences among treatments for average DM intake. Average DM intake of lambs receiving ground barley grains was on average 1.39 kg and was lower than lambs feeding whole barley grains. This finding is supported by result of experiment 1 and is consistent with the traditional belief that finely ground grains produce dust and depress DM intake (Mathison, 1996).

Lower ruminal pH of lambs fed ground barley grain compared to lambs fed whole barley grain decreases cellulolytic bacteria count (Russell and Wilson, 1996), resulting in decreased DM intake of lambs fed ground barley grain with 1 mm screen.

Overall, feeding lambs with ground barley grain with a 1 mm screen decreased ADG of lambs compared to lambs fed whole barley grain. Our results are not consistent with Sormunen-Cristian (2013) and Salo et al. (2016). Salo et al. (2016) reported that feeding malted and cracked barley grain had no positive effect on ADG of Highland lambs. However, Voia et al. (2009) reported that total weight gain and ADG of ground barley-fed lambs was higher than whole barley–fed lambs. The higher ADG lambs fed whole barley could be due the higher DM intake on this diet compared to ground barley-fed lambs.

The FCR of the lambs ranged from 8.95 to 11.59 and these values were considerably higher than those reported by Petit (2000) and Sormunen-Cristian (2013), probably due to the higher initial weight of the lambs used in the present study. Lambs fed ground barley grain with 5 mm screen had better (P<0.05) FCR than lambs fed ground barley grain with 1 mm screen. However, Crane et al. (2014) reported that FCR of feedlot lambs was not affected by corn particle size. The lowest cost of feed per kg live weight was achieved on the diet based on ground barley with 5 mm screen.

There were no differences between the treatments in terms of the cold carcass weights and dressing percentage (Table 3). The dressing percentage in our study was similar to those obtained by Fozooni and Zamiri (2007). Similar results were reported by Crane et al. (2014) and Salo et al. (2016). Carcass weight and dressing percentage of Highland sheep were not affected by processing of barley grain (Salo et al. 2016).

Lambs fed diet containing ground barley with a 3 mm of screen had (P<0.05) higher longissimus muscle area compared to lambs fed whole barley grain. The pattern of fat deposition and increase in longissimus area was altered by changing screen size.

Table 1 Effects of barley grain particle size on rumen pH and ruminal ammonia concentration

Items	Basal diet[*] + whole barley grain	Basal diet + ground barley with a 5 mm screen	Basal diet + ground barley with a 3 mm screen	Basal diet + ground barley with a 1 mm screen	SEM
Average pH	6.21[a]	6.17[ab]	6.15[ab]	6.05[b]	0.06
Average NH_3-N (mg/dL)	14.08	13.98	13.87	13.74	0.23

* Basal diet on DM basis consisted of 365 g/kg corn silage, 10 g/kg limestone, 10 g/kg urea, 5 g/kg salt, 10 g/kg a vitamin-mineral premix and 600 g/kg barley.
The means within the same row with at least one common letter, do not have significant difference (P>0.05).
SEM: standard error of the means.

Table 2 Effects of barley grains particle size on performance of male lambs

Items	Basal diet[*] + whole barley grain	Basal diet + ground barley with a 5 mm screen	Basal diet + ground barley with a 3 mm screen	Basal diet + ground barley with a 1 mm screen	SEM
Final body weight (kg)	54.3	54.9	52.3	51.5	1.77
Average dry matter intake (kg)	1.46[a]	1.41[ab]	1.39[ab]	1.39[b]	0.023
Average daily gain (g)	165.7	163.9[a]	141.7[ab]	121.3[b]	12.48
Feed conversion ratio	9.19[ab]	8.95[b]	10.16[ab]	11.59[a]	0.901
Cost of feed for kg live weight (Rial)	65610[bc]	63903[c]	72542[b]	82753[a]	534.1

* Basal diet on DM basis consisted of 365 g/kg corn silage, 10 g/kg limestone, 10 g/kg urea, 5 g/kg salt, 10 g/kg a vitamin-mineral premix and 600 g/kg barley.
The means within the same row with at least one common letter, do not have significant difference (P>0.05).
SEM: standard error of the means.

Table 3 Effects of barley grains particle size on carcass characteristics of male lambs

Items	Basal diet[*] + whole barley grain	Basal diet + ground barley with a 5 mm screen	Basal diet + ground barley with a 3 mm screen	Basal diet + ground barley with a 1 mm screen	SEM
Cold carcass weight (kg)	27.64	27.92	26.92	26.1	0.813
Dressing percentage (%)	50.90	50.86	51.47	50.68	1.07
Longissimus muscle area (cm²)	14.7[b]	15.9[ab]	16.4[a]	16.2[ab]	0.543
Leg weight (kg)	6.68	6.62	6.64	6.26	0.285
Shoulder weight (kg)	4.38	4.63	4.43	4.16	0.163
Back weight (kg)	5.59	5.40	5.23	5.14	0.147
Brisket weight (kg)	3.35[a]	3.32[ab]	3.29[ab]	3.14[b]	0.123
Neck weight (kg)	1.25	1.20	1.30	1.22	0.051
Flap weight (kg)	1.41[b]	1.81[a]	1.50[ab]	1.43[b]	0.107
Head weight (kg)	2.75	2.65	2.65	2.69	0.064
Back fat thickness (cm)	0.600	0.505	0.512	0.498	0.063
Abdominal fat (kg)	0.890[a]	0.785[ab]	0.535[b]	0.527[b]	0.901
Pelvic fat (kg)	0.205[a]	0.197[ab]	0.120[b]	0.138[ab]	0.023
Kidney fat (kg)	0.178	0.192	0.145	0.168	0.019

* Basal diet on DM basis consisted of 365 g/kg corn silage, 10 g/kg limestone, 10 g/kg urea, 5 g/kg salt, 10 g/kg a vitamin-mineral premix and 600 g/kg barley.
The means within the same row with at least one common letter, do not have significant difference (P>0.05).
SEM: standard error of the means.

Feeding ground barley grain with a 3 mm screen could provide the energy and protein needs in synchronization, which may stimulate the microbial growth in the rumen (Witt *et al.* 1999). Crane *et al.* (2014) reported that feeding ground corn grain increased loin eye area of feedlot lambs fed high crude protein diet. The improved microbial growth in the rumen led to improving of longissimus muscle area. Treatments had no significant effect on leg, shoulder, back and neck weights of lambs (P>0.05). There was no difference in back fat thickness and kidney fat of lambs fed whole and ground barley grains as previously reported by Petit (2000). Moreover, Crane *et al.* (2014) reported that grinding of corn grain had not significant effect on dressing percentage and fat depth of feedlot lambs. Lambs fed whole barley grain also had higher abdominal and pelvic fats, which is in line with results of Sormunen-Cristian (2013)

who, reported that lambs receiving whole barley grains had a higher kidney fat than those receiving crushed or ground barley grains. Feeding whole compared to ground barley decreases synchronization of energy and protein in rumen and increases acetate: propionate ratio, overall resulting in increased abdominal and pelvic fats (Ørskov *et al.* 1974).

Effects of barley grains particle size on internal organs of lambs are shown in Table 4. Kidney, heart, lungs and gastrointestinal tract weights of lambs were not affected by treatments (P>0.05). However, liver weight of lambs was statistically different among treatments. The highest liver weight was obtained from lambs fed whole barley grains. While data on the effects of grain particle size on internal organs of lambs is limited, there is high correlation between live body weight and internal organs weights (Abegaz *et al.* 1996).

Table 4 Effects of barley grains particle size on internal organs weights (kg) of male lambs

Organs weights	Basal diet[*] + whole barley grain	Basal diet + ground barley with a 5 mm screen	Basal diet + ground barley with a 3 mm screen	Basal diet + ground barley with a 1 mm screen	SEM
Kidney	0.138	0.132	0.127	0.133	0.004
Liver	0.783[b]	0.862[a]	0.665[b]	0.672[b]	0.037
Heart	0.227	0.223	0.210	0.212	0.007
Lungs	0.577	0.568	0.608	0.570	0.024
Gastrointestinal tract	2.60	2.58	2.55	2.42	0.047

* Basal diet on DM basis consisted of 365 g/kg corn silage, 10 g/kg limestone, 10 g/kg urea, 5 g/kg salt, 10 g/kg a vitamin-mineral premix and 600 g/kg barley.
The means within the same row with at least one common letter, do not have significant difference (P>0.05).
SEM: standard error of the means.

According to the non -significant effect of barley particle size on live body weight, this result is predictable.

CONCLUSION

Barley grain particle size affects the performance of young feedlot lambs. Grinding the grains to pass through 5 and 3 mm sieves reduced feed cost of gain, abdominal and pelvic fat deposition. Our results suggest ground barley grain with a 5 mm or a 3 mm screen for feeding 7-8 months lambs fed high urea diet. Further research on particle size effects on ruminal nutrient degradability and microbial crude protein synthesis is needed.

ACKNOWLEDGEMENT

Authors are grateful to Shahr-e-Qods Branch of the Islamic Azad University for financial supports.

REFERENCES

Abegaz S., Tiyo D. and Gizachew C. (1996). Compensatory in Horro lambs of Ethiopia. Pp. Pp. 209-213 in Proc. 3rd Biennial Conf. African Small Rumin. Network. Kampala, Uganda.

AFRC. (1992). Energy and Protein Requirements of Ruminants. CAB International, Wallingford, UK.

Callison S.L., Firkins J.L., Eastridge M.L. and Hull B.L. (2001). Site of nutrient digestion by dairy cows fed corn of different particle sizes or steam-rolled. J. Dairy Sci. 84, 1458-1467.

Canpolat O. and Karabulut A. (2010). Effect of urea and oregano oil supplementation on growth performance and carcass characteristics of lamb fed diets containing different amounts of energy and protein. Turkish J. Vet. Anim. Sci. 34, 119-128.

Crane A.R., Redden R.R., Berg P.B. and Schauer C.S. (2014). Effects of diet particle size and Lasalocid on growth, carcass traits, and N balance in feedlot lambs. Sheep. Goat. Res. J. 29, 17-23.

Crooke W.M. and Simpson W.E. (1971). Determination of ammonia in kjeldahl digests of crops by an automated procedure. J. Sci. Food Agric. 22, 903-916.

Ebrahimi-Mahmoudabad S.R. and Taghinejad-Roudbaneh M. (2015). Effects of feeding whole and ground barley grains on its ruminal dry matter and crude protein degradability and performance of male lambs. Int. J. Biosci. 6(1), 201-209.

Economides S., Kumas A., Georghiades E. and Hadjipanayioto M. (1990). The effects of barley-sorghum grain processing and form of concentrate mixture on performance of lambs, kids and calves. Anim. Feed Sci. Technol. 31, 105-116.

Farid A. (1989). Direct, maternal and hetrosis effects for slaughter and carcass characteristics in three breeds of fat-tailed sheep. Livest. Prod. Sci. 23, 137-162.

Fozooni R. and Zamiri M.J. (2007). Relationships between chemical composition of meat from carcass cuts and the whole carcass in Iranian fat-tailed sheep as affected by breed and feeding level. Iranian J. Vet. Res. 8, 304-312.

Khalil J.K., Sawya W.N. and Hyder S.Z. (1986). Nutrient composition of atriplex leaves grown in Saudi Arabia. J. Range. Manage. 39, 104-107.

Mathison G.W. (1996). Effects of processing on the utilization of grain by cattle. Anim. Feed Sci. Technol. 58, 113-125.

McAllister T.A., Bae H.D., Jones G.A. and Cheng K.J. (1994). Microbial attachment and feed digestion in the rumen. J. Anim. Sci. 72, 3004-3018.

NRC. (1985). Nutrient Requirements of Sheep. 6th Ed. National Academy Press, Washington, DC, USA.

Ørskov E.R., Fraser C. and Gordon J.G. (1974). Effect of processing of cereals on rumen fermentation, digestibility, rumination time and firmness of subcutaneous fat in lambs. Br. J. Nutr. 32, 59-68.

Petit H.V. (2000). Effect of whole and rolled corn or barley on growth and carcass quality of lambs. Small Rumin. Res. 37, 293-297.

Reynal S.M., Ipharraguerre I.R., Lineiro M., Brito A.F., Broderick G.A. and Clark J.H. (2007). Omasal flow of soluble proteins, peptides, and free amino acids in dairy cows fed diets supplemented with proteins of varying ruminal degradabilities. J. Dairy Sci. 90, 1887-1903.

Russell J.B. and Willson D.B. (1996). Why are ruminal cellulolytic bacteria unable to digest cellulose at low pH? J. Dairy Sci. 79(8), 1503-1509.

Salo S., Urge M. and Animut G. (2016). Effects of supplementation with different forms of barley on feed intake, digestibility, live weight change and carcass characteristics of Highland sheep fed natural pasture. J. Food Proc. Technol. 7(3), 1-5.

SAS Institute. (1996). SAS®/STAT Software, Release 6.11. SAS Institute, Inc., Cary, NC. USA.

Satter L.D. and Slyter L.L. (1974). Effect of ammonia concentration on rumen microbial protein production in vitro. Br. J. Nutr. 32, 199-205.

Sinclair L.A., Garnsworthy P.C., Newbold J.R. and Buttery P.J. (1993). Effect of synchronizing the rate of dietary energy and nitrogen release on rumen fermentation and microbial protein synthesis in sheep. *J. Agric. Sci.* **120,** 463-472.

Sormunen-Cristian R. (2013). Effect of barley and oats on feed intake, live weight gain and some carcass characteristics of fattening lambs. *Small Rumin. Res.* **109,** 22-27.

Stanto T.L. and Whittie J. (2006). Urea and NPN for cattle and sheep. Service in Action. *Colorado State Univ.* **1,** 608.

Tagawa S.I., Holtshausen L., McAllister T.A., Yang W.Z. and Beauchemin K. (2016). Effects of particle size of processed barley grains, enzyme addition and microwave treatment on *in vitro* disappearance and gas production for feedlot cattle. *Asian Australasian J. Anim. Sci.* **30(4),** 479-485.

Trevaskis L.M., Fulkerson W.J. and Gooden J.M. (2001). Provision of certain carbohydrate-based supplements to pasture-fed sheep, as well as time of harvesting of the pasture, influences pH, ammonia concentration and microbial protein synthesis in the rumen. *Australian J. Exp. Agric.* **41,** 21-27.

Vanhatalo A., Varvikko T. and Huhtanen P. (2003). Effects of casein and glucose on responses of cows fed diets based on restrictively grass silage. *J. Dairy Sci.* **86,** 3260-3270.

Van Soest P.J. (1994). Nutritional Ecology of the Ruminants. Cornell University Press, New York, USA.

Voia S., Bogdan G.H., Drinceanu D., Padeanu I. and Daraban S. (2009). Effects of barley processing on the bio productive indices in fattening the weaned lambs (note II small particles). *Zootech. Biotehnol.* **42,** 511-515.

Witt M.W., Sinclair L.A., Wilkinson R.G. and Buttery P.J. (1999). The effects of synchronizing the rate of dietary energy and nitrogen supply to the rumen on the production and metabolism of sheep: food characterization and growth and metabolism of ewe lambs given food *ad libitum*. *Anim. Sci.* **69,** 223-235.

Zamiri M.J. and Izadifard J. (1997). Relationships of fat-tail weight with fat-tail measurements and carcass characteristics of Mehraban and Ghezel rams. *Small Rumin. Res.* **26,** 261-266.

Genetic Diversity and Molecular Phylogeny of Iranian Sheep based on Cytochrome b Gene Sequences

S. Savar Sofla[1], H.R. Seyedabadi[1*], A. Javanrouh Aliabad and R. Seyed Sharifi[2]

[1] Animal Science Research Institute of Iran (ASRI), Agricultural Research Education and Extension Organization (AREEO), Karaj, Iran
[2] Department of Animal Science, Faculty of Agricultural and Natural Resources, University of Mohaghegh Ardabili, Ardabil, Iran

*Correspondence E-mail: h_seyedabadi@asri.ir

ABSTRACT

Phylogenetic relationships and genetic variation between two Iranian sheep breeds were analyzed using cytochrome b (cyt-b) gene sequences. The genomic DNA was isolated by salting out method and amplified cytochrome b gene using polymerase chain reaction restriction (PCR) method with a pair of primer. A partial sequence of cyt-b gene of Iranian sheep is 780 bp and contained 13 variable sites and 11 haplotypes. Phylogenetic analysis of haplotype in the combination with the sheep from GenBank showed that Iranian sheep made a separated cluster. This study is provided useful information for understanding relationships between breeds from different parts of the world. This study may simplify the future researchers and breeders for better understanding the genetic structure and breed differentiation for designing future breeding strategies to the conservation of animal genetic resources.

KEY WORDS genetic diversity, Iranian sheep, mitochondrial cytochrome b gene, phylogenetic analysis.

INTRODUCTION

Sheep and goats are two important livestock species in Iranian rural areas. More than 57% of the available animal units in the country are sheep and goats. More than 27 breeds of sheep with a variety of sizes, shapes, types and color have been recognized in Iran (Mobini, 2013). All Iranian native sheep, except the Zel are fat-tailed breeds (Mobini, 2013). Ghezel and Shal are of the predominant sheep breeds in Iran and being very well adapted to harsh environmental conditions. They are fat tailed sheep used mainly for meat production (Atashi and Izadifar, 2012). Animal genetic resources are mainly facing two challenges. On one side, the demand for livestock products are increasing in developing countries as estimated by Food Agricul-

ture Organization (FAO, 1993) and the demand for milk and meat from livestock have increased twice than usual. On the other hand, animal genetic resources are menaced because of the aimless development (Ruane *et al.* 2006). Strategies for genetic progress of domestic animals mainly involve the use of the genetic variation. Genetic diversity studies in livestock aim at evaluating genetic diversity within and between breeds, since the breed is the management unit for which factors such as inbreeding are controlled (Tolonea *et al.* 2012). Therefore, a molecular genetics study of the population diversity may improve the comprehension of the genetic resources (Tolonea *et al.* 2012). Mitochondrial DNA (mtDNA) is the genetic material that exists outside the nucleus in eukaryotic cells (Burgstaller *et al.* 2015). mtDNA is highly conserved and its relatively slow mutation rates (compared to other DNA regions such as microsatellites) make it useful for studying the evolutionary, relationships, phylogeny of organisms. It has multiple copies, has a rapid evolutionary rate and follows maternal inheritance. The cytochrome b (cyt-b) gene is one of the important coding genes in mtDNA; it is about 1.2 kb in length (Sawaimul et al. 2014). Because of its maternal inheritance, its well-known gene structure and sequence, the occurrence of low recombination and other characteristics, the cyt-b gene has been widely used for phylogenetic evolution of several animal species (Patwardhan et al. 2014). The purpose of this study was to investigate the genetic diversity and phylogenetic evolution of two Iranian sheep (Ghezel and Shal) based on the analysis of the partial sequence of the cyt-b gene. This investigation will be helpful for the conservation, utilization, and exploitation of the genetic resources of the indigenous Iranian sheep.

MATERIALS AND METHODS

Population sampling
Blood samples from two Iranian sheep breeds (Ghezel and Shal) were considered for the study. Samples were collected from sheep that were judged to be true to type with the phenotypic characteristics of that breed. The animals selected had unrelated parents based on the information provided by the owners. A total of 50 individuals from different locations were sampled and the blood was stored at 4 °C up to 21 days. Genomic DNA was extracted from fresh blood according to standard procedures (Javanrouh et al. 2006) and was quantitated by spectrophotometry (Nanodrop ND1000).

PCR amplification and sequencing
At the first step, cyt-b of the mtDNA was amplified and sequenced. To amplify the cyt-b region of sheep mtDNA, a pair of primers was designed using the known sheep mtDNA sequence (GenBank Accession No NC_001941.1). The primers cyt-b-F 5'-CATTCTCCTCTGTAACCCACATCTG-3' and cyt-b-R 5'-GTCCAATAATGATGTAGGGGTGTTC-3' were used to amplify an 870 bp DNA fragment. PCR amplifications were conducted in a 30 µL volume containing 5 µL of 10x reaction buffer, 1.5 mM MgCl$_2$, 0.2 mM dNTPs, 0.2 uM each primer, 1U Taq DNA polymerase (TaKaRa Biosystems) and approximately 150 ng genomic DNA. The PCR mixture underwent 4 min at 95 °C, 35 cycles 50 s at 94 °C, 1 min at 60 °C and 1 min at 72 °C and 5 min at 72 °C. PCR products were purified by using PCR Purification Kit (Watson BioTechnologies, Shanghai) and then sequenced using ABI PRISM BigDyeTM Terminator Cycle Sequencing Ready Reaction Kit and ABI PRISM 3130 Geneti Analyzer (Applied Biosystems, Foster City, USA).

Phylogenetic reconstruction
The quality of the 780 bp cyt-b gene sequence for individuals was firstly evaluated on the basis of sequeing peak value and then these sequences were manually edited using program Chromas version 2.23. Then sequences were arranged using the BioEdit program and were aligned using CLUSTALW (http://ebi.ac.uk/clustalw) software. These results were compared with other sequences obtained from GenBank. To investigate genetic relationship between mitochondrial sequences, phylogenetic tree unweighted pair group method with arithmetic mean (UPGMA) and neighbor joining (NJ) were constructed using the Tamura–Nei distance method (Tamura and Nei, 1993). The phylogenetic tree construction is incorporated in the MEGA version 6.1 (Tamura et al. 2013). DnaSP 5.0 (Librado and Rozas, 2009) was used to analyze the diversity parameters including haplotype diversity (HD), nucleotide diversity (π) and the average number of nucleotide differences.

RESULTS AND DISCUSSION

A total of 780 base pairs (bp) of the cyt-b region (from np 14410 to np 15190) were obtained for 50 samples. There were no insertions/deletions in 50 sequences of cyt-b region. The average percentage of nucleotides T, C, A and G were 26.3, 28.7, 29.71 and 13.65%, respectively. Percentage of nucleotide pairs A + T and C + G was 56% and 44%, respectively, suggesting that A + T nucleotides were higher in the cyt-b region of mtDNA Iranian sheep breeds. Because of the well-known gene structure and lack of recombination, the cyt-b gene has been generally used alone or in combination with other mtDNA encoding genes and hyper variable regions for phylogenetic studies between species (Chen et al. 2006). Generally, the AT content is always higher than the GC content in cyt-b (Sawaimul et al. 2014) which is consistent with our results. However, the result was different from that of Sawaimul et al. (2014), probably because of the differences for the sheep breeds and the length of the sequences that were studied. The cyt-b sequences were polymorphic. Fifty sequences rendered 11 divergent haplotypes with 13 variable sites defined. The largest haplotype group consisted of 5 individuals. The number of haplotypes detected in each breed ranged from 4 in Ghezel to 7 in Shal (Table 1). As an encoding gene of mtDNA, the incidence of mutation of the cyt-b gene is medium compared to mutation in the D-loop and other encoding genes (Chen et al. 2006). The nucleotide sequence of cyt-b genes revealed several nucleotides differences with Iranian sheep (Table 2).

These variable sites showed similarities among of the Iranian sheep breeds, but clearly were different with other sheep breeds.

Table 1 Haplotypes, parsimony informative sites, singleton and polymorphic sites for each breed

Breed	n	Haplotypes	PSI	Singleton sites	Polymorphic sites
Shal	25	7	11	1	13
Ghezel	25	4	10	0	12

PSI: parsimony informative sites.

Table 2 Analysis of genetic variations based on mtDNA cytochrome b gene refers to the other sheep breeds

Population	Site												
	67	81	86	151	154	253	271	454	471	493	565	571	748
Iran (Shal)	A	C	T	C	C	A	T	C	T	T	G	G	T
Iran (Ghezel)	-	T	-	C	T	-	-	-	-	C	-	A	-
AF010406 (Germany)	C	T	C	T	T	G	C	T	-	C	A	A	C
NC001941 (Germany)	T	T	C	T	T	G	C	T	-	C	A	A	C
AY858379 (Korea)	T	T	C	T	T	-	C	T	-	C	A	A	C
EF490451 (Austria)	T	T	C	T	T	G	C	T	C	C	A	A	C
HE577849 (Israel)	C	T	C	T	T	G	C	T	-	C	A	A	C
HM236179 (Australia)	C	T	C	-	-	-	C	-	-	-	A	-	-
JX235837 (Pakistan)	T	T	C	T	T	-	C	T	-	C	A	A	C
JX567831 (China)	T	T	C	T	T	G	C	T	-	C	A	A	C
KF938345 (China)	T	T	C	T	T	-	C	T	-	C	A	A	C
KF229236 (China)	T	T	C	T	T	-	C	T	-	C	A	A	C
KU899150 (China)	T	T	C	T	T	-	C	T	-	C	A	A	C
KF302446 (Italy)	T	T	C	T	T	-	C	T	-	C	-	A	C

Transversions occurred only at one position 67 (A/C) and in all the other positions, transitions occurred (G/A, 3 and T/C, 9). Haplotypes diversity values were moderate in two populations. Values ranged from 0.791 ± 0.021 in Shal to 0.623 ± 0.014 in Ghezel. As can be seen from Table 3, synchronous with haplotype diversity enhancement, the nucleotide diversity of mtDNA and polymorphism of the population were increased. The nucleotide diversity (π) ranged from 0.013 ± 0.011 (Ghezel) to 0.014 ± 0.002 (Shal). The average number of nucleotide differences (k) was quite relevant and the highest was for the Shal breed (6.641) (Table 3). Nucleotide diversity and haplotype diversity of mtDNA cyt-b region are the important indices for assessing population polymorphism and genetic differentiation.

Table 3 Values of haplotypes diversity (HD), nucleotide diversity (π) and average number of nucleotide differences (k) for each breed

Breed	(HD±SD)	(π±SD)	k
Shal	0.791±0.021	0.014±0.002	6.641
Ghezel	0.623±0.014	0.013±0.011	6.157

SD: standard deviation.

It was far lower than that of the D-loop region (Javanrouh et al. 2016) indicating that the cyt-b gene is relatively conserved and that most base substitutions did not change the coding of the amino acid. The extent of gene differentiation of these sheep breeds was in accordance with that obtained from microsatellites (Molaee et al. 2009). Molaee et al. (2009), studied six Iranian indigenous sheep populations by investigating their nuclear DNA using microsatellite markers and the result showed that the mean polymorphism information content of the six breeds were moderate.

The cyt-b gene has been used to study other aspects such as intra or interspecific relationships and gene flow as well (Alves et al. 2003). It is generally recognized that the domestic animals experience a bottleneck effect after domestication (Xin et al. 2006). But in this study, none of the sheep population expansion events irrespective of the size of the population. Out of the 11 haplotypes observed in this study, only 4 haplotypes are common to these breeds, suggesting that a moderate level of genetic diversity was present within each of these breeds. This unique pattern of haplotype distribution may also be attributed to reproductive isolation due to harsh geographical structure of the country and unique husbandry practices (migratory farming system) associated with to this specific region. We identified 1 singleton sites and 10 parsimony informative sites. A singleton site contains at least two types of nucleotides (or amino acids) with, at most, one occurring multiple times. DNAsp identifies a site as a singleton site if at least three sequences contain unambiguous nucleotides or amino acids.

A site is parsimony-informative if it contains at least two types of nucleotides (or amino acids) and at least two of them occur with a minimum frequency of two (http://www.megasoftware.net).

Phylogenetic relationship on Iranian sheep breeds

The phylogenetic trees of Shal and Ghezel sequences were constructed using UPGMA method with reported sheep sequences from Italy (KF302446), China (KP229236, KF938345, KU899150, JX567831), Korea (AY858379), Austria (EF490451), Australia (HM236179), Pakistan (JX235837) and Germany (NC001941 and AF010406), as in groups and with goat (AB004070.1) and cattle (AB074964.1) sequences as out groups (Figure 1).

Phylogeny tree of cyt-b gene nucleotide showed that Iranian sheep made a separated cluster.

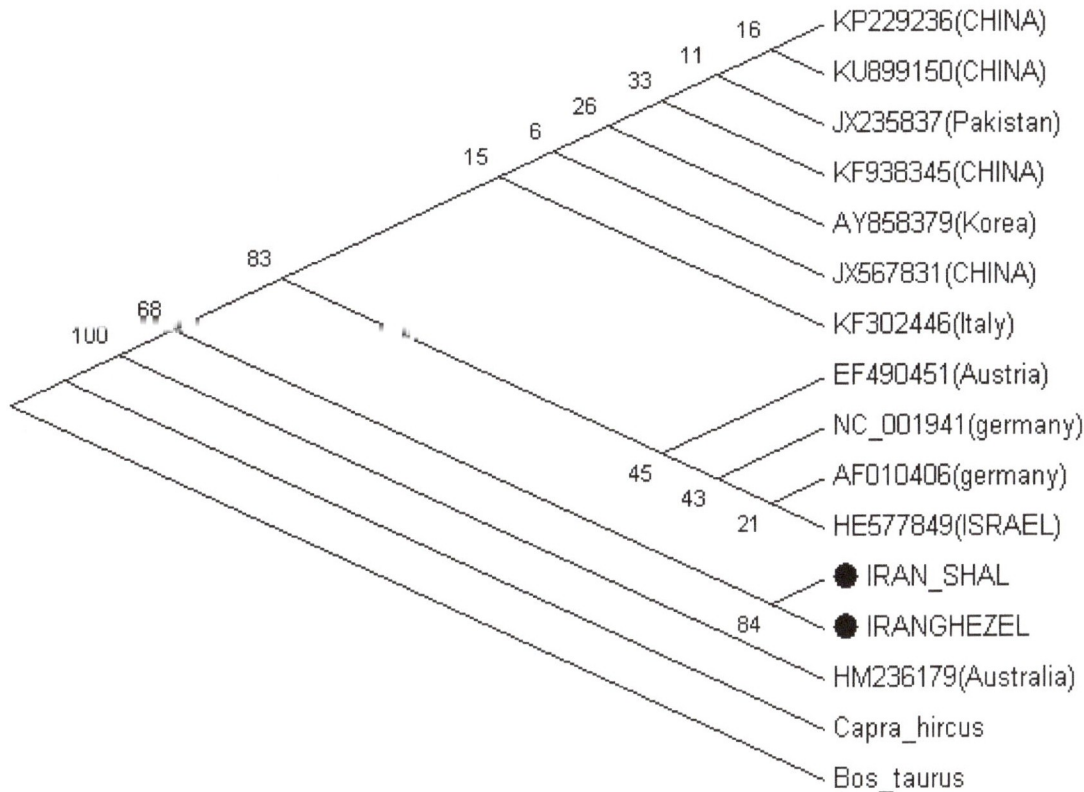

Figure 1 UPGMA phylogenetic tree constructed for Iranian sheep mtDNA sequences with the 12 reference sequences

	1	2	3	4	5	6	7	8	9	10	11
1. IRAN											
2. GERMANY	0.009										
3. KOREA	0.007	0.003									
4. CATTLE	0.185	0.183	0.182								
5. GOAT	0.122	0.120	0.118	0.153							
6. AUSTRIA	0.010	0.001	0.004	0.182	0.121						
7. ISRAIL	0.009	0.000	0.003	0.183	0.120	0.001					
8. AUSTRALIA	0.011	0.013	0.011	0.176	0.118	0.014	0.013				
9. PAKISTAN	0.007	0.003	0.001	0.180	0.117	0.004	0.003	0.010			
10. CHINA	0.007	0.002	0.002	0.180	0.117	0.003	0.002	0.011	0.000		
11. ITALYA	0.008	0.004	0.003	0.182	0.118	0.005	0.004	0.011	0.001	0.002	

Figure 2 Estimates of mean distance over sequence pairs between groups

This result is supported by the bootstrap value of 100%. Bootstrap value is a criterion to determine the level of accuracy of phylogeny tree. The estimated genetic distances between populations also indicated that the Iranian and Australia sheep populations are far away (0.011) and Iranian sheep are closely related to Pakistani, Korea and China (0.007) sheep populations (Figure 2). Clustering the different sheep breeds within one branch of phylogeny is because of the low sequence substitutions in cyt-b gene (Sultana *et al.* 2003). Clustering also occurred in other comparator groups because of nucleotide substitutions in cyt-b gene. The present information could be used to strengthen the monitoring, characterization and conservation of animal genetic resources towards the sustainable rearing of the autochthonous sheep breeds. However, further studies involve the existing knowledge from microsatellite marker will help to unravel the history of domestication of Iranian sheep.

CONCLUSION

In the present study, we investigated the diversity and the organization of cyt-b region in Iranian sheep breeds. The cyt-b region of mtDNA using sequencing techniques was suitable tool for analyzing genetic variability, phylogenetic relationship and time of divergence between the Ghezel and Shal sheep breeds. The evolutionary divergence into distinct entities of Iranian sheep breeds based on cytochrome b sequence appear to closely follow their geographical distribution in Iran and this could have implications for management, improvement and conservation strategies in Iranian sheep.

ACKNOWLEDGEMENT

The authors acknowledged the three reviewers for constructive comments on the manuscript. We gratefully acknowledge all farmers who took part in the present study, giving access to the animals.

REFERENCES

Alves P.C., Ferrand N., Suchentrunk F. and Harris D.J. (2003). Ancient introgression of *Lepus timidus* mtDNA into *Lepus granatensis* and *Lepus europaeus* in the Iberian Peninsula. *Mol. Phylogenet. Evol.* **27,** 70-80.

Atashi H. and Izadifar J. (2012). Estimation of individual heterosis for lamb growth in Ghezal and Mehraban sheep. *Iranian J. Appl. Anim. Sci.* **2,** 127-130.

Burgstaller M., Iain G., Johnston L. and Poulton J. (2015). Mitochondrial DNA disease and developmental implications for reproductive strategies Joerg Patrick. *Mol. Hum. Reprod.* **21,** 11-22.

Chen S., Fan B., Liu B., Yu M., Zhao S., Zhu M., Xiong T. and Li K. (2006). Genetic variations of 13 indigenous Chinese goat breeds based on cytochrome b gene sequences. *Biochem. Gen.* **44,** 87-95.

FAO. (1993). Food and Agriculture Organization of the United Nations (FAO), Rome, Italy.

Javanrouh A., Banabazi M.H., Esmaeilkhanian S., Amirinia C., Seyedabadi H.R. and Emrani H. (2006). Optimization on salting out method for DNA extraction from animal and poultry blood cells. Pp. 103 in Proc. 57th Ann. Meet. European Assoc. Anim. Prod. Antalya. Turkey.

Javanrouh Aliabad A., Khodamoradi S. and Seyedabadi H.R. (2016). Analysis of the genetic diversity and the phylogenetic evolution of Iranian sheep based on D-loop region sequences. *J. Vet. Res.* **20,** 353-361.

Librado P. and Rozas J. (2009). Dnasp V5: a software for comprehensive analysis of DNA polymorphism data. *Bioinformatics.* **25,** 1451-1452.

Mobini B. (2013). A quantitative evaluation of different regions of skin in adult Iranian native sheep. Vet. Med. **58,** 260-263.

Molaee V., Osfoori R., Eskandari Nasab M.P. and Qanbari S. (2009). Genetic relationships among six Iranian indigenous sheep populations based on microsatellite analysis. *Small rumin. Res.* **84(1),** 121-124.

Patwardhan A., Ray S. and Roy A. (2014). Molecular markers in phylogenetic studies. A review. *J. Phylogen. Evol. Biol.* **2,** 1-9.

Ruane P., Lang J., DeJesus E., Berger D.S. and Dretler R. (2006). Pilot study of once-daily simplification therapy with abacavir/lamivudine/zidovudine and efavirenz for treatment of HIV-1 infection. *HIV Clin. Trials.* **7,** 229-236.

Sawaimul A.D., Sahare M.G., Ali S.Z., Sirothia A.R. and Kumar S. (2014). Assessment of genetic variability among Indian sheep breeds using mitochondrial DNA cytochrome b region. *Vet. World.* **7,** 852-855.

Sultana S., Mannen H. and Tsuji S. (2003). Mitochondrial DNA diversity of Pakistani goats. *J. Anim. Genet.* **34,** 417-421.

Tamura K. and Nei M. (1993). Estimation of the number of nucleotide substitutions in the control region of mitochondrial DNA in humans and chimpanzees. *Mol. Biol. Evol.* **10,** 512-526.

Tamura K., Stecher G., Peterson D., Filipski A. and Kumar S. (2013). MEGA6: molecular evolutionary genetics analysis version 6.0. *Mol. Biol. Evol.* **30,** 2725-2729.

Tolonea M., Mastrangeloa S., Rosac A.J.M. and Portolanoa B. (2012). Genetic diversity and population structure of Sicilian sheep breeds using microsatellite markers. *Small Rumin. Res.* **102,** 18-25.

Xin W., Yue-Hui M.A. and Hong C. (2006). Analysis of the genetic diversity and the phylogenetic evolution of Chinese sheep based on cyt-b gene sequences. *Acta Genet. Sinica.* **33,** 1081-1086.

23

Effects of Specific Gravity and Particle Size of Passage Marker on Particulate Rumen Turnover in Holestine Dairy Cattle

A. Teimouri Yansari[1*]

[1] Department of Animal Science, Faculty of Agricultural Science, Sari University of Agricultural Sciences and Natural Resources, Sari, Iran

*Correspondence E-mail: astymori@yahoo.com

ABSTRACT

This experiment was carried out to evaluate the effect of specific gravity (SG) and particle size (PS) of Chromium-mordanated alfalfa neutral detergent fiber (NDF) markers on estimation of particulate rumen turnover. Thirty-two multiparous, mid lactation Holstein dairy cows (body weight=654±24 kg) were allotted to a completely randomized design experiment with two replications in a 4×4 factorial method. The experiment was accomplished over 21 d (adaptation, 14 d; sample collection, 7 d). Cows were *ad libitum* fed twice daily at 09:00 and 21:00 with similar total mixed rations. To prepare the marker, alfalfa in the individual bales was chopped for theoretical cut length 19, 10 and 5 mm. The fine alfalfa was prepared by grinding the 5 mm cut alfalfa through a 2 mm screen. The marker preparation followed a modified procedure of Uden *et al.* (1980) by different PS (8.45, 4.38, 3.00 and 1.10 mm) and $K_2Cr_2O_7$ concentration in mordanting solution (50, 300, 500 and 900 g/kg $K_2Cr_2O_7$ instead of 300 g/kg). Fecal grab samples were taken at 0, 6, 12, 16, 20, 24, 30, 36, 48, 60, 72, 96, 120 and 144 h after dosing. Reduction of PS and enhancement of $K_2Cr_2O_7$ concentration increased SG of marker, ruminal passage rate, but decreased ruminal mean retention time, time delay and total mean retention time. The coefficient of correlation between SG and PS of marker, ruminal passage rate, time delay and ruminal mean retention time were -0.62, -0.70 and 0.65; 0.45, 0.75 and 0.12, respectively. The SG of themarker was the most influential factor affecting ruminal passage rate and mean retention time.

KEY WORDS chromium-mordanated marker, particle size, ruminal particulate turn over, specific gravity.

INTRODUCTION

According to Van Soest's "Hotel Theory" feed particles in rumen compete on retention for digestion or passage (Van Soest, 1994). Ruminants consuming forage diets need to balance the advantages of rapid transit of digesta through the gut to maintain high feed intake, against allowing sufficient time for fermentation by rumen microbes of fiber to maximize the digestion and yield of energy. In order to predict the quantitative importance of rumen digestion on fiber, information is needed on the flow rate of feed particles from rumen. The K_p can be measured by tagging the feed with various indigestible markers. An ideal marker should: 1) be inert without toxic physiological effects on the animal or microflora, 2) not be absorbed or metabolized within the gastrointestinal tract, 3) not influence gastrointestinal secretion, digestion, absorption, or motility, 4) have physiochemical properties that allow for precise, quantitative analysis, and it must not interfere with other analyses and finally and 5) flow parallel with or be physically similar to or intimately associated with the material it is to mark. Rare earth-labeled particles and chromium (Cr)-

mordanated fiber often are used to estimate the K_p of particle-associated components, whereas polyethylene glycol and EDTA chelates of Cr and cobalt often are used to measure fluid K_p (Galyean, 1987). The Cr mordanating fulfils most of desirable criteria for particulate marking. It yields a stable marker of solids forming hexacoordinate ligands with hydroxyl groups that are very difficult to hydrolyze (Udén et al. 1980). The Cr mordanated fiber is the most tenaciously bound of the particulate markers, which have been examined. Concentrations of Cr in excess of 80 mg/g dry matter (DM) render NDF essentially indigestible (Colucci et al. 1982); as the content of Cr on NDF decreases, the digestibility of NDF increased. However, the single dose procedure provides estimates of passage rate, mean retention time and gastrointestinal trace fill. The Cr mordanating of forages increase their density, reduce their digestibility and alter their chemical composition (Ehle, 1984; Lindberg, 1985; Ramanzin et al. 1991; Udén et al. 1980). Therefore, all of those alterations should be considered on estimation of turnover rate parameters.

Many experiments showed that physicochemical properties of mordanated markers change with marking. Passage rates measured with Cr mordanated particles can give a fairly accurate estimation of indigestible cell wall components (Bosch and Bruining, 1995). The method can seriously overestimate the passage rates of digestible cell wall components as measured by evacuation techniques (Aitchison et al. 1986). These differences indicate that mordanated particles do not readily simulate the passage of natural feed components. This may be because the markers have a different PS distribution and functional specific gravity (FSG) compared to the digesta component and the rate of PS reduction for the marker particles and feeds may be different. Mordanated hay particles are rendered indigestible by the preparation method, and are broken down only by physical mechanisms (Offer and Dixon, 2000) and it seems their rate of breakdown is vary greater than the feeds. Without an accurate description of the PS range and FSG of the markers, it is difficult to interpret differences in outflow rates measured in different experiments.

Bruining and Bosch (1992) found that PS of the Cr-NDF has a great influence on the calculated fractional passage rate from the rumen. Bruining and Bosch (1992) reported that passage rate of mordanated hay particles is dependent on their size and particles lower than 0.3 mm have twice passage rate (4.1±1.0 %/h) compared to that for those of size 0.6 - 1.0 mm (2.1±0.5 %/h). Chopped particles were found to have a slower rate than ground particles, and these in turn, had a slower rate than ground particles in which the NDF had been removed prior to mordanating (Ramanzin et al. 1991).

Ehle and Stern (1986) reported that labeling of feed particles with high Cr concentrations alters density of markers, which changes the availability of particles to rumination and passage. Therefore, for making prediction of quantitative importance of rumen digestion the use of suitable marker is critical.

The current experiment was carried out to test effects of PS distributions and density of a Cr-mordanated marker on the passage rate, ruminal mean retention time (RMRT), total mean retention time (TMRT) and time delay (TD) of markers.

MATERIALS AND METHODS

Particle size measurements

Alfalfa harvested at early flowering on one day from a single field without rain was cut and dried. Individual bales were chopped with a forage field harvester (Jaguar#62, Class Company, Germany) to 19, 10 and 5 mm theoretical cut length (TCL) for preparation of three different PS distributions. The fine alfalfa was prepared by grinding the 5 mm cut alfalfa using a farm grinder (2 mm screen size; Behsaz company#11.02 Jahadeh-e-Daneshgahie Mashhad, Iran).

The GM and the standard deviation of GM in each type of alfalfa were determined as ASAE (2002) (Table 1). Also, the GM and the standard deviation of GM of corn silage and alfalfa used total mixed ration (TMR) and four sizes of alfalfa used to marking and four sizes of markers were determined as ASAE Standard (2002) S424.1 (Table 1).

The four types of alfalfa analyzed for dry matter (DM), organic matter (OM), Kjeldahl N, ether extracts (EE) (Feldsine et al. 2002), NDF, acid detergent fiber (ADF) (Van Soest et al. 1991) and ash at 605 °C. Nonfiber carbohydrates (NFC) was calculated by 1000 - (CP(g/kg of DM) + NDF (g/kgof DM) + Ash (g/kgof DM) + EE (g/kgof DM)) (Table 1).

Marker Preparation

Cr-mordanated alfalfa NDF was prepared using a modified procedure of Udén et al. (1980). Basic modifications were to vary $K_2Cr_2O_7$ concentration in mordanating solution (50, 300, 500 and 900 g/kg) and feeds PS. Dichromate potassium ($K_2Cr_2O_7$; 231-906-6, solid crystallize, Merck Co.) was used for preparation of markers. The concentration of H_2O in $K_2Cr_2O_7$ was determined by titration with acidic 0.1 N cerium sulphate using ferroin as an indicator. Markers were analyzed for DM, OM, Kjeldahl N, ether extract NDF, ADF and ash as described above. NFC was calculated as described in equation 1 (Table 3). Also, the GM of markers determined according to ASAE (2002); Table 3.

Table 1 Chemical composition, functional specific gravity and particle size distribution of alfalfa, total mixed ration and four sizes of alfalfa used for marker preparation according to chop length

Screen size	TMR	Theoretical length cut of alfalfa			
		19 mm	10 mm	5 mm	Fine
Dry matter (g/kg)	625	938.3	938.3	938.3	938.3
Neutral detergent fiber (g/kg)	334	450.0	450.6	450.9	451.5
Acid detergent fiber (g/kg)	178	313.3	313.5	313.6	313.9
Nonfiber carbohydrates (g/kg)	432	326.7	324.7	324.1	323.0
Crude protein (g/kg)	170	166.2	167.5	167.9	168.1
Ether extracts (g/kg)	25	15.3	15.1	15.2	15.2
Ash (g/kg)	49	41.8	42.1	41.9	42.2
FSG	-	0.890^d	1.003^c	1.121^b	1.208^a
19 mm	12.51	27.19	12.40	11.19	0.00
12.7 mm	8.52	23.51	8.51	9.26	0.00
6.3 mm	16.50	14.20	16.30	6.68	0.00
3.96 mm	0.00	14.00	23.70	21.02	5.90
1.18 mm	15.12	15.30	23.00	22.93	37.00
Pan	16.14	5.30	16.0	28.92	57.10
Geometric mean	4.75	9.13^a	4.51^b	3.34^c	1.20^d
Standard deviation of GM	3.02	3.01^{de}	3.84^c	2.95^b	1.66^f

TMR: total mixed ration and FSG: functional specific gravity.
The means within the same row with at least one common letter, do not have significant difference (P>0.05).

Table 2 Chemical composition of feeds used in the total mixed ration

Feeds	Chemical composition									
	DM (g/kg)	OM (g/kg)	CP (g/kg)	NDF (g/kg)	ADF (g/kg)	ADL (g/kg)	EE (g/kg)	ASH (g/kg)	NFC (g/kg)	Chromium (ng/g DM)
Alfalfa	938.3^a	938.7^{ef}	164.3^d	442.1^g	314.9^{bc}	106.0^b	14.7^h	61.3^{ef}	326^e	204^b
Corn silage	316.7^e	929.3^{hg}	89^g	463.3^f	298.4^c	37.3^e	32.5^c	70.7^{cd}	345.6^d	318^a
Barely	925.3^{bc}	975.9^a	142.5^e	210.8^i	80.8^f	22.7^g	23.4^e	24.1^j	599.1^b	104^d
Beet pulp	943.6^a	926.9^{hi}	96.1^f	396.7^h	238.8^d	16.6^h	7.8^i	73.1^{bc}	426.3^c	116^d
Soybean meal	912.3^c	923.3^{hi}	444.7^a	183.8^j	110.9^f	3.7^i	18.7^g	76.7^{bc}	266.2^g	16^c
Wheat bran	916.1^c	942.9^e	164.9^d	498.9^e	168.5^e	29.6^f	46.4^b	57.1^f	232.6^h	81^e

The means within the same row with at least one common letter, do not have significant difference (P>0.05).
DM: dry matter; OM: organic matter; CP: crud protein; NDF: neutral detergent fiber; ADF: acid detergent fiber; ADL: acid detergent lignin; EE: ether extract and NFC: non-fiber carbohydrate.

Table 3 Density and geometric mean of types of alfalfa used for marker preparation and markers and mordanated chromium % of alfalfa NDF, according to chop length

Item	Theoretical length cut of alfalfa															
	19 mm				10 mm				5 mm				Fine			
K$_2$Cr$_2$O$_7$ Concentration (g/kg of DM)	50	300	500	900	50	300	500	900	50	300	500	900	50	300	500	900
Alfalfa	0.89^d				1.003^c				1.121^b				1.208^a			
FSG	9.13 mm^a				4.51 mm^b				3.34 mm^c				1.20 mm^d			
GM (mm)																
Markers																
Composition of marker (g/kg of DM)																
Mordanated Cr	22.4^j	137^h	202^f	375^c	225^j	133^i	204^f	376^c	23.7^j	136^{hi}	208^e	3802^b	2.43^j	14.1^g	22.0^d	40.3^a
NDF	955^a	841^b	775^d	603^g	955^a	844^b	773^d	601^g	954^a	866^b	756^f	596^h	953^a	837^c	758^f	574^i
ADF	691^a	609^b	562^d	437^g	691^a	611^b	560^d	436^g	690^a	610^b	557^e	432^h	689^a	606^c	548^f	417^i
CP	16.2	16.4	16.8	16.6	16.5	16.9	16.4	16.5	16.6	16.2	16.4	16.8	16.8	16.6	16.4	16.5
EE	0.14	1.5	1.5	1.4	1.5	1.5	1.5	1.5	1.5	1.4	1.5	1.5	1.5	1.5	1.4	1.5
Ash	27.4^j	141^h	207^f	379^c	270^j	138^i	209^f	381^c	27.9^j	116^{hi}	226^e	386^b	28.7^j	145^g	224^d	408^a
Physical characteristics																
Initial FSG	11^k	12.6^i	15.8^f	19^c	11.6^j	13.1^h	16^{ef}	19.2^c	12.5^i	13.2^h	16.2^e	19.6^b	12.9^h	14.2^g	16.7^d	20.1^a
FSG12	12.1^j	13.1^h	15.6^f	19^c	12.1^j	13.3^h	16^{ef}	19.1^c	12.6^i	13.3^h	16.2^e	19.5^b	13.1^h	14.3^g	16.5^d	19.9^a
FSG24	12.5^i	13.2^h	15.6^f	18.9^c	12.4^i	13.4^h	15.9^{ef}	19.1^c	12.6^i	13.4^h	16.3^e	19.6^b	13.1^h	14.2^g	16.5^d	20^a
GM (mm)	84.2^a	84.6^a	84.5^a	84.6^a	43.6^b	43.5^b	43.9^b	44.2^b	30.2^c	30^c	30^c	30.1^c	11.1^d	11^d	10.6^d	11.4^d

FSG: functional specific gravity; FSG12: functional specific gravity of materials after 12 h incubation and FSG24: functional specific gravity of materials after 24 h incubation.
DM: dry matter; NDF: neutral detergent fiber; ADF: acid detergent fiber; CP: crud protein and EE: ether extract.
The means within the same row with at least one common letter, do not have significant difference (P>0.05).

Functional specific gravity measurements

The functional specific gravity (FSG) of four alfalfa types, markers and TMRs were measured using 100 mL pycnometer (Wattiaux, 1990).

The dry samples (1.5 g) were incubated for 24 h in a 50 mL pycnometers with thermometer (Ambala Cantt, Ambala- 133001, Haryana) pycnometer and their FSG were measured at 12 and 24 h after incubation (Table 3). All the measurements for the kinetics of hydration were made at 39.0 ± 0.5 °C. The mixed rumen fluid from two steers fed only alfalfa were collected prior to feeding via a cannula and rinsed through 8 layers of cheese cloth, centrifuged at $30000 \times g$, for 10 min and the supernatant with density 1.0039 ± 0.0024 g/mL was used as the hydration solution. Sodium azide (0.50 g/L) and penicillin G (25000 units/L) were added to the hydration solution to prevent microbial growth.

Animal and diets

All procedures used in this study were approved based on Proposing a National Ethical Framework for Animal Research in Iran (16). Thirty-two multiparous, mid lactation Holstein dairy cows (body weight=654±24 kg) were allotted to a completely randomized design with two replications in a 4×4 factorial for evaluation of four size of markers with 8.45, 4.38, 3.00 and 1.10 mm of the GM and four $K_2Cr_2O_7$ concentration (50, 300, 500 and 900 g/kg of $K_2Cr_2O_7$ per DM were used for preparation of Cr-mordanated alfalfa NDF markers in different specific gravity). The experiment was undertaken over 21 d (adaptation, 14 d; sample collection, 7 d). During the experiment, cows were housed in tie-stalls and fed twice daily at 09:00 and 21:00.

The TMRs were offered ad libitum, allowing for at least 100 g resiual (as-fed basis). All diets had a 40 forage:60 concentrate ratio and contained 200, 200, 350, 70, 75, 100, 3, 1 and 1 g/kg of DM alfalfa, corn silage, barely grain, soybean meal, beet pulp, wheat bran, DCP, vitamin premix and salt, respectively. Water and mineralized salt were available for all cows over the experiment. The TMRs were formulated using the NRC (2001). The TMRs had similar chemical component and contained 334, 178, 432, 170, 5 and 5 (g/kg of DM) NDF, ADF, NFC, CP, EE, calcium and phosphors, respectively.

Markers were mixed with 400 g concentrate and 200 g molasses, fed to all cows at d 15, at the time of the morning feeding. Fecal grab samples were taken at 0, 6, 10, 12, 14, 18, 22, 26, 30, 36, 42, 48, 54, 60, 72, 84, 96, 120 and 144 h after dosing to determine the rate of passage, RMRT, TMRT and TD. Feeds, fecal samples and markers were dry-ashed and concentrations of Cr were determined by direct current plasma emission spectroscopy (Feldsine et al. 2002).

Sample collection

Body weights were recorded weekly. Dry matter intake was measured daily for all cows. Samples of forage, concentrates, and residual samples were collected each day. Samples were dried at 55 °C, ground through a Wiley mill (1-mm screen), and composited by animal within period. Total collection of feces was carried out for all cows over d 14 to 20. The feces were dried at 55 °C and ground through a Wiley mill (1-mm screen). Feeds and TMRs were analyzed as descried above and digestibility was calculated for DM, OM and all nutrients (Table 4).

Statistical analysis

The data were analyzed as a completely randomized design in a 4×4 factorial with two replicates using the following the model:

$$Y_{ijk} = \mu + PS_i + SG_j + PS \times SG_{ij} + e_{ijk}$$

Where:
Y_{ijk}: depended variable.
μ: overall mean.
PS_i: random effect of PS (i=1, 2, 3 and 4).
SG_j: random effect of specific gravity (j=1, 2, 3 and 4).
e_{ijk}: experimental error.

PROC GLM in SAS (2002) was used for analyses. In the first analysis, dry matter intake (DMI) was significantly different between the treatments; therefore DMI was considered as a covariate in future analysis. The data of PS were analyzed as a completely randomized design by using the REML variance component and PROC MIXEDof SAS (2002). Mean separation was determined using the PDIFF procedure and significance was declared at P < 0.05.

Fecal Cr excretion curves were fitted to the double compartment model represented by two exponential constants and a time delay (Grovum and Williams, 1973):

$$Y_t = Ae^{-k_1(t-TD)} - Ae^{-k_2(t-TD)}, k_1 = k_2 \text{ for } t \geq T,$$
$$Y = 0 \text{ for } t = TD,$$

Where:
Y_t: marker concentration (ppm).
A: scale parameter.
k_1: ruminal rate of passage (%/h).
k_2: lower digestive tract rate of passage (%/h).
t: sampling time post dosing (h).
TD: time delay.

The TMRT in the digestive tract was calculated as the sum of RMRT ($1/k_1$) and in the lower digestive tract ($1/k_2$) plus the TD (Table 4). Data were analyzed by nonlinear regression using the SAS (2002) (PROC NLIN®, iterative Marquardt method). The model in SAS® programming language (G4G1) that was used for fitting the excretion curves of markers are listed in appendix 1. Values used in the PARAMS statement were as follows: C-zero, 100, 500, 1200, 2400; Lambda, 0.1, 0.4, 0.8, 2; K, 0.03, 0.05, 0.07 and TD, 1, 10 (Moore *et al*. 1992). All of estimated parameters were analyzed as a completely randomized design in 4×4 factorial with two replicates. The orthogonal methods were used to compare means of two factors and all treatments (Table 4).

Analysis of correlation between PS and FSG of markers and ruminal turnover parameters was done using the PROC CORR of SAS (2002); (Table 5 and 6). The concentration of $K_2Cr_2O_7$ in mordanating solutions and the GM of alfalfa were used in analysis of regression for describing the relationship with CR concentrations in CR mordanated alfalfa NDF.

RESULTS AND DISCUSSION

The Cr concentrations of feeds rarely are measured, but all feeds used in the TMR had Cr concentrations in the normal range. Chemical composition and FSG of four sizes of alfalfa used for marker preparation are shown in Table 2. There were no significant differences between the chemical compositions of four sizes of alfalfa. But FSG of alfalfa types were significantly different and increased as PS of alfalfa was reduced because reduction of PS decreases the volume of feeds. Also, as theoretical cut length (TLC) of alfalfa reduced, the GM of alfalfa particles reduced. The GM of particle was different for four sizes of alfalfa. The GM of alfalfa and corn silage particles that used in TMRs were 4.43, 4.42 and 3.94 mm, respectively (Table 3). All of TMRs had the same composition and the same distribution of PS.

Particle size of alfalfa ($P<0.0001$) and $K_2Cr_2O_7$ concentration as a percentage of DM of materials ($P<0.0001$) and their interaction ($P=0.0045$) had significantly effect on the density of markers. Density of marker increased as $K_2Cr_2O_7$ concentration as a percentage of DM of materials was increased and PS was reduced (Table 4). Regardless the size, when $K_2Cr_2O_7$ concentration as 900 g/kg of DM of alfalfa was used, all markers had density more than 1.90 g/ml. However, alfalfa with TLC 19 mm produced the lowest dense markers, but fine alfalfa produced the highest dense markers. As ratios of concentration of $K_2Cr_2O_7$ in mordanating solutions to DM were increased, there was a concomitant increase of bounded Cr to alfalfa NDF.

The concentration of $K_2Cr_2O_7$ in mordanating solutions and the GM of alfalfa were used in analysis of regression for describing the relationship with Cr concentrations in Cr mordanated alfalfa NDF.

The equations describing this relationship are:

$Y= 0.38631 + 0.42099X_1$, $R^2= 99.6$ (Equation 1)
$Y= 19.21436 – 0.05762X_2$, $R^2= 4.0$ (Equation 2)
$Y= 0.33958 + 0.42119X_1 + 0.00537X_2$, $R^2= 99.6$ (Equation 3)

Where:

Y: Cr concentrations in Cr mordanated alfalfa NDF (mg/g cell wall).
X_1: concentration of $K_2Cr_2O_7$ in mordanating solutions.
X_2: GM of alfalfa that used for marker preparation (mm).

Ehle (1984) used 20, 40, 80, 160 and 320 g/kg Cr per cell wall and found that as ratios of Cr to fiber DM were increased, Cr bound to alfalfa fiber was increased (24.23, 28.17, 31.95, 38.99 and 41.13 mg/g cell wall in each Cr concentration; respectively) and the equation described this relationship was:

$Y= 0.533X + 26.28$ with $R^2= 91$

Where:

Y: Cr concentration (mg/g cell wall).
X: Cr concentration in mordanating solutions.

However, the PS of mordanated alfalfa in Ehle's study was 2.07 mm, but in the current study, PS of alfalfa that used for marking was significantly different.

As ratios of concentration of $K_2Cr_2O_7$ in mordanating solutions to DM were increased, the density of Cr mordanated alfalfa NDF was increased. The results were similar to Ehle (1984), Ramanzin *et al*. (1991). We used the concentration of Cr mandated alfalfa NDF and GM in the regression analyses to describe the relationship between density and Cr concentration and marker PS in the current study. The equations describing this relationship are:

$Y= 1.1124 + 0.00217X_1$, $R^2= 94.0$ (Equation 4)
$Y= 1.597 – 0.0179X_2$, $R^2= 3.3$ (Equation 5)
$Y= 1.1772 + 0.0216X_1 – 0.0149X_2$, $R^2= 96.0$ (Equation 6)

Where:

Y: SG of markers.
X_1: Cr concentration (mg/g cell wall).
X_2: GM of Cr mordanated alfalfa NDF.

The recommended $K_2Cr_2O_7$ or $Na_2Cr_2O_7$ application by Udén *et al*. (1980) is between 300 to 330 g/kg of fiber weight to get a Cr concentration equivalent to 12-14 g/kg (average 13 g/kg) in the cell wall. Based on our current study using the different concentrations of $K_2Cr_2O_7$ resulted in Cr concentration between 22.4 to 403.1 g/kg of Cr mordanated NDF.

The data of FSG of markers after 12 and 24 h incubation in pycnometer are shown in Table 4. Particle size (P=0.0241) had a significant effect on FSG changes over the incubation time, but, concentrations of $K_2Cr_2O_7$ had no significant effect (P=0.4563). Over the incubation time, only FSG of markers that originated with alfalfa in 19 and 10 mm TCL and mordanated by 50 g/kg concentrations of $K_2Cr_2O_7$ were increased.

The GM of alfalfa types and markers were significantly different. Preparation of markers reduced the GM of alfalfa types and amount of reduction were 8, 3, 11 and 9% for 19, 10 and 5 mm TLC and fine alfalfa, respectively. Markers that prepared with alfalfa with TLC 19 mm had the largest GM but markers that prepared with fine alfalfa had the lowest GM (Table 4). There were no significant differences between the GM of prepared markers in each TLC size of alfalfa. The geometric means of markers were 8.45, 4.38, 3.00 and 1.10 mm that originated from 19, 10, 10 mm TLC and fine alfalfa, respectively. Therefore, it can be expected that different digestion parameters result in different density of markers.

Chemical compositions of markers are shown in Table 2. There were no significant differences for CP and EE between the markers. There were significant different for NDF, ADF, ash and mordanated Cr between the markers. As ratios of concentration of $K_2Cr_2O_7$ in mordanating solutions to DM were increased, ash and mordanated Cr were increased and NDF and ADF were decreased. It seems that the concentration of $K_2Cr_2O_7$ in mordanating solutions was the most influence factor affected on chemical compositions and physical properties of markers.

BW, DMI and digestibility of nutrients used in the ration are shown in Table 4. There were no significantly difference in BW between the cows. DMI was significantly different between cows in different treatments. Therefore, DMI was considered as a covariate in the final model of ANOVA. However, digestibility of DM, NDF, ADF and CP were not significantly different between the treatments (Table 4). The passage rate and digestibility are correlated and digestibility can be easily related to the digestive mechanism and it is a function of the kinetic of digestion and passage (Colucci et al. 1982; Grovum and Williams, 1973; Allen and Mertens, 1988; Van Soest, 1994). Also, DMI is related to fiber digestion because it is limited by the rate of disappearance of material from the digestive tract (Allen and Mertens, 1988). There is clear evidence that the major dietary factors affecting rumen outflow of liquid and

particle components are BW (body size), level of feeding, ration composition and form, forage:concentrate ratio and digestibility of nutrient.

In current experiment, BW, level of feeding, forage: concentrate ratio and digestibility of nutrient, especially fiber components (NDF and ADF) were not significantly different between treatments and all cows used in experiment were multiparous, therefore, it seems significant difference between the cows in DMI may be result of different ruminal capacity.

Digestion kinetic parameters are shown in Table 4. Marker PS (P=0.0003) and concentration of $K_2Cr_2O_7$ in mordanating solutions (P<0.0001) and interaction of PS and concentration of $K_2Cr_2O_7$ (P<0.0001) had significant effects on Kp. Regardless the size, as concentration of $K_2Cr_2O_7$ was increased or density of markers were increased, Kp significantly were decreased. The highest Kp was observed when marker was prepared with 300 g/kg of DM concentration of $K_2Cr_2O_7$ by using alfalfa in 19 mm TLC. In all size markers, markers that were prepared with 5% of DM concentration of $K_2Cr_2O_7$ had higher Kp than the others.

Ehle (1984) used 20, 40, 80,160 and 320 g/kg Cr per cell wall (with density 1.126, 1.165, 1.242, 1.396 and 1.703 in each Cr concentration; respectively) and found that rumen turnover rates increased linearly with increased Cr concentration (0.0107, 0.0072, 0.0191, 0.0228 and 0.0194 per h in each Cr concentration; respectively). Lirette and Milligan (1989) used 0.1 and 5 g/kg DM Cr to NDF of stem in 1-2 or 10 mm in length and found that Cr concentration had no significant effect on turnover rate parameters, but PS had a significant effect. However, in current study, we increased Cr concentration to 400 g/kg of cell wall and obtained greater density than previous studies (we obtained markers with density greater than 1.90 with 900 g/kg of DM concentration of $K_2Cr_2O_7$ in all sizes). The Kp was increased when concentration of $K_2Cr_2O_7$ increased from 50 to 300 g/kg but decreased when concentration of $K_2Cr_2O_7$ more than 300 g/kg were used for marking.

Marker PS (P=0.0381) and concentration of $K_2Cr_2O_7$ in mordanating solutions (P=0.0034) had a significant effect on RMRT, but interaction of PS and concentration of $K_2Cr_2O_7$ (P=0.3148) did not have a significant effect on RMRT.

Lirette and Milligan (1989) used 0.1 and 5 g/kg DM Cr to NDF of stem in 1-2 or 10 mm lengths and found that Cr concentration had not significantly effect on RMRT.

In normal rations the coarsest material floats in an upper layer and forms aruminal mat. The mat is one of the main sorting mechanisms and elimination allows the escape of larger particles (Van Soest, 1994). Rumen contents are stratified into several layers. PS and FSG of the mat contents are important in the consistency of ruminal mat.

Table 4 Dry matter intake, digestible nutrients of rations and kinetic of digestion

	Theoretical length cut of alfalfa															
$K_2Cr_2O_7$	19 mm				10 mm				5 mm				Fine			
Concentration (g/kg of DM)	50	300	500	900	50	300	500	900	50	300	500	900	50	300	500	900
BW (kg)	650.5	657.5	656.5	656	660	655	655	657	661.5	656.5	661	657.5	655	655	655	653.5
DMI (Kg/d)	24.9[abcd]	23.8[abcd]	24.9[d]	24.8[abcd]	24.6[abcd]	24.4[cd]	25.7[abc]	25.0[abcd]	24.5[bcd]	24.7[abcd]	25.3[abc]	24.9[abcd]	26.0[ab]	23.7[d]	23.8[d]	26.0[a]
Digestible nutrients (g/kg)																
DM	763	773	774	765	770	753	762	764	760	771	770	764	763	764	763	767
NDF	566	568	572	559	568	572	563	557	559	560	563	571	568	558	556	562
ADF	515	521	522	513	501	516	498	503	496	490	521	506	512	496	505	514
CP	753	760	764	753	751	750	761	753	746	753	754	748	762	754	752	750
Kinetic of digestion																
Kp (%/h)	4.5[bc]	5.45[a]	5.0[ab]	3.1[e]	4.2[cd]	5.0[ab]	3.4[e]	3.6[de]	5.0[ab]	3.6[de]	3.1[e]	3.1[e]	5.1[ab]	3.7[de]	3.4[e]	3.1[e]
K_2 (%/h)	-	-	-	*					*	-	-	-	-	-	-	-
RMRT (h)	22.2[de]	18.4[e]	19.9[e]	32.2[ab]	25.2[cd]	19.8[e]	29.4[abc]	27.9[bc]	19.9[e]	28.0[bc]	32.8a	32.8[a]	19.7[d]	27.6[ii]	31.1	0.7[b]
FMRT (h)	11.9[ab]	10.9[bc]	10.9[bc]	10.2[bcd]	10.3[bcd]	10.9[bc]	11.6[abc]	9.88[bcd]	10.2[bcd]	9.2[cd]	13.6[a]	10.9[bc]	9.38[bcd]	10.2[bcd]	10.5[bc]	7.79[d]
TMRT (h)	53.2[a]	46.3[b]	45.5[b]	51.6[a]	52.2[a]	45.8[b]	51[a]	45.5[b]	39.4[dc]	42.3[bc]	53[a]	52[a]	37.8[d]	44.7[b]	44.6[b]	43.8[b]
TD (h)	20.0[a]	17.0[b]	14.7[c]	9.3[de]	16.7[b]	15.1[c]	10.0[d]	7.7[fg]	9.3[de]	5.1[i]	6.6[h]	8.3[ef]	8.7[e]	6.8[hg]	4.7[ij]	3.8[j]
Recovery rate (%)	97.55	97.6	97.0	96.8	97.3	97.4	97.2	96.75	97.4	96.8	97.0	97.1	97.3	97.4	97.5	96.9

BW: body weight and DMI: dry matter intake.
Kp: ruminal passage rate; K_2: lower compartments passage rate; RMRT: ruminal mean retention time; FMRT: mean retention time in lower compartments; TMRT: total mean retention time and TD: tTime delay.
DM: dry matter; NDF: neutral detergent fiber; ADF: acid detergent fiber and CP: crud protein.
The means within the same row with at least one common letter, do not have significant difference (P>0.05).

Table 5 The effects of particle size and $K_2Cr_2O_7$ concentration on ruminal turnover

Item	P-value		
	Particle size	$K_2Cr_2O_7$ concentration	Interaction
Kp (%/h)	0.0003	< 0.0001	< 0.0001
K_2 (%/h)	0.0012	< 0.0001	0.0704
RMRT (h)	0.0004	< 0.0001	< 0.0001
FMRT (h)	0.0381	0.0034	0.3148
TMRT (h)	< 0.0001	0.0297	< 0.0001
Time delay (h)	< 0.0001	< 0.0001	< 0.0001

Kp: ruminal passage rate; K_2: lower compartments passage rate; RMRT: ruminal mean retention time; FMRT: mean retention time in lower compartments and TMRT: total mean retention time.
The means within the same row with at least one common letter, do not have significant difference (P>0.05).

Table 6 Correlation between particle size and density of markers and ruminal turn over parameters (%, above diagonal) and their P-value (below diagonal)

Item	DMI (Kg/d)	Kp (%/h)	K_2 (%/h)	RMRT (h)	FMRT (h)	TMRT (h)	Time delay (h)	Particle size	Density of markers
DMI (kg/d)		-0.13	0.18	-0.03	0.16	0.06	-0.11	-0.09	0.16
Kp (%/h)	0.4825	-	-0.16	0.07	-0.96	-0.43	0.63	0.37	-0.55
K_2 (%/h)	0.3267	0.3814	-	-0.95	0.14	-0.52	-0.41	-0.40	0.40
RMRT (h)	0.8923	0.6956	< 0.0001	-	-0.04	0.54	0.30	0.29	-0.33
FMRT (h)	0.3866	< 0.0001	0.4510	0.8186	-	0.47	-0.64	-0.33	0.59
TMRT (h)	0.7604	0.0124	0.0024	0.0016	0.0064	-	0.35	0.43	-0.12
Time delay (h)	0.5312	0.0001	0.0196	0.0918	< 0.0001	0.0518	-	0.74	-0.70
Particle size	0.3789	0.0364	0.0248	0.1057	0.0695	0.0133	< 0.0001	-	-0.16
Density of markers	0.6124	0.0010	0.0241	0.0636	0.0004	0.5224	< 0.0001	0.3692	-

DMI: dry matter intake; Kp: ruminal passage rate; K_2: lower compartments passage rate; RMRT: ruminal mean retention time; FMRT: mean retention time in lower compartments and TMRT: total mean retention time.
The means within the same row with at least one common letter, do not have significant difference (P>0.05).

Evans *et al.* (1973) reported that FSG is a major factor in the separation of ruminal mat contents. Fine and heavy materials cause cessation of rumination and the relative elimination of the floating mat (Van Soest, 1994).

Heavy material sinks through the rumen pack whereas light material floats and forms the upper part of the fibrous mass. Material with SG less than 1 was heavily ruminated and slowly passed (DesBordes and Welch, 1984).

Intermediate SG (1.17 and 1.42) passed most rapidly. The densest particles (1.77 and 2.14) passed more slowly with little evidence of remastication. Apparently, few of the heavy particles were incorporated in the active moving materials in the region of the cardiac, where rumination boluses are formed. Particle lengths from 1 to 10 mm showed the same trends: light SG passed very slowly and the 1.17 to 1.42 range passed most rapidly.

Marker PS (P<0.0001) and concentration of $K_2Cr_2O_7$ in mordanating solutions (P=0.0297) and interaction of PS and interaction of PS and concentration of $K_2Cr_2O_7$ (P<0.0001) had significant effects on TMRT (Table 4). Marker PS (P<0.0001) and concentration of $K_2Cr_2O_7$ in mordanating solutions (P<0.0001) and interaction of PS and interaction of PS and concentration of $K_2Cr_2O_7$ (P<0.0001) had significantly effect on transit time of markers. As TCL was decreased and concentration of $K_2Cr_2O_7$ in mordanating solutions was increased, time delay was decreased. Regardless the density, the markers that originated with TCL 19 mm and fine alfalfa had highest and lowest time delay, respectively. Marker PS (P=0.3452) and concentration of $K_2Cr_2O_7$ in mordanating solutions (P=0.6542) and interaction of PS and interaction of PS and concentration of $K_2Cr_2O_7$ (P=0.5874) had not significantly effect on recovery rate of markers (Table 4).

CONCLUSION

The results obtained in this experiment showed that the Cr mordanting procedure can greatly affect the physicochemical properties of feed that use for marking, thus altering their passage rate estimation. Calculated outflow rates are highly dependent on marker properties such as density and PS distribution, which should be properly, defined in future experiments. The results of the current experiment indicate that outflow from the rumen is greatly affected by PS and FSG and FSG is the most important factor that influenced the ruminal turn over rate parameters. The Cr mordanating of forages increase their density, reduce their digestibility and alter their chemical composition. Therefore, all of those alterations should be considered on estimation of turnover rate parameters. Without an accurate description of the PS range and FSG of the markers, it is difficult to interpret differences in outflow rates measured in different experiments. Thus the passage rate of mordanated alfalfa NDF is dependent on their FSG and PS distributions (geometric mean).

ACKNOWLEDGEMENT

This work was supported by the research grant of Sari Agricultural and Natural Resources University (SANRU), Mazandaran, Iran. The author thank from Mrs. Maassoumeh Vahmian and Rohangiz Vahmian for preparation of Cr-Mordanated marker of forage particles. Also, the authors thank Mr. Mousa and Esmaiel Teimouri, Ali Mahdavi and Abolfazl Hydarian, the staff of the dairy unit for care of the cows and sample collection, for their assistance.

REFERENCES

Aitchison E., Gill M., France J. and Dhanoa M.S. (1986). Comparison of methods to describe the kinetics of digestion and passage of fibre in sheep. *J. Sci. Food Agric.* **37,** 1065-1072.

Allen M. and Mertens D.R. (1988). Evaluating constraints on fiber digestion by rumen microbes. *J. Nutr.* **118(2),** 261-270.

Bosch M.W. and Bruining M. (1995). Passage rate and total clearance rate from the rumen of cows fed on grass silages differing in cell-wall content. *British J. Nutr.* **73,** 41-49.

Bruining M. and Bosch M. (1992). Ruminal passage rate as affected by CrNDF particle size. *Anim. Feed Sci. Technol.* **37,** 193-200.

Colucci P., Chase L. and Van Soest P. (1982). Feed intake, apparent diet digestibility, and rate of particulate passage in dairy cattle. *J. Dairy Sci.* **65,** 1445-1456.

DesBordes C. and Welch J. (1984). Influence of specific gravity on rumination and passage of indigestible particles. *J. Anim. Sci.* **59,** 470-475.

Ehle F. (1984). Influence of particle size on determination of fibrous feed components. *J. Dairy Sci.* **67,** 1482-1488.

Ehle F. and Stern M. (1986). Influence of particle size and density on particulate passage through alimentary tract of Holstein heifers. *J. Dairy Sci.* **69,** 564-568.

Evans E., Pearce G., Burnett J. and Pillinger S.L. (1973). Changes in some physical characteristics of the digesta in the reticulo-rumen of cows fed once daily. *British J. Nutr.* **29,** 357-376.

Feldsine P., Abeyta C. and Andrews W.H. (2002). AOAC international methods committee guidelines for validation of qualitative and quantitative food microbiological official methods of analysis. *J. AOAC. Int.* **85,** 1187-1200.

Galyean M. (1987). Factors influencing digesta flow in grazing ruminants. Pp. 24-27 in Proc., Grazing Livest. Nutr. Conf. Colorado, USA.

Grovum W. and Williams V. (1973). Rate of passage of digesta in sheep. *British J. Nutr.* **30,** 313-329.

Lindberg J. (1985). Retention time of chromium-labelled feed particles and of water in the gut of sheep given hay and concentrate at maintenance. *British J. Nutr.* **53,** 559-567.

Lirette A. and Milligan L. (1989). A quantitative model of reticulo-rumen particle degradation and passage. *British J. Nutr.* **62,** 465-479.

Moore J., Pond K., Poore M.H. and Goodwin T. (1992). Influence of model and marker on digesta kinetic estimates for sheep. *J. Anim. Sci.* **70,** 3528-3540.

NRC. (2001). Nutrient Requirements of Dairy Cattle. 7th Ed. National Academy Press, Washington, DC, USA.

Offer N. and Dixon J. (2000). Factors affecting outflow rate from the reticulo-rumen. Pp. 833-844 in Proc. Nutr. Abst. Rev. Livest. Feeds and Feeding. CAB International, Wallingford, Oxfordshire, United Kingdom.

Ramanzin M., Bittante G. and Bailoni L. (1991). Evaluation of different chromium-mordanted wheat straws for passage rate studies. *J. Dairy Sci.* **74,** 2989-2996.

SAS Institute. (2002). SAS®/STAT Software, Release 9.1. SAS Institute, Inc., Cary, NC. USA.

Standard. (1998). Method of determining and expressing particle size of chopped forage material by screening. Published by the American Society of Agricultural and Biological Engineers, St. Joseph, Michigan.

Udén P., Colucci P.E. and Van Soest P.J. (1980). Investigation of chromium, cerium and cobalt as markers in digesta. Rate of passage studies. *J. Sci. Food Agric.* **31,** 625-632.

Van Soest P.J. (1994). Nutritional ecology of the ruminant. Cornell University Press, Ithaca, New York.

Van Soest P.J., Robertson J.B. and Lewis B. (1991). Methods for dietary fiber, neutral detergent fiber and nonstarch polysaccharides in relation to animal nutrition. *J. Dairy Sci.* **74,** 3583-3597.

Wattiaux M.A. (1990). A mechanism influencing passage of forage particles through the reticulo-rumen: change in specific gravity during hydration and digestion. Ph D. Thesis. University of Wisconsin-Madison, Madison, USA.

24

Evaluation of Total Antioxidant, Total Calcium, Selenium, Insulin, Free Triiodothyronine and Free Thyroxine Levels in Cows with Ketosis

S. Kozat[1*] **and N. Yüksek**[1]

[1] Department of Internal Medicine, Faculty of Veterinary, Yuzuncu Yıl University, Zeve Campus, Van, Turkey

*Correspondence E-mail: skozat@hotmail.com

ABSTRACT

Ketosis is an important metabolic disease of high milk-producing cows. There are significant changes in many metabolite and hormonal concentrations in metabolic diseases. This study was carried out to assess the concentrations calcium (Ca), selenium (Se), total antioxidant (TAOC), insulin, free triiodothyronine (fT_3) and free thyroxine (fT_4) in cows with subclinical and clinical ketosis. This study included 20 dairy cows within the first two months of lactation, aged between 4-8 years. Cows with ß-hydroxybutyrate acid (BHBA) concentrations 1.20 mmol/L were considered healthy, whereas 1.20 and 1.50 mmol/L were considered subclinical and 1.60-2.20 mmol/L were classified as clinically ketotic. Serum aspartate aminotransferase (AST), alanine aminotransferase (ALT), low density lipoprotein (LDH), glucose, Ca, plasma TAOC capacity and BHBA concentrations were performed spectrophotometrically. Serum insulin, free triiodothyronine and free thyroxine concentrations were measured using the chemi-luminescence method. Serum Se concentrations were measured using an Inductively Coupled Plasma-Mass Spectrometry (ICP-MS). In conclusion, significant changes were noted in decreased concentrations of TAOC, Ca, Se, fT_3, fT_4 and insulin in cows with subclinical and clinical ketosis. The study identified important parameters, changes in the levels of these parameters will be important in determining the treatment and prognosis of the disease. Their use may also help reduce the economic losses suffered by dairy farmers as a result of the disease.

KEY WORDS calcium, dairy cows, insulin, ketosis, selenium, thyroid.

INTRODUCTION

Ketosis is a metabolic disease of cows with high milk yields, which occurs in the last stage of gestation and in the two months after parturition (Kennerman, 2004). Late gestation and early lactation, more energy is required than is consumed resulting in mobilization of body reserves (Djoković et al. 2009; Sahinduran et al. 2010). Triglycerides are mobilised from fat reserves and decompose into fat acids and glycerol (Kennerman, 2004). While glycerol participates directly in glucose synthesis, fat acids complexed with serum albumin and are transported to the liver.

After the transport of free fat acids to the mitochondria and their decomposition as a result of oxidation, acetyl coenzyme A is formed (Katoh, 2002). Energy deficiency causes inadequate oxaloacetate; as a result, acetyl coenzyme A cannot participate in the tricarboxylic acid (TCA) cycles, which leads to an increased amount of ketone bodies (Kennerman, 2004; Katoh, 2002). Excessive lipid metabolism leads to the production of acetoacetate, acetone and ß-hydroxybutyrate acid (BHBA). Detecting increased levels of these in blood, urine or milk is used to diagnose the disease. At the same time the animal develops hyperketonaemia, she also undergoes hypoglycaemia, low thyroxine and high non-esterified plasma fat acids are reported. In addition, an increase in the levels of aspartate aminotransferase (AST), alanine aminotransferase (ALT) and alkaline phosphatase activities due to liver failure is reported (Sahinduran *et al.* 2010). Changes in calcium (Ca) concentrations are also reported in cows with ketosis (Sahinduran *et al.* 2010; Saldago Hernández *et al.* 2009). In addition to metabolic changes, hormonal changes (Katoh, 2002) have also been reported in ketosis as well as lower blood insulin concentrations than is found in healthy cows at the first stage of lactation (Djoković *et al.* 2009). Thyroid hormones have a great impact on the basic metabolic rate and are body weight and energy expenditure (Mullur *et al.* 2014). Working in conjunction with growth hormone and insulin, protein synthesis is stimulated and nitrogen excretion is reduced. In this way, growth, metabolism and their production may be affected (Kozat, 2007). Changes in free triiodothyronine (fT$_3$) and free thyroxine (fT$_4$) have been reported in a number of studies (Kennerman, 2004; Saldago Hernández *et al.* 2009; Sahinduran *et al.* 2010). Other research suggests that Se levels should be at a specific for the transformation of serum fT$_4$ to fT$_3$ (Kozat, 2007).

In metabolic diseases, free radical out bursts caused by oxidative stress lead to a decrease in neutralising antioxidant capacity (Sahoo *et al.* 2009). In studies on cows with high triglycerides and non-esterified fat acids, high levels of reactive oxygen metabolites and a low level of total antioxidant capacity have been observed (Katoh, 2002; Haces *et al.* 2008; Wang *et al.* 2010). Free fat acids not only cause oxidative stress but also lead to deterioration in endogenous antioxidant defence by lowering the level of intracellular glutathione peroxidase (GSH-Px) (Sahoo *et al.* 2009). Selenium (Se) is a key component of various selenoproteins, which play a role in the enzymatic functions required for oxidation reduction and thyroid hormone metabolism in order to maintain homeostasis (Stranges *et al.* 2010). Se is present in many selenoproteins. There is a positive relationship between plasma Se concentration and GSH-Px activity (Kozat, 2007; Pavlata *et al.* 2002).

The present study was undertaken to investigate possible associations between levels of serum Ca, Se, insulin, fT$_3$, fT$_4$, TAOC capacity and BHHA concentrations in dairy cows and blood concentrations of glucose and some other metabolites.

MATERIALS AND METHODS

Animal material

A total of 86 Simmental cows (4-8 years old) from one commercial dairy herd were included in the study. This research was conducted with a total of 30 dairy cows, including 10 healthy cows, 8 with subclinical and 12 with clinical ketosis. All cows enrolled in study were within the first 2 months after parturition. All cows aged between 4-8 years. The cows that had had subclinical ketosis, with milk yield ranging from 20 to 32.5 kg/day of milk at the time of diagnosis, whereas cows with clinical ketosis were producing from 12 to 22.5 kg/day at diagnosis. This study was carried out during the early spring housing period in April. All animals were routinely treated against endoparasites and no parasite eggs were observed during faecal examination.

Systemic examination of the cows was carried out blood and urine samples were obtained from the animals thought to have the disease. The findings were then examined clinically and systematically. Rothera and spin react 100 tests (test strips–Combur [10]Test®M, Roche, İstanbul, Turkey) were performed. Cows that were positive for the Rothera and spin react 100 tests were taken into study. Animals that were positive according to Rothera and spin react 100 tests, but whose BHHA concentrations were low in laboratory data (plasma D-3-Hydroxy butyrate concentrations) were excluded from the study. These tests yielded positive results and clinical ketosis were defined as plasma BHHA concentrations ≥ 1.60 mmol/L. In addition, eight cows with no ketosis symptoms were diagnosed with subclinical ketosis according to evaluation of plasma D-3-Hydroxy butyrate concentrations which are 1.20 mmol/L and serum glucose concentrations which are 35.63 ± 0.73 mg/dL. Threshold values of plasma D-3-Hydroxy butyrate concentrations in subclinical ketosis in this study classified as to on the basis of threshold values to separate healthy cows from cows with SCK are reported by researchers (Voyvoda and Erdogan, 2010; Sakha *et al.* 2007).

Cows with subclinical ketosis were administered 500 mL of 30% serum dextrose solution (Dekstrovet 30%, İ.E. Ulagay Pharmaceutical Trading Co, İstanbul, Turkey), administered intravenously. According to body weight, 0.5 IU/kg insulin (Humulin®M 70/30 100 IU Flakon, Lilly Pharmaceutical Trading Co, İstanbul, Turkey) was administered via intramuscular injection. For cows with clinical ketosis, the amounts were, respectively, 1000 mL of 30% serum dextrose solution, administered intravenously, and 0.5 IU/kg insulin, administered via intramuscular injection for 2 consecutive days.

Blood samples were obtained from all the animals in order to evaluate biochemical parameters. Samples were taken from the jugular vein and placed in anticoagulated and coagulant-free blood tubes. The serum and plasma from the blood samples was extracted after centrifuging at 3000 rpm. Biochemical parameters from the serum and plasma were then measured. Blood samples were taken from all cows before treatment and after treatment (3rd day) for biochemical parameters. Control cows should have been monitored and sampled more frequently to be sure that they did not experience ketosis.

Serum AST, ALT, LDH, glucose concentrations were performed spectrophotometrically (Photometer 5010®Boehringer Mannheim Gmbh, Germany) according to the test kit methods. Plasma TAOC capacity (Rel Assay Diagnostics® Research& Clinical Chemistry, United Kingdom) and plasma D-3-Hydroxy butyrate (Randox Laboratories Ltd., United Kingdom) were also analysed spectrophotometrically (Photometer 5010®Boehringer Mannheim Gmbh, Germany) according to the test kit procedures. Serum insulin, free triiodothyronine and free thyroxine levels in healthy cows and cows with ketosis were examined using the chemi-luminescence method by hormone device (Architect i2000-USA), according to ABBOTT commercial test kit procedures. Similarly, Se concentrations were measured using an Inductively Coupled Plasma-Mass Spectrometry (ICP-MS) device (Thermo Scientific X2, Switzerland).

Statistical analysis

Definitive statistics for healthy cows and cows with subclinical and clinical ketosis were stated at mean, standard deviation, minimum and maximum values. A comparison of biochemical parameter groups was performed with the Kruskal-Wallis test. In order to determine different groups, the Duncan Multiple Range Test was used. Spearman correlation multipliers were calculated in order to determine the association between these variables. The level of statistical significance was set at 5% (SPSS, 2011).

RESULTS AND DISCUSSION

The clinical examination of cows with clinical ketosis revealed a 50-60% decrease in milk yield, loss of appetite, decrease in ruminal movements and weakness. In dairy cows with subclinical ketosis, indigestion symptoms were found less often but milk yield also decreased. When the biochemical parameters of cows with subclinical ketosis were compared with the control group before treatment, AST, ALT and LDH activities and plasma BHBA concentrations were high ($P<0.001$, $P<0.05$, $P<0.001$ and $P<0.001$, respectively). Serum glucose levels, Ca values, fT_3, fT_4, Se and plasma TAOC amounts were low ($P<0.001$, $P<0.001$, $P<0.01$, $P<0.001$, $P<0.001$ and $P<0.001$, respectively). Insulin values did not change and were closer to those in the control group ($P>0.05$). On the third day after treatment, when the biochemical parameters of the cows with subclinical ketosis were compared with the control group, although Se values were lower than in the control group ($P<0.05$), other parameters were closer to those in the control group.

When the biochemical parameters in cows with clinical ketosis were compared with the control group before treatment, AST, ALT and LDH activities and plasma BHBA concentration were higher ($P<0.001$, $P<0.01$, $P<0.001$ and $P<0.001$, respectively). Serum glucose levels, Ca values, fT_3, fT_4 and TAOC levels and Se concentrations were lower ($P<0.001$, $P<0.01$, $P<0.001$, $P<0.001$, $P<0.001$ and $P<0.001$, respectively). However, insulin levels did not change. On the third day after treatment, when the biochemical parameters of cows with clinical ketosis were compared with the control group, plasma BHBA concentration had returned to normal levels ($P>0.05$). Although glucose levels had increased, they remained lower than in the control group ($P<0.001$). AST, ALT and LDH activities remained high and these increases were statistically significant ($P<0.01$, $P<0.05$ and $P<0.001$, respectively). Although serum Ca, fT_3, fT_4 and TAOC levels were higher, these levels were lower than in the control group ($P<0.05$, $P<0.01$, $P<0.001$ and $P<0.001$, respectively). Again, although serum insulin levels in cows with clinical ketosis were closer to those in the control group ($P>0.05$), serum Se concentration levels remained lower ($P<0.001$) (Table 1).

We then analysed the pre-treatment parameters of cows with subclinical ketosis. The results showed a positive correlation between BHBA and LDH concentration ($P<0.05$, r=0.835) and a negative correlation between fT_3 and LDH ($P<0.01$, r=−0.919). A negative correlation was shown between glucose and LDH ($P<0.05$, r=−0.719), with a positive correlation between glucose and Se ($P<0.01$, r=0.839). The results showed a positive correlation between AST and LDH ($P<0.01$, r=0.966) and a negative correlation between AST and selenium ($P<0.05$, r=−0.796). There was a negative correlation between LDH, Ca, fT_3 and Se ($P<0.01$, r=−0.954; $P<0.01$, r=−0.922; $P<0.05$, r=−0.791). The results showed a positive correlation between Ca, fT_3 and Se ($P<0.01$, r=0.843; $P<0.05$, r=0.734, respectively). Correlation between fT_3 and Se was also deemed positive ($P<0.05$, r=0.780). Biochemical parameters for cows with subclinical ketosis were also taken on the third day after treatment. Analysis of these parameters showed a negative correlation between serum BHBA and Ca concentration ($P<0.01$, r=−0.870). An increase in Ca concentration was shown along with a decrease in BHBA.

When the parameters in cows with clinical ketosis were analysed, a positive correlation was found between serum AST and serum BHB ($P<0.05$, r=0.675). A negative correlation was found between serum Se and BHB ($P<0.05$, r=−0.699). In addition, the findings showed a negative correlation between serum glucose levels and serum AST levels ($P<0.05$, r=−0.673).

Table 1 Biochemical parameters in cows with ketosis and healthy cows

Variable	Control (n=10) (Mean±SE)	Subclinical ketosis (n=8) (Mean±SE)		Clinical ketosis (n=12) (Mean±SE)	
		BT	AT (3rd day)	BT	AT (3rd day)
BHBA (mmol/L)	0.22±0.03[a]	1.25±0.14[b]	0.25±0.04[a]	1.76±0.21[b]	0.28±0.03[a]
Glucose (mg/dL)	69.80±3.55[a]	35.63±0.73[b]	68.00±2.65	27.75±1.07[b]	54.92±0.57[b]
AST (U/L)	51.86±1.92[a]	74.9±0.52[b]	59.38±0.58[d]	85.39±1.57[b]	68.03±0.72[c]
ALT (U/L)	23.43±3.07[a]	27.91±2.39[d]	25.34±1.28	35.50±4.62[c]	28.96±3.22[d]
LDH (U/L)	1404±27[a]	1907±19[b]	1515±17[d]	2068±52[b]	1674±26[b]
Calcium (mg/dL)	9.37±0.06[a]	7.81±0.10[c]	9.04±0.11[d]	6.78±0.13[c]	8.64±0.09[d]
fT3 (pg/dL)	2.52±0.04[a]	2.03±0.01[c]	2.38±0.01[d]	1.71±0.04[b]	2.16±0.02[c]
fT4 (ng/dL)	0.86±0.03[a]	0.67±0.10[b]	0.82±0.03	0.69±0.05[b]	0.70±0.10[b]
TAOC (mmol/L)	4.78±0.07[a]	3.13±0.04[b]	4.55±0.10	2.59±0.06[b]	3.53±0.05[b]
Insulin (pmol/L)	3.54±0.21	2.95±0.46	3.28±0.34	2.75±0.26	2.93±0.41[a]
Se (µg/L)	69.82±1.17[a]	54.75±0.59[b]	63.63±1.03[d]	43.83±1.34[b]	57.58±0.50[b]

BT: before treatment; AT: after treatment; BHBA: β hydroxybutyrate acid; AST: aspartate aminotransferase; ALT: alanine aminotransferase; fT3: free triiodothyronine; fT4: free thyroxine and TAOC: total antioxidant.
The means within the same row with at least one common letter, do not have significant difference (P>0.05).
SE: standard error.

There was a positive correlation between serum glucose levels, fT$_3$ and TAOC levels (P<0.01, r=−0.793; P<0.01, r=−0.807). The findings also showed an increase in fT3 and TAOC levels, in parallel with the increase in serum glucose levels. There was a negative correlation between serum AST activities, fT$_3$ and TAOC levels (P<0.05, r=−0.651; P<0.05, r=−0.697, respectively) and an increase in fT$_3$ and TAOC levels, in parallel with the decrease in serum AST activities. There was a positive correlation between serum Ca values, fT$_3$ and TAOC levels (P<0.01, r=−0.769; P<0.01, r =−0.803, respectively) and an increase in fT$_3$ and TAOC levels, in parallel with the increase in serum Ca values. The findings also showed a positive correlation between serum fT$_3$ and TAOC levels (P<0.01, r=−0.788) and an increase in fT$_3$ levels, in parallel with the increase in TAOC levels. On the third day after treatment, the analysis of parameters from cows with clinical ketosis showed a positive correlation between serum BHB and serum AST (P<0.05, r=0.637) and a negative correlation between serum BHB and serum fT$_3$ (P<0.01, r=−0.798). There was a negative correlation between serum AST activities, serum Ca values, fT3 and TAOC levels (P<0.05, r=−0.587; P<0.01, r=−0.782 and P<0.05, r=−0.629, respectively). A decrease in serum AST activities levels was found, in parallel with the increase in serum Ca values, fT$_3$ and TAOC levels. Although a negative correlation was found between ALT and serum Ca values (P<0.01, r=−0.716), a positive correlation was found between fT$_3$ and plasma TAOC values (P<0.05, r=0.701).

Energy deficiency occurs in dairy cows with high milk yields between the second and sixth weeks of lactation as the animals need to maintain tissue functions and because the required energy is more than the amount of energy obtained from feed (Kennerman, 2004; Saldago Hernández et al. 2009). To obtain the required energy when a negative energy balance occurs, triglycerides are broken down to fat

acids and glycerine by mobilising fat reserves (Kennerman, 2004; Sahoo et al. 2009). Hepatic ketogenesis increases as a result of fatty acid oxidation and gluconeogenesis. BHBA levels increase due to inadequate oxidation of fatty acids to the TCA cycles and an increase in ketone bodies (Kennerman, 2004). An abnormal increase in plasma ketone bodies may lead to clinical and subclinical ketosis (Goldhawk et al. 2009; Zhang et al. 2013). In dairy cows with subclinical ketosis, the required energy amount to produce milk two weeks after calving is approximately 30% higher (Goldhawk et al. 2009). It is important to pay attention to the subclinical form of the disease because lack of appetite, body weight loss and a sudden decrease in milk production are significant issues from an economic perspective (Sahoo et al. 2009). Tests are performed to diagnosis SCK which are evaluate concentrations of BHBA, acetoacetate, and acetone. But among of them the measurement of BHBA concentration level plays importance role for diagnosis of subclinical ketosis (Zhang et al. 2013). In another studies, a diagnosis of subclinical ketosis is reached after measuring plasma glucose, non-esterified fatty acids and the concentration of ketone bodies in blood, milk and urine (Kennerman, 2004; Sahinduran et al. 2010; Goldhawk et al. 2009; Gartner et al. 2009). Findings have shown that serum glucose levels are low and BHBA levels high in cows with subclinical ketosis compared with healthy cows (Zhang et al. 2011), while evaluation of glucose concentration is not a very good indicator for the energy status of dairy cows. In another study of cows with subclinical ketosis, serum glucose concentrations in cows with subclinical ketosis were significantly lower (P<0.01) and BHBA concentrations significantly higher (P<0.01) (Zhang et al. 2013; Xu and Wang, 2008). In the current study, no distinctive symptoms were found. However, according to the laboratory data, serum glucose levels were measured as 35.63 ± 0.73 mg/dL and BHBA levels as 1.25 ± 0.14 mmol/L. Glucose levels were lower than those in healthy cows, whereas BHBA levels were higher. In the light of these findings, the animals were diagnosed with ketosis. In parallel with the decrease in serum glucose levels in the clinical ketosis group, a significant increase in serum BHBA levels was found. In this study, although the glucose levels in the control group were physiological, glucose levels in cows with subclinical and clinical ketosis were lower than the values reported for the healthy cows. An increase in serum glucose levels in both groups of cows with ketosis was found after the third day of glucose administration. In parallel with this increase, serum BHBA levels had decreased. In cows with both subclinical and clinical ketosis, serum glucose levels were lower and serum BHBA levels were higher before treatment. This occurred by ketogenesis due to energy deficit.

In ketosis, free radical damage to living cells has a significant effect on the antioxidant systems of dairy cows (Spears and Weiss, 2008). At the same time, ketone body metabolism is an important source of free oxygen radicals (Çelik and Karagul, 2005). In studies showing the role of antioxidants on erythrocyte oxidative stress, antioxidants have been used as an additional treatment. In this study, too, oxidative stress was decreased in conjunction with conventional treatment. The erythrocyte lipid peroxide levels of cows in the group that received both conventional and antioxidant treatment were significantly lower (Sahoo et al. 2009). In this study, the TAOC levels of both the subclinical and clinical ketosis groups were shown to be lower than those in the control group (P<0.001). On the third day after treatment, although TAOC levels of cows with subclinical ketosis were closer to the control group's TAOC levels (P>0.05), the TAOC levels of cows with clinical ketosis remained low (P<0.01). The lower levels of TAOC in the group with ketosis when compared with the control group supports the data of other researchers (Sahoo et al. 2009; Çelik and Karagul, 2005).

In healthy cows, Ca concentrations decrease several days before parturition as Ca is used to synthesise colostrum. Usually, Ca does not reach normal levels until some days after parturition (Arslan and Tufan, 2010). Some studies have reported low insulin concentrations in cows with hypocalcaemia (Forslund et al. 2010). Hypocalcaemia occurs in cows with subclinical ketosis and has been reported as being more severe in cows with clinical ketosis (Saldago Hernández et al. 2009). Toxaemia occurs due to energy deficit in pregnant sheep; in contrast to increases in BHBA levels, a decrease has been found in their glucose and Ca concentrations (Moghaddam and Hassanpour, 2008). Similar to the serum Ca levels of groups with subclinical and clinical ketosis before treatment were recorded as 7.81 ± 0.10 and 6.78 ± 0.13 mg/dL, respectively. In both groups,

serum Ca levels were lower than in the control group (P<0.01). A positive correlation was found between the decrease in serum Ca levels and serum glucose levels before the treatment of cows with clinical ketosis (P<0.05, r=0.624).

In dairy cows, physiological signs can change rapidly. Metabolic profile analysis shows that AST and ALT activities increase at a greater rate at the time of parturition than before parturition (Avcı and Kızıl, 2013). The reason for the increase is thought to result from cellular damage caused by lipid mobilisation related to the energy deficit that occurs close to the time of parturition (Avcı and Kızıl, 2013; Elitok et al. 2006). In this study, the increase in serum AST, ALT and LDH levels in cows with subclinical and clinical ketosis before treatment in comparison with the control group was statistically significant (P<0.01). This is in line with the increase in serum AST, ALT and LDH activities reported by other researchers (Avcı and Kızıl, 2013; Elitok et al. 2006).

Insulin is a peptide hormone secreted from β cells in the Langerhans islet and it has an anabolic effect (Kennerman, 2004). Insulin deficiency leads to acceleration in lipolysis, an increase in ketogenesis and a decrease in the usage of ketone bodies in muscles (Henze et al. 1998). The endocrine system, especially the pancreas, may play a role in the development of ketosis in ruminants. The decrease in plasma insulin concentrations triggers lipolysis and increases plasma-volatile fat acid concentration (Brockman, 1979). It has been reported that the serum insulin level of cows with ketosis is lower than the serum insulin level of healthy cows (Djoković et al. 2009). In a different study, in order to treat ketosis, the administration of 500 IU insulin as an addition to 30 % dextrose solution resulted in more effective treatment (Saldago Hernández et al. 2009). Similar data were obtained in a study performed on sheep with ketosis (Henze et al. 1998). In this study, the serum insulin levels of cows with clinical ketosis were lower than in the control group. Insulin data in the study match the data from other studies and support the findings of other researchers (Saldago Hernández et al. 2009; Henze et al. 1998; Brockman, 1979).

Thyroid hormones play a crucial role in the growth, development and function of most vertebrate tissues such as brain, bone, adipose tissue, skeletal muscles (Cassar-Malek et al. 2007), and thermoregulation (Kennerman, 2004). Although circulatory T_4 is higher, it is accepted as a prohormon (Kennerman, 2004; Cassar-Malek et al. 2007). T_3 hormone, which has biological activity in peripheral tissues, is formed at about 80% during transformation of T_4 (Brockman, 1979). Circulatory T_3 is formed through type 1 deiodinase enzyme by deionisation of T_4 (Kennerman, 2004; Haces et al. 2008; Cassar-Malek et al. 2007). Type 1 deiodinase enzyme is one of the seleno-enzymes (Cassar-Malek *et al.* 2007), which requires the Se element for activation (Haces *et al.* 2008; Cassar-Malek *et al.* 2007; Forrer *et al.* 1998). In lambs with Se deficiencies, serum T_3 levels are lower than serum T_3 levels in healthy lambs (Haces *et al.* 2008). In oxen with Se deficiency, transformation of T_4 to T_3 has been reported as corrupted (Rowntree *et al.* 2004). In cases of weight gain in humans, an increase in serum T_3 can be seen. This is lowered where weight decreases (Rosenbaum *et al.* 2000). In another study, low circulatory thyroid hormone levels have been reported in cows with energy deficiency (Blum *et al.* 1983). In this study, the serum T_3 and T_4 levels of cows with subclinical and clinical ketosis were lower than in the control group (P<0.001). The transformation level of T_4 to T_3 before treatment was particularly low in cows with clinical ketosis, where T_3 and T_4 values were higher than the values before treatment. The T_3:T_4 ratio increased in cows with clinical ketosis. The reason for the increase in T_3:T_4 ratio levels results from increased Se serum levels in cows after treatment. Further thorough investigation should be carried out to confirm these results (Kennerman, 2004; Haces *et al.* 2008; Cassar-Malek *et al.* 2007; Contreras *et al.* 2002; Forrer *et al.* 1998; Rowntree *et al.* 2004; Rosenbaum *et al.* 2000; Blum *et al.* 1983).

Several antioxidants and trace minerals play important roles in the immune function. These antioxidants affect the health of pregnant dairy cows in the peripartum period (Spears and Weiss, 2008). Se participates in the GSH-Px enzyme structure, which metabolises the hydrogen peroxide and lipoperoxydes formed during normal cell metabolism. It also protects the cell from the harmful effects of free radicals (Haces *et al.* 2008; Spears and Weiss, 2008). Se is an essential trace element for antioxidant and thyroid hormone processes (Rowntree *et al.* 2004). In order to determine serum Se concentrations in cattle, a number of studies have been carried out (Gerloff, 1992; Rowntree *et al.* 2004). These have found 80-300 µg/L to be normal, 30-70 µg/L to be critical l and 2-25 µg/L to be deficient (Blum *et al.* 1983). Another study suggests that 40 µg/L of serum Se level indicates deficiency, that 40-70 µg/L is critical and that concentrations higher than 70 µg/L are adequate (Gerloff, 1992). In our study, serum Se levels in healthy cows were measured as 69.82 ± 1.17 µg/L. Before treatment, serum Se levels in cows with subclinical ketosis were 54.75 ± 0.59 µg/L. In cows with clinical ketosis, the levels were 43.83 µg/L. On the third day after treatment, serum Se levels in cows with subclinical ketosis were 63.63 ± 1.03 µg/L.

For cows with clinical ketosis, this was 57.58 ± 0.50 µg/L. Serum Se levels in groups with both clinical and subclinical ketosis were lower than the values found in healthy

cattle. It can be concluded, therefore, that ketosis leads to Se deficiency in dairy cows, according to these data. In the same time, the Se levels of cows with subclinical and clinical ketosis were lower than in the control group (P<0.001).

BHBA levels increased in parallel with the low Se levels of the group with ketosis. A negative correlation was found between Se and BHBA (P<0.001 r=0.919). There was a positive correlation (P<0.05, r=0.780) with serum Se concentration in cows with subclinical and clinical ketosis before treatment. Low levels of serum fT_3 were found as a result of low Se concentration. This situation suggests an increase in oxidative stress and a decrease in antioxidants in ketosis.

CONCLUSION

Serum insulin, thyroid hormone and Ca levels in subclinical and clinical ketosis cow were lower than in the control group on the third day after the treatment. The evaluation of insulin, thyroid hormones, Se and Ca parameters in addition to the current routine parameters will help reduce yield losses caused by ketosis. As these parameters are directly associated with yields, regulating these parameters will help prevent losses. Examining Se, insulin, thyroid hormone and Ca levels together with total antioxidant levels will help prevent yield losses and increase the chance of treatment and prognosis. In addition, we think that this study will provide an insight for future research into ketosis.

ACKNOWLEDGEMENT

This research was carried out in the Department of Internal Medicine at Faculty of Veterinary Medicine at Yuzuncu Yil University. All authors interpreted the data and revised the manuscript and approved the final version. The research was supported by the Yuzuncu Yil University Scientific Research Projects Directorate as 2012-VF-B028 project.

REFERENCES

Arslan C. and Tufan T. (2010). Feeding the transition dairy cow I. Physiologic, hormonal, metabolic and immunogical changes and nutrient requirement of dairy cow during this period. *Kafkas Univ. Vet. Fak. Derg.* **16**, 151-158.

Avcıl C. and Kızıl O. (2013). The effects of injectable trace elements on metabolic parameters in transition cow. *Kafkas Univ. Vet. Fak. Derg.* **19**, 73-78.

Blum J.W., Kunz P., Leuenberger H., Gautschi K. and Keller M. (1983). Thyroid hormones, blood plasma metabolites and haematological parameters in relationship to milk yield in dairy cows. *Anim. Sci.* **36**, 93-104.

Brockman R.P. (1979). Roles for insulin and glucagon in the development of ruminant ketosis: a review. *Canadian Vet. J.* **20**, 121-126.

Cassar-Malek I., Picard B., Kahl S. and Hocquette J.F. (2007). Relationships between thyroid status, tissue oxidative metabolism, and muscle differentiation in bovine foetuses. *Domest. Anim. Endocrinol.* **33**, 91-106.

Çelik S. and Karagul H. (2005). The red blood cell membrane proteins in rabbits with experimental ketosis. *Turkian J. Vet. Anim. Sci.* **29**, 151-155.

Contreras P.A., Matamoros R., Monroy R., Kruze J., Leyan V., Andaur M. and Wittwer F. (2002). Effect of a selenium-deficient diet on blood values of T_3 and T_4 in cows. *Comp. Clin. Pathol.* **11(2)**, 65-70.

Djoković R., Šamanc H., Ilić Z. and Kurćubić V. (2009). Blood glucose, insulin and inorganic phosphorus in healthy and ketotic dairy cows after intravenous infusion of glucose solution. *Acta. Vet. Brno.* **78**, 449-453.

Elitok B., Kabu M. and Elitok Ö.M. (2006). Evaluation of liver function test in cows during periparturient period. *Fırat Univ. Vet. J. Health. Sci.* **20**, 205-209.

Forrer R., Gautschi K., Stroh A. and Lutz H. (1998). Direct determination of selenium and other trace elements in serum samples by ICP-MS. *J. Trace Elem. Med. Biol.* **12**, 240-247.

Forslund K.B., Ljungvall Ö.A. and Jones B.B. (2010). Low cortisol levels in blood from dairy cows with ketosis: a field study. *Acta. Vet. Scandinavica.* **52**, 1-6.

Gartner V., Potočnik K. and Jovanovac S. (2009). Test-day records as a tool for subclinical ketosis detection. *Acta. Vet.* **59**, 185-191.

Gerloff B.J. (1992). Effects of selenium supplementation on dairy cattle. *J. Anim. Sci.* **70**, 3934-3940.

Goldhawk C., Chapinal N., Veira D.M., Weary D.M. and Von Keyserlingk M.A.G. (2009). Prepartum feedding behavior is an early indicator of subclinical ketosis. *J. Dairy Sci.* **92**, 4971-4977.

Haces M.L., Hernández-Fonseca K., Medina-Campos O.N., Montiel T., Pedraza-Chaverri J. and Massieu L. (2008). Antioxidant capacity contributes to protection of ketone bodies against oxidative damage induced during hypoglycemic conditions. *Exp. Neurol.* **211(1)**, 85-96.

Henze P., Bickhard K., Fuhrmann H. and Sallmann H.P. (1998). Spontaneous pregnancy toxaemia (ketosis) in sheep and the role of insulin. *J. Vet. Med. A.* **45**, 255-266.

Katoh N. (2002). Relevance apolipoproteins in the development of fatty liver and fatty liver-related peripartum disease in dairy cows. *J. Vet. Med. Sci.* **64**, 293-307.

Kennerman E. (2004). Serum insulin, triidothyronine (T_3), and thyroxine (T_4) levels in cows with ketosis. *J. Vet. Surg.* **10**, 34-37.

Kozat S. (2007). Serum T_3 and T_4 concentrations in lambs with nutritional myodegeneration. *J. Vet. Int. Med.* **21**, 1135-1137.

Moghaddam J.G. and Hassanpour A. (2008). Comparison of blood serum glucose, beta hydroxybutyric acid, blood urea nitrogen and calcium concentrations in pregnant and lambed ewes. *J. Anim. Vet. Adv.* **7**, 308-311.

Mullur R., Liu Y.Y. and Brent G.A. (2014). Thyroid hormone regulation of metabolism. *Physiol. Rev.* **94(2)**, 355-382.

Pavlata L., Illek J., Pechova A. and Matějíček M. (2002). Selenium status of cattle in the Czech Republic. *Acta. Vet. Brno.* **71(1)**, 3-8.

Rosenbaum M., Hirsch J., Murphy E. and Leibel R.L. (2000). Effects of changes in body weight on carbonhydrate metabolism, catecholamine excretion, and thyroid function. *Am. J. Clin. Nutr.* **71**, 1421-1432.

Rowntree J.E., Hill G.M., Hawkins D.R., Link J.E., Rincker M.J., Bednar G.W. and Kreft R.A. (2004). Effect of Se on selenoprotein activity and thyroid hormone metabolism in beef and dairy cows and calves. *J. Anim. Sci.* **82(10)**, 2995-3005.

Sahinduran S., Sezer K., Buyukoğlu T., Albay M.K. and Karakurum M.C. (2010). Evaluation some haematological and biochemical parameters before and after treatment in cows with ketosis and comparison of different treatment methods. *J. Anim. Vet. Adv.* **9**, 266-271.

Sahoo S.S., Patra R.C., Behera P.C. and Swarup D. (2009). Oxidative stress indices in the erythrocytes from lactating cows after treatment for subclinical ketosis with antioxidant incorporated in the therapeutic regime. *Vet. Res. Commun.* **33**, 281-290.

Sakha M., Ameri M., Sharifi H. and Taheri I. (2007). Bovine subclinical ketosis in dairy herds in Iran. *Vet. Res. Commun.* **31(6)**, 673-679.

Saldago Hernández E.G., Bouda J., Avila García J. and Navarro Hernández J.A. (2009). Effect of postpartum administration of calcium salts and glucose precursors on serum calcium and ketone bodies in dairy cows. *Vet. Méx.* **40**, 17-26.

Spears J.W. and Weiss W.P. (2008). Role of antioxidants and trace elements in health and immunity of transition dairy cows. *Vet. J.* **176**, 70-76.

SPSS Inc. (2011). Statistical Package for Social Sciences Study. SPSS for Windows, Version 20. Chicago SPSS Inc.

Stranges S., Navas-Acien A., Rayman M.P. and Guallar E. (2010). Selenium status and cardiometabolic health: state of the evidence. *Nutr. Metab. Cardiovasc.* **20**, 754-760.

Voyvoda H. and Erdogan H. (2010). Use of a hand-held meter for detecting subclinical ketosis in dairy cows. *Res. Vet. Sci.* **9(3)**, 344-351.

Wang Y.M., Wang J.H., Wang C., Wang J.K., Chen B., Liu J.X. and Guo F.C. (2010). Effect of dietary antioxidant and energy density on performance and anti-oxidative status of transition cows. *Asian-Australasian J. Anim. Sci.* **23(10)**, 1299-1307.

Xu C. and Wang Z. (2008). Comparative proteomic analysis of livers from ketotic cows. *Vet. Res. Commun.* **32**, 263-273.

Zhang Z.G., Li X.B., Gao L., Li Y.F., Liu G.W., Wang H.B. and Wang Z. (2011). Serum antioxidant capacity of dairy cows with subclinical ketosis. *Vet. Record.* **168(1)**, 22-27.

Zhang Z.G., Xue J.D., Gao R.F., Liu J.Y., Wang J.G., Yao C.Y. and Wang Z. (2013). Evaluation of the difference of L-selectin, tumor necrosis factor-α and sialic acid concentration in dairy cows with subclinical ketosis and without subclinical ketosis. *Pakistan Vet. J.* **33(2)**, 225-228.

Effect of Different Levels of Milk Thistle (*Silybum Marianum*) in Diets Containing Cereal Grains with Different Ruminal Degradation Rate on Rumen Bacteria of Khuzestan Buffalo

Z. Nikzad[1], M. Chaji[1*], K. Mirzadeh[1], T. Mohammadabadi[1] and M. Sari[1]

[1] Department of Animal Science, Ramin Agricultural and Natural Resources University, Mollasani, Ahvaz, Iran

*Correspondence E-mail: chaji@ramin.ac.ir

ABSTRACT

The objective of this study was to investigate the effects of diets containing different levels of milk thistle (0, 100 and 200 g/kg dry matter) and grains (barley and corn) with different rumen degradation rates on rumen bacteria and whole rumen microorganisms (WRM) of Khuzestan buffalo. The gas production (GP) technique, two steps digestion and specific bacteria culture medium methods were used for this purpose. The rumen fluid was taken from two fistulated buffaloes. The results of GP potential of experimental diets by WRM were not significantly different, however in both basal diets GP increased by increasing the level of milk thistle (P>0.05). The highest amount of GP in diet based on barley and corn was for diets containing 200 and 100 g milk thistle, respectively. Rate of GP by WRM was significantly different (P<0.05), so that in both basal diets at the level of 100 g milk thistle had the highest GP rates. Potential and rate of GP of diets by buffalo rumen bacteria was not significantly different (P>0.05). In two-steps digestion method using of different levels of milk thistle in diets (based on barley and corn) had no negative effect on the digestibility of nutrients by WRM (P>0.05). In barley-based diet adding milk thistle was numerically increased dry matter and neutral detergent fiber (NDF) digestibility compared with the control (P>0.05), while in the corn-based diet dry matter and NDF digestibility were reduced (P>0.05). The digestibility of dry matter by bacteria in corn-based diet was significantly reduced compared with the control (P<0.05). Nutrient digestibility of experimental diets by bacteria in the specific bacteria culture did not be affected by milk thistle in diets (P>0.05). Therefore, results provided here suggest that milk thistle could be used up to 20% of buffalo's diet without any negative effect on digestion and fermentation characteristics by WRM and bacteria.

KEY WORDS bacteria, barley grain, corn grain, whole rumen microorganisms.

INTRODUCTION

Milk thistle (*Silybum marianum*) or Maritighal is an annual grass with medicinal characteristics and is grown in different parts of Iran. The medicinal properties of milk thistle are due to presence of silymarin in its seed which is primarily consists of an isomeric mixture of 6 phenolic compounds, silydianin, silychristin, diastereoisomers of silybin (silybin A and B), and diastereoisomers of isosilybin (isosilybin A and B) (Lee *et al.* 2007). Silymarin has been used to treat liver disorders such as unwell of alcoholic and various hepatitis (Carmen Tamayo, 2007; Gazak *et al.* 2007). It also inhibits chemically induced carcinogenesis and shows direct anti-carcinogenic activity against several human carcinoma cells. The silymarin has antidiabetic, hypolipidaemic, anti-inflammatory, cardioprotective, neurotrophic and neuroprotective effects (Kren and Walterova, 2005; Abascal *et al.* 2003). The seed of this plant also contains betaine, tri-methyl glycine and a large amount of oil that play an important role in anti-inflammatory and anti-hepatitis effects of the extract (Hadolin *et al.* 2001).

In was reported that by using milk thistle no adverse effects have been found but nitrate poisoning maybe occurred (Macadam, 1966). Nitrate is not always toxic to animals but it is converted to nitrite and then to ammonia, by rumen microorganisms when animals consume feed containing nitrate. When high levels of nitrites accumulate in the gastrointestinal tract, they are absorbed into the bloodstream and it changes haemoglobin to methaemoglobin (Robson, 2007). If enough energy (such as grain) is available in the rumen, nitrite easily converted into ammonia, and ammonia ultimately can be used for microbial protein synthesis. Since the converting of nitrate to nitrite and ammonia is an endergonic process, therefore the degradation rate of carbohydrate source may influence on ruminants which consumed the plants containing nitrates, such as milk thistle (Kamra, 2005).

Rumen microorganisms are comprised predominantly of cellulolytic bacterial populations, anaerobic fungi and protozoa. In recent years, it shown the rumen cellulolytic bacteria have a primary role in fiber digestion and so animal performance (Denmen and McSweeney, 2006).

Tannins are phenolic compounds, which are able to precipitate proteins, alkaloids and gelatin. High concentrations of the tannins in fodder plants inhibit gastrointestinal bacteria and reduce ruminant performance for usage the nitrogen and protein sources (Smith *et al.* 2005).

In an early experiments on nutritional value of milk thistle in sheep, both positive and negative effects on fiber and other nutrients digestibility were observed. Also, in a gas production (GP) experiment, milk thistle showed extremely negative effect on the digestion and fermentation activities of microorganisms (Mojadam, 2012). It is likely that some of these results may be due to the effects of active ingredients of *Silybum marianum* on rumen microorganisms, including bacteria. In Khuzestan province, milk thistle is being consumed by animals such as sheep, camels and cows, but there is no information about its effects on rumen microorganisms. Therefore, the aim of the present experiment was to investigate the effects of diets containing *Silybum marianum* and grains with different degradation rates on rumen bacteria of Khuzestan buffalos.

MATERIALS AND METHODS

This experiment was carried out at Ramin Agricultural and Natural Resources University of Khuzestan. Whole milk thistle plant (leaves, stems and flowers) was collected from Mollasani. After drying and powdering, its dried powder was added to the basal diet containing corn or barley (fast fermentation rate than corn) grains in 3 levels of 0, 100 and 200 g/kg DM as top dress.

Experimental animal

Rumen fluid was taken before morning feeding from 2 fistulated water buffalo (live weight around 450 kg). Buffalo diets were prepared based on the standard requirements (NRC, 2001) and included 50% forage (30% alfalfa hay and 20% wheat straw) and 50% concentrate (36.9% barley or corn, 12% wheat bran, 0.03% fish meal [for corn base diet only], 0.3% salt, 0.7% calcium carbonate, 0.1% mineral and vitamin supplement). The diets were offered to the animals about 6 kg per day on DM base. The effect of milk thistle on bacteria and whole rumen microorganisms (WRM) of the buffalo was studied using the following techniques.

Gas production by WRM of buffalo

Gas production of the experimental treatments were measured in 100 mL glass syringes containing 300 mg of ground sample, 20 mL of artificial saliva and 10 mL of rumen liquid (Menke and Stingass, 1988). Artificial saliva was supplied by McDougall method (McDougall, 1948). Rumen fluid was collected through the rumen fistula of 2 buffalo that were fed at maintenance level for 6 weeks before the start of experiment. Collected rumen fluid was strained through 4 layers of cheese cloth and mixed with artificial saliva.

Gas production by rumen bacteria of buffalo

The method used for determining GP of rumen bacteria was the same that used for WRM of rumen, but for isolation of rumen bacteria, after collection and strain of liquor, rumen fluid was centrifuged (1000 g, 10 min) for removing protozoa. Then bacteria were isolated from non-protozoa strained rumen fluid using antifungal agent, benomyle (500 mg/L of medium culture; Sigma-Aldrich Co., Taufkirchen, Germany) and metalaxyle (10 mg/L; Sigma-Aldrich Co., Taufkirchen, Germany) (Mohammadabadi *et al.* 2012). The GP on 39 degrees of centigrade was measured at 2, 4, 6, 8, 10, 12, 16, 24, 48, 72 and 96 h. Cumulative GP data were analyzed with the exponential equation (Orskov and McDonald, 1979).

Two steps digestion

Rumen fluid was obtained from buffalos and purified by the methods described in the previous section (Mohammadabadi *et al.* 2012). Then, *in vitro* digestibility of experimental treatments by whole microorganisms and buffalo rumen bacteria were measured in the 100 mL test tubes containing 0.5 g sample, 40 mL artificial saliva (McDougall, 1948) and 10 mL of rumen fluid (1/4 ratio). Digestibility of dry matter and neutral detergent fiber (NDF) was calculated based on the differences of raw material and material remaining at end of incubation (Tilly and Terry, 1963). Neutral detergent fiber was measured by Van Soest et al. (1991) method.

Specific rumen bacteria culture medium

The experimental treatments were cultured in specific culture medium of rumen bacteria (Galdwell and Bryant, 1966). The culturing glasses, containing 1 g sample, were sterile by autoclave for 15 min at 120 °C, and then, 25 mL of a mixture of centrifuged rumen fluid of buffalos with cellobiose, sodium sulfide, cysteine-HCL, sodium carbonate (Merck Co., Darmstadt, Germany), fungicides (Benomyl and metalaxyl; Sigma-Aldrich Co., Taufkirchen, Germany), trypticase and yeast extract (Merck Co., Darmstadt, Germany) was added to them. After that, 3 mL from sucrose solution was added to each glass. In order to create anaerobic conditions, carbon dioxide injected to test tubes. Then, 4 mL of rumen fluid was added to each bottle. The culturing glasses were incubated at a temperature of 39 °C for 96 h. Six replicates for each treatment were considered, after drying and weighing, the disappearance of DM and NDF samples of bacteria were calculated.

Statistical analysis

Data was analysis by split-plot design (main plot was basal diet containing barley or corn and subplot diets include different value of milk thistle), using the General Linear Model procedure of SAS (2004) Duncan's multiple range tests was used to compare the significant means (Mohammadabadi et al. 2012).

RESULTS AND DISCUSSION

Gas production by WRM

The potential of GP of experimental diets (Table 1) did not show significant differences (P>0.05). Numerically, in diets based on barley and corn grain, diet containing 200 and 100 g milk thistle had the highest potential of GP, respectively. Compared with the control, GP rate of experimental diets was significantly different (P<0.05). In both basal diet, the highest rate of GP was for diet containing 100 g milk thistle. In total, the highest GP rate was belonged to barley diet containing 100 g milk thistle.

Gas production by rumen bacteria

The trends of results related to potential of GP of experimental diets by rumen bacteria were the same as WRM (Table 1).

The potential and rate of GP of experimental diets by rumen bacteria of buffalo was not significantly different (Table 1). In diets based on barley and corn, the highest potential of GP was for diets containing 100 and 200 g milk thistle, respectively. Therefore, adding milk thistle to either barley or corn based diets, had no significant effect on GP by rumen bacteria and WRM (P>0.05). However, GP increased numerically as the amount of milk thistle was increased in the diets.

Regardless the level of milk thistle (Table 2), GP potential by WRM for the corn-based diet was higher than barley. Rate of GP by WRM in barley-based diets was significantly higher than diets containing corn (P<0.05). The basal diets had no significant effects on the potential and the rate of GP by rumen bacteria (P>0.05).

Regardless the type of basal diet (Table 3), potential of GP by WRM for the different levels of milk thistle was similar (P>0.05), but GP rate was significantly different (P<0.05). Adding milk thistle to the diets significantly increased GP rate, and the 100 g milk thistle had the highest GP rate (P<0.05), which was not significant compared with the diet containing 200 g milk thistle.

The potential and GP rate of buffalo rumen bacteria for different levels of milk thistle were not significant (P>0.05). Compared with the control diet, 100 g milk thistle increased the potential of GP (P>0.05).

Digestibility of experimental diets
WRM

The milk thistle inclusion in diets (based on barley and corn) had no significant effect on nutrients digestibility (Table 4) (P>0.05). In barley-based diets, increasing milk thistle increased numerically DM digestibility compared with the control group, but had no significant effect on NDF digestibility (P>0.05). In diets based on corn grain, , adding milk thistle to diet caused to slightly decrease in DM and NDF digestibilities compared with the control diet. Diet without milk thistle showed the highest DM and NDF digestibilities (P>0.05). Within corn-based diets (Table 4), the control had the highest DM and NDF digestibilities (P>0.05).

Bacteria

DM digestibility (Table 4) was higher in barley-based diets with milk thistle compared to the control, but there was no significant difference between 2 levels of milk thistle (P>0.05). In diet based on corn grain, milk thistle decreased DM digestibility significantly compared to the control diet (P<0.05), and there was significant difference between 2 levels of milk thistle. Increasing the amounts of milk thistle in both diets, had no significant impact on NDF digestibility by bacteria (P>0.05).

Table 1 Effect of milk thistle on gas production parameters of the experimental diets by whole rumen microorganisms (WRM) and bacteria of buffalos

Basal diet	Milk thistle in diet, (g/kg DM)	WRM		Bacteria	
		b, (mL)	c, (mL/h)	b, (mL)	c, (mL/h)
Barley	0	70.53	0.039[ab]	39.99	0.041
	100	70.75	0.047[a]	38.17	0.048
	200	80.31	0.042[ab]	43.01	0.045
Corn	0	79.40	0.034[b]	44.58	0.046
	100	81.32	0.040[ab]	49.58	0.037
	200	76.21	0.037[b]	37.23	0.041
SEM		2.99	0.0024	3.010	0.0038
P-value		0.084	0.0377	0.104	0.387

b: potential of gas production and c: gas production rate.
The means within the same row with at least one common letter, do not have significant difference (P>0.05).
SEM: standard error of the means.

Table 2 Effect of the type of basal diet on gas production parameters by whole rumen microorganisms (WRM) and bacteria of buffalos (regardless the level of milk thistle)

Basal diet	WRM		Bacteria	
	b (mL)	c (mL/h)	b (mL)	c (mL/h)
Barley	73.89[b]	0.043[a]	40.39	0.045
Corn	79.25[a]	0.037[b]	43.80	0.041
SEM	1.727	0.0014	1.738	0.0022
P-value	0.049	0.0105	0.19	0.298

b: potential of gas production and c: gas production rate.
The means within the same row with at least one common letter, do not have significant difference (P>0.05).
SEM: standard error of the means.

Table 3 Effect of milk thistle on gas production parameters by whole rumen microorganisms (WRM) and bacteria of buffalos (regardless of the basal diet)

Level of milk thistle in diet, g/kg (DM)	WRM		Bacteria	
	b (mL)	c (mL/h)	b (mL)	c (mL/h)
0	75.03	0.037[b]	42.29	0.0433
100	76.09	0.043[a]	43.87	0.0426
200	78.60	0.039[ab]	40.12	0.0430
SEM	2.11	0.0017	2.13	0.0027
P-value	0.49	0.0422	0.48	0.983

b: potential of gas production and c: gas production rate.
The means within the same row with at least one common letter, do not have significant difference (P>0.05).
SEM: standard error of the means.

Table 4 Nutrients digestibility of the experimental diets by whole rumen microorganisms (WRM) and bacteria of buffalos

Basal diet	Level of milk thistle in diet, g/kg DM	WRM		Bacteria	
		Dry mater (%)	NDF (%)	Dry mater (%)	NDF (%)
Barley	0	67.51	79.10	66.72[b]	78.27
	100	73.84	79.15	69.31[b]	78.55
	200	71.80	78.74	70.42[b]	82.09
Corn	0	78.79	84.71	81.09[a]	84.09
	100	72.82	78.28	62.09[b]	75.47
	200	75.13	76.03	64.98[b]	78.69
SEM		3.50	3.21	2.66	1.98
P-value		0.43	0.28	0.024	0.16

NDF: neutral detergent fiber.
The means within the same row with at least one common letter, do not have significant difference (P>0.05).
SEM: standard error of the means.

In diet based on barley grain, diet containing 200 g milk thistle had the highest NDF digestibility; while in corn-based diets, use of milk thistle decreased numerically NDF digestibility (P>0.05). Regardless the levels of milk thistle in the diets (Table 5), the differences between nutrients digestibilities of barley- or corn- based diets by WRM and bacteria were not significant (P>0.05).

The DM and NDF digestibilities (Table 6) of the diets were also not affected by the levels of milk thistle (P>0.05). However, DM (P<0.05) and NDF digestibility by bacteria decreased significantly (P>0.05) with increasing the level of milk thistle in the diets (Table 6).

Table 5 Nutrients digestibility of the basal diets by whole rumen microorganisms (WRM) and bacteria of buffalos (regardless the level of milk thistle) using two-step digestion method

Basal diet	WRM		Bacteria	
	Dry mater (%)	NDF (%)	Dry mater (%)	NDF (%)
Barley	71.05	78.99	68.82	79.63
Corn	75.58	79.67	69.39	79.42
SEM	2.02	1.86	1.53	1.14
P-value	0.16	0.80	0.80	0.89

NDF: neutral detergent fiber.
The means within the same row with at least one common letter, do not have significant difference (P>0.05).
SEM: standard error of the means.

Table 6 Effect of milk thistle on nutrients digestibility by whole rumen microorganisms (WRM) and bacteria of buffalos (regardless of the basal diet)

Level of milk thistle in diet, g/kg DM	WRM		Bacteria	
	Dry mater (%)	NDF (%)	Dry mater (%)	NDF (%)
0	73.15	81.90	73.90[a]	81.18
100	73.33	78.71	65.70[b]	77.00
200	73.46	77.38	67.70[ab]	80.39
SEM	2.48	2.27	1.88	1.40
P-value	0.996	0.41	0.049	0.160

NDF: neutral detergent fiber.
The means within the same row with at least one common letter, do not have significant difference (P>0.05).
SEM: standard error of the means.

Nutrients digestibilities by specific bacteria culture method

In Table 7 the effect of milk thistle on digestibility of DM and NDF was not significant in 2 basal diets (P>0.05). Regardless the levels of milk thistle (Table 8) there was no significant difference between basal diets in digestibility of DM and NDF (P>0.05). As Table 9 demonstrated, regardless the type of basal diet, differences in nutrient digestibility of various levels of milk thistle was not significant (P>0.05). The DM and NDF digestibility decreased with increasing milk thistle (P>0.05), such that the zero level of milk thistle shown the highest digestibility (P>0.05).

GP by WRM and bacteria

Numeral increasing of GP (20 and 10% for barley- and corn- based diets, respectively) may be due to sugar compounds existing in milk thistle. Compatible with the results of the present experiment, Dong et al. (2010) in their study on goats, despite of a reduction in fermentative and microbial activities for high levels of phytogenic products, did not found any significant effect for phytogenic products on in vitro rumen fermentation and methane emission. Broudiscou et al. (2000) screened action of 13 plant extracts with high flavonoid content on fermentation and rumen microbes. Their result indicated that GP was increased by L. officinalis and S. virgaurea (from family of milk thistle). However, opposite to the results of the present study; a significant decrease of methane emission was reported by Sun et al. (2011) in evaluating the effects of Cichorium intybus on GP in sheep. In the present study may be the numerically reduced of GP (levels 20% in corn diet) was related with effects of tannin in milk thistle, tannins readily

complex with carbohydrates and proteins (Hagerman et al. 1992).

Condensed tannin extracted from quebracho trees reduced methane emissions from cattle (Beauchemin et al. 2007) which is compatible with our results. They observed that feeding up to 2% of the dietary DM quebracho tannin extract failed to reduce enteric methane emissions from growing cattle.

Although, Carulla et al. (2005) reported that feeding Acacia mearnsii extract (2.5% of dry matter intake (DMI)) to sheep (as source of tannin) decreased methane per kilogram of DM intake by approximately 12%. In general, the addition of milk thistle to the diets in the current study did not have any negative effect on GP by WRM and bacteria of buffalo, perhaps due to the rapid adaptation of rumen microorganisms of buffalo to the tannin content of in milk thistle.

Maldar et al. (2010) reported, which habit of goats Alamut to oak leaf (rich source of tannin), reduce the negative impact of tannins and change conditions of rumen fermentation and digestibility of the byproduct was increased. Increasing the proportion of tannin-resistant bacteria in the rumen by adaptation protects ruminants from antinutritional effects (Smith et al. 2005). The higher GP potential in diets based on corn compared with the barley (Table 2) is probably due to the high fiber content in barley which is mostly hemicellulos; but non fiber carbohydrates of barley and corn were 53.9% and 60.6%, respectively (Parnian Khaje Dizaj, 2011). On the other hand, in feed sources with lower NDF potential of GP are high and increasing proportion of the lignin in cell wall reduced GP, and lead to reduction of digestibility (Sommart et al. 2000).

Table 7 Nutrients digestibility of the experimental diets by rumen bacteria of buffalos after 96 h of incubation

Basal diet	Level of milk thistle in diet, g/kg DM	Nutrients	
		Dry mater (%)	NDF (%)
Barley	0	42.60	22.97
	100	40.51	21.47
	200	40.65	18.04
Corn	0	43.58	22.60
	100	43.04	18.47
	200	37.79	78.69
SEM		1.78	1.98
P-value		0.32	0.16

NDF: neutral detergent fiber.
The means within the same row with at least one common letter, do not have significant difference (P>0.05).
SEM: standard error of the means.

Table 8 Nutrients digestibility of the experimental diet by rumen bacteria of buffalos (regardless the level of milk thistle) after 96 h of incubation

Basal diet	Nutrients	
	Dry mater (%)	NDF (%)
Barley	41.26	20.83
Corn	41.47	20.78
SEM	1.029	1.007
P-value	0.80	0.97

NDF: neutral detergent fiber.
The means within the same row with at least one common letter, do not have significant difference (P>0.05).
SEM: standard error of the means.

Table 9 Effect of milk thistle on nutrients digestibility of experimental diets by rumen bacteria of buffalos after 96 h incubation (regardless of the basal diet)

Level of milk thistle in diet, g/kg DM	Nutrients	
	Dry mater (%)	NDF (%)
0	43.09	22.79
100	41.78	19.94
200	39.22	19.69
SEM	1.26	1.23
P-value	0.17	0.22

NDF: neutral detergent fiber.
The means within the same row with at least one common letter, do not have significant difference (P>0.05).
SEM: standard error of the means.

The higher GP rate of diets based on barley compared to diets based on corn perhaps was due to the difference between corn and barley starch degradation rate. The *in situ* experiments (Sadeghi and Shawrang, 2006; Sadeghi and Shawrang, 2008) shown that barley, unlike corn starch have a high effective degradability and its rumen degradability was faster and results in higher rates of GP from barley than corn grain (Herrera-Saldana *et al.* 1990; Nocek and Tamminga, 1991). Also, Khorasani *et al.* (2001) were examined the effects of replacing barley grain with corn on ruminal fermentation characteristics of Holstein dairy cows, and their results showed that barley starch is digested faster than corn starch. However, Garcia *et al.* (2000) reported that apparent digestibility of starch in the rumen of Holstein heifers was not different for rations with barley and corn. Perhaps one reason for slight decrease in GP by bacteria (level of 200 g milk thistle) was declines the population due to the presence of tannins in milk thistle, Zargari (1996) reported that various parts of milk thistle plant have tannin;

and condensed tannins reduce methanogens variation (Tan *et al.* 2001). Feeding tannin-containing forages to ruminants may reduce methane emissions (Pinares-Patino *et al.* 2003).

Ruminal bacteria play a particularly important role in the biological degradation of dietary fiber because of their much larger biomass (Koike *et al.* 2003). Tannins are most effective against the fiber degrading bacteria. The reports have indicated that in animals fed tannin rich *Calliandra calothyrsus*, the population of *Ruminococcus* spp. and *Fibrobacter* spp. was reduced considerably, but fungi, protozoa and proteolytic bacteria were less affected by this diet (McSweeny *et al.* 2000). Sotohy *et al.* (1997) reported that the number of total bacteria in the rumen of goats decreased significantly when the animals were fed tannin-rich plant (*Acacia nilotica*), and this decrease had direct relation with the level of this feed in the diet. Disagree with above researches Brooker *et al.* (1994) isolated facultative anaerobe gram-positive bacteria (cellulolytic strains) from the rumen liquor of feral goats browsing on *Acacia* spp. leaves, which they could grow in a medium containing 2.5% tannic acid or condensed tannins. Also Odenyo and Osuji (1998) isolated three strains of tannin-tolerant bacteria from sheep, goat and an antelope and observed that the isolates could tolerate tannins up to 8 g/L of the medium. Therefore, in present experiment, adding milk thistle to diet had no negative effect on bacteria and WRM.

Nutrient digestibility of experimental diets
WRM

Within diets based on corn (Table 4), control diet had the highest DM and NDF digestibility. The numerical reduce of DM and NDF digestibility probably was because of unsaturated fatty acids found in milk thistle and corn seed. Milk thistle contains 28.9% oil and unsaturated fatty acid and linoleic acid is the predominant unsaturated fatty acid of its oil (Malekzadeh et al. 2011). As Palmquist and Jenkins (1980) stated, feeding unprotected fats, especially unsaturated, resulted to lower ruminal fiber degradability. Agree with the results of present experiment, Mojadam (2012) reported that DM and NDF digestibility of soybean meal and wheat straw, which were incubated by rumen fluid of sheep that fed with the diets containing milk thistle (barely and corn in basal diets) not affected by the treatments. Also Szumacher-Strabel et al. (2009) studied the effect of oils rich in linoleic acid (like oil of milk thistle) on in vitro rumen fermentation parameters of sheep, goats and dairy cows, there, unsaturated fatty acid does not disrupt ruminal fermentation; but Harvatine and Allen (2006) represented that using of fat sources in diet was effective on ruminal digestion.

Totally, adding milk thistle (as source of unsaturated fatty acid) to diet, particularly in higher levels and associated with diet contains corn, caused more decrease of nutrient digestibility; which one reason for that maybe was decrease digestibility of structural carbohydrates (White et al. 1992; Zinn, 1989; Zinn and Shen, 1996). Therefore, since in present experiment reduction of digestibility of DM and NDF was not significant, it can be said that the negative effects of tannin, unsaturated fatty acids and other compounds of milk thistle on digestibility were negligible.

Bacteria

Decline rumen digestibility of nutrients by bacterial in diets based on corn (Table 4) probably was due to the additive effect of unsaturated fatty acids of milk thistle and corn grain. Cellulolytic bacteria constitute 5% to 7% of the total bacterial population of rumen (Denmen and McSweeney, 2006). In beef cattle diets when fat added at the end of the finishing period, organic matter and fiber digestibility decreased.

This reduction was due to negative effect of fat on fiber digestibility in the rumen. The fats have directly inhibitory effects on fibrolytic bacteria and protozoa, also because of the physical coverage of the fibers and thus reduce rumen fermentation by microorganisms and reduce the absorption of cations, had a negative effect on digestibility (Lundy et al. 2004). Agrees with the present study, Harvatine and Allen (2006) reported that supplementation of saturated and unsaturated fatty acids reduced apparent digestibility of DM organic matter and NDF, because unsaturated fatty acids in oil with covering particles of fiber in rumen probably impaired cellulolytic bacteria function and ultimately reduce fiber digestion in the rumen, which is consistent with the findings of the present analysis.

The cell wall of barley is higher than that of corn (Parnian Khaje Dizaj, 2011), so probably, the higher digestibility (numerically) of corn diet (Table 5) was due to more cell wall of barley. Agrees with the results of present experiment, there was a trend for decreased total tract NDF digestibility when barley replaced by corn in diets of cattle (McCarthy et al. 1989; Overton et al. 1995).

Regardless the type of basal diet, DM and NDF digestibility by bacteria decreased with increasing level of milk thistle in diet (Table 6). The results were shown that negative effect of milk thistle on the bacteria was more than WRM, thus the hypothesis about interference of fat or oil on bacteria is supported. Totally, the digestibility results showed that using of milk thistle up to 20% in buffalo diets have not adverse effect on rumen bacteria and WRM.

Nutrients digestibility of experimental diets by rumen bacteria (with specific rumen bacteria culture method)
Probably the reason of numerical decrease in bacterial digestibility of the nutrients (Table 7) was unsaturated fatty acids (mainly linoleic acid) of milk thistle (Malekzadeh et al. 2011). Fatty acids are toxic to bacteria in pure culture experiments (Maczulak et al. 1981), but once in the rumen, fatty acids predominately are associated with feed particles their effects may satisfied (Harfoot, 1978).

Regardless the type of basal diet (Table 9), the milk thistle had no significant effect on DM and NDF digestibility. Thus, we may conclude from specific bacteria culture test that different levels of milk thistle have no negative effect on rumen bacteria functions.

CONCLUSION

In the present research the effects of using different levels of milk thistle in diet of buffalo on bacteria a whole rumen microorganism digestion function was investigated, and almost, the negative effects were not observed, even at most case the digestibility was improved. There was no signify-cant difference between two basal diets in nutrient digesti-bility. Therefore, results suggest that milk thistle could be used up to 20% in diets of buffalo without any negative effect on digestion and fermentation characteristics of WRM and bacteria.

ACKNOWLEDGEMENT

The authors are grateful to Ramin Agriculture and Natural Resources University of Khuzestan for financial support and preparation of experiment conditions.

REFERENCES

Abascal K., Herbalist J.D. and Yarnell E. (2003). The many faces of *Silybum marianum* (milk thistle). *J. Altern. Complement. Med.* **9(4),** 170-175.

Beauchemin K.A., McGinn S.M., Martinez T.F. and McAllister T.A. (2007). Use of condensed tannin extract from quebracho trees to reduce methane emissions from cattle. *J. Anim. Sci.* **85,** 1990-1996.

Brooker J.D., O'Donovan L.A., Skene I., Clarke K., Blackall L. and Muslera P. (1994). *Streptococcus caprinus* SP-Nov., a tannin resistant ruminal bacterium from feral goats. *Lett. Appl. Microbiol.* **18,** 313-318.

Broudiscou L.P., Papon Y. and Broudiscou A.F. (2000). Effects of dry plant extracts on fermentation and methanogenesis in con-tinuous culture of rumen microbes. *Anim. Feed Sci. Technol.* **87(3),** 263-277.

Carmen Tamayo M.D. (2007). Review of clinical trials evaluating safety and efficacy of milk thistle (*Silybum marianum*). *Integr. Cancer. Ther.* **6(2),** 146-157.

Carulla J.E., Kreuzer M., Machmller A. and Hess H.D. (2005). Supplementation of *Acacia mearnsii* tannins decreases methanogenesis and urinary nitrogen in forage-fed sheep. *Aus-tralian J. Agric. Res.* **56,** 961-970.

Denmen S.E. and McSweeney C.S. (2006). Development of a real-time PCR assay for monitoring anaerobic fungal and cel-lulolytic bacterial populations within the rumen. *FEMS Mi-crobiol. Ecol.* **58,** 572-582.

Dong G.Z., Wang X.J., Liu Z.B. and Wang F. (2010). Effects of phytogenic products on *in vitro* rumen fermentation and meth-ane emission in goats. *J. Anim. Feed Sci.* **19,** 218-229.

Galdwell D.R. and Bryant M.P. (1966). Medium without rumen fluid for non-selective enumeration and isolation of rumen bacteria. *Appl. Microbiol.* **14,** 794-801.

Garcia S.C., Santini F.J. and Elizalde J.C. (2000). Sites of diges-tion and bacterial protein synthesis in dairy heifers fed fresh oats with or without corn or barley grain. *J. Dairy Sci.* **83(4),** 746-755.

Gazak R., Walterova D. and Kren V. (2007). Silybin and sily-marin, new and emerging applications in medicine. *Curr. Med. Chem.* **14(3),** 315-324.

Hadolin M., Skerget M., Knez Z. and Bauman D. (2001). High pressure extraction of vitamin E rich oil from *Silybum mari-anum. Food Chem.* **74,** 355-364.

Hagerman A.E., Robbins C.T., Weerasuriya Y., Wilson T.C. and McArthur C. (1992). Tannin chemistry in relation to digestion. *J. Range Manage.* **45(1),** 57-62.

Harfoot C.G. (1978). Lipid metabolism in the rumen. *Prog. Lipid. Res.* **17,** 21-27.

Harvatine J. and Allen S. (2006). Fat supplements affect fractional rates of ruminal fatty acid biohydrogenation and passage in dairy cows. *J. Nutr.* **136,** 677-683.

Herrera-Saldana R.E., Huber J.T. and Poore M.H. (1990). Dry matter, crude protein and starch degradability of five cereal grains. *J. Dairy Sci.* **73,** 2386-2393

Kamra D.N. (2005). Rumen microbial ecosystem. *Curr. Sci.* **89(1),** 124-135.

Khorasani G.R., Okine E.K. and Kennelly J.J. (2001). Effects of substituting barley grain with corn on ruminal fermentation characteristics, milk yield, and milk composition of Holstein cows. *J. Dairy Sci.* **84,** 2760-2769.

Koike S., Pan J., Kobayashi Y. and Tanaka K. (2003). Kinetics of *in sacco* fiber-attachment of representative ruminal cellulolytic bacteria monitored by competitive PCR. *J. Dairy Sci.* **86,** 1429-1435.

Kren V. and Walterova D. (2005). Silybin and silymarin new ef-fects and applications. *Biomed. Pap.* **149,** 29-41.

Lee L., Narayan M. and Barrett J.S. (2007). Analysis and com-parison of active constituents in commercial standardize sily-marin extract by liquid chromatography-electrospray ioniza-tion mass spectrometry. *J. Chromatogr. B.* **845,** 95-103.

Lundy F.P., Block E., Bridges W.C., Bertrand J.A. and Jenkins T.C. (2004). Ruminal biohydrogenation in Holstein cows fed soybean fatty acid as amides or calcium salts. *J. Dairy Sci.* **87,** 1038-1046.

Macadam J.F. (1966). Some poisonous plants in the northwest. *Agric. Gaz. New South Wales.* **77(2),** 73-78.

Maczulak A.E., Dehority B.A. and Palmquist D.L. (1981). Effects of long-chain fatty acids on growth of tureen bacteria. *Appl. Environ. Microbiol.* **42,** 856-862.

Maldar S.M., Roozbehan Y. and Alipour D. (2010). The effect of adaptation to oak leaves on digestibility (*in vitro*) and ruminal parameters in Alamout goat. *Iranian J. Anim. Sci.* **41(3),** 243-252.

Malekzadeh M., Mirmazloum S.I., Rabbi Anguorani H., Morta zavi S.N. and Panahi M. (2011). The physicochemical proper-ties and oil constituents of milk thistle (*Silybum marianum* Gaertn. cv. Budakalászi) under drought stress. *J. Med. Plants Res.* **5(8),** 1485-1488.

McCarthy Jr.R.D., Klusmeyer T.H., Vicini J.L., Clark J.H. and Nelson D.R. (1989). Effects of source of protein and carbohy-drate on ruminal fermentation and passage of nutrients to the small intestine of lactating cows. *J. Dairy Sci.* **72,** 2002-2016.

McDougall E.L. (1948). Studies on ruminant saliva. 1. The com-position and output of sheep's saliva. *J. Biochem.* **43,** 99-106.

McSweeny C., Palmer B., Krause D.O. and Brooker J.D. (2000). Rumen microbial ecology and physiology in sheep and goats fed a tannin-containing diet. Pp. 140-145 in Proc. Tannins Livest. Hum. Nutr, Adelaide, Australia.

Menk K.H. and Stingass H. (1988). Estimation of the energetic feed value obtained from chemical analysis and *in vitro* gas production using rumen fluid. *Anim. Res. Dev.* **28**, 7-55.

Mojadam A. (2012). Evaluation of chemical composition, and digestibility and fermentation milk thistle (*Silybum marianum*) and its effects on rumen fermentation and digestion of fiber and protein in Arabic sheep. MS Thesis. Ramin Agriculture and Natural Resources Univ., Ahvaz, Iran.

Mohammadabadi T., Danesh Mesgaran M., Chaji M. and Tahmasebi R. (2012). Evaluation of the effect of fat content of sunflower meal on rumen fungi growth and population by direct (quantitative competitive polymerase chain reaction) and indirect (dry matter and neutral detergent fiber disappearance) methods. *African J. Biotechnol.* **11(1)**, 179-183.

Nocek J.E. and Tamminga S. (1991). The prediction of nutrient supply to dairy cows from rate and extent of ruminal degradation of ration components. *J. Dairy Sci.* **74**, 3598-3629.

NRC. (2001). Nutrient Requirements of Dairy Cattle. 7th Ed. National Academy Press, Washington, DC, USA.

Odenyo A.A. and Osuji P.O. (1998). Tannin-tolerant ruminal bacteria from east African ruminants. *Canadian. J. Microbiol.* **44**, 905-909.

Orskov E.R. and McDonald I. (1979). The estimation of protein degradability in the rumen from incubation measurements weighted according to rate of passage. *J. Agric. Sci.* **92**, 499-503.

Overton T.R., Cameron M.R., Elliott J.P., Clark J.H. and Nelson D.R. (1995). Ruminal fermentation and passage of nutrients to the duodenum of lactating cows fed mixtures of corn and barley. *J. Dairy Sci.* **78**, 1981-1998.

Palmquist D.L. and Jenkins T.C. (1980). Fat in lactation rations: review. *J. Dairy Sci.* **63**, 1-14.

Parnian Khaje Dizaj F., Taghizadeh A., Moghaddam G.A. and Janmohammadi H. (2011). Use of *in vitro* gas production technique for evaluation of nutritive parameters of barley and corn grain treated by different microwave irradiation times. *Anim. Sci. Res.* **21(1)**, 15-27.

Pinares-Patino C.S., Ulyatt M.J., Waghorn G.C., Lassey K.R., Barry T.N., Holmes C.W. and Johnson D.E. (2003). Methane emission by alpaca and sheep fed on lucerne hay or grazed on pastures of perennial ryegrass/white clover or birdsfoot trefoil. *J. Agric. Sci.* **140**, 215-226.

Robson S. (2007). Nitrate and Nitrite Poisoning in Livestock. Available at: www.dpi.nsw.gov.au/primefacts. Accessed Feb. 2007.

Sadeghi A.A. and Shawrang P. (2006). Effects of microwave irradiation on ruminal protein and starch degradation of corn grain. *Anim. Feed Sci. Technol.* **127**, 113-123.

Sadeghi A.A. and Shawrang P. (2008). Effects of microwave irradiation on ruminal dry matter, protein and starch degradation characteristics of barley grain. *Anim. Feed Sci. Technol.* **141**, 184-194.

SAS Institute. (2004). SAS®/STAT Software, Release 9.1. SAS Institute, Inc., Cary, NC. USA.

Smith A.H., Zoetendal E. and Mackie R.I. (2005). Bacterial mechanisms to overcome inhibitory effects of dietary tannins. *Microbial. Ecol.* **50**, 197-205.

Sommart K., Parker D.S., Rowlinson P. and Wanapat M. (2000). Fermentation characteristics and microbial protein synthesis in an *in vitro* system using cassava, rice straw and dried Ruzi grass as substrates. *Asian-Australas J. Anim. Sci.* **13**, 1084-1093.

Sotohy S.A., Sayed A.N. and Ahmed M.M. (1997). Effect of tannin- rich plant (*Acacia nilotica*) on some nutritional and bacteriological parameters in goats. *Deut. Tierarztl. Woch.* **104**, 432-435.

Sun X.Z., Hoskin S.O., Muetzel S., Molan G. and Clark H. (2011). Effects of forage chicory (*Cichorium intybus*) and perennial ryegrass *(Lolium perenne)* on methane emissions *in vitro* and from sheep. *Anim. Feed Sci. Technol.* **166**, 391-397.

Szumacher-Strabel M., Cieślak A. and Nowakowska A. (2009). Effect of oils rich in linoleic acid on *in vitro* rumen fermentation parameters of sheep, goats and dairy cows. *J. Anim. Feed Sci.* **18**, 440-452.

Tan H.Y., Sieo C.C., Lee C.M., Abdullah N., Liang J.B. and Ho Y.W. (2011). Diversity of bovine rumen methanogens *in vitro* in the presence of condensed tannins, as determined by sequence analysis of 16S rRNA gene library. *J. Microbiol.* **49(3)**, 492-498.

Tilly J.M.A. and Terry R.A. (1963). A two stage technique for the indigestion of forage crops. *Grass. Forage. Sci.* **18**, 104-111.

Van Soest P.J., Robertson J.B. and Lewis B.A. (1991). Methods of dietary fiber, neutral detergent fiber, and nonstarch polysaccharides in relation to animal nutrition. *J. Dairy Sci.* **74**, 3583-3597.

White T.W., Bunting L.D., Sticker L.S., Hembry F.G. and Saxton A.M. (1992). Influence of fish meal and supplemental fat on performance of finishing steers exposed to moderate or high ambient temperatures. *J. Anim. Sci.* **70**, 3289-3292.

Zargari A. (1996). Medicinal Plants. Institute Publications and Printing of Tehran University, Tehran, Iran.

Zinn R.A. (1989). Influence of level and source of dietary fat on its comparative feeding value in finishing diets for steers, feedlot cattle growth and performance. *J. Anim. Sci.* **67**, 1029-1037.

Zinn R.A. and Shen Y. (1996). Interaction off dietary calcium and supplemental fat on digestive function and growth performance in Feedlot steers. *J. Anim. Sci.* **74**, 2303-2309.

Effects of Sodium Bentonite on Blood Parameters, Feed Digestibility and Rumen Fermentation Parameters of Male Balouchi Sheep Fed Diet Contaminated by Diazinon, an Organophosphate Pesticide

M.H. Aazami[1*], A.M. Tahmasbi[1], V. Forouhar[1] and A.A. Naserian[1]

[1] Department of Animal Science, Faculty of Agriculture, Ferdowsi University of Mashhad, Mashhad, Iran

*Correspondence E-mail: M.H.Aazami@gmail.com

ABSTRACT

The remnants of pesticides in livestock feeds have been increased by excessive using of these pesticides so as to meet extreme demands for more feeds. Finding a new strategy for reducing pesticides negative effects is absolutely necessary. Therefore, evaluation of the effects of sodium bentonite on blood parameters, feed digestibility and rumen fermentation parameters in sheep fed diets contaminated by diazinon, an organo-phosphate pesticide, was the aim of this study. Eight canulated male Balouchi sheep (40±2 kg) assigned to a 2 × 2 factorial arrangement in four 21-day-period. Treatments were: 1) control group, 2) control + 4% so-dium bentonite, 3) control + 21 ppm diazinon and 4) control + 4% sodium bentonite + 21 ppm diazinon. Dry matter, organic matter, crude protein, neutral detergent fiber (NDF) and acid detergent fiber (ADF) digestibility were not affected by treatments. Sodium bentonite decreased pH variation after feeding. Am-monia nitrogen of rumen liquor was the same among treatments before feeding however 3 and 6 hours after feeding, amount of ammonia nitrogen was significantly higher in groups fed pesticide contaminated diets (P<0.05). Hemoglobin, white blood cells and hematocrit were not affected by treatments but red blood cells and acetylcholinesterase activity were significantly reduced in groups fed pesticide contaminated diets (P<0.05). Results of present study showed that using sodium bentonite as a binder was effective to reduce negative effects of diazinon on pH variation and acetylcholinesterase activity and it has no effect on hemo-globin, white blood cells, hematocrit and nutrient digestibility. Therefore, sodium bentonite can be used as an effective diazinon binder in sheep diet.

KEY WORDS blood parameter, diazinon, male Balouchi sheep, nutrient digestibility, ru-men parameter, sodium bentonite.

INTRODUCTION

Most of animal feeds are grow n using pesticides. Chemical control of weeds, insects and other pests has increased agri-cultural productivity. However, these economic benefits are not without risks to animal health and environmental da-mage. Feed and fodder offered to animals are often conta-minated with pesticide residues (Raikwar and Nag, 2003)

and after feeding, these residues assimilated into the body systems of the animals (Prasad and Chhabra, 2001). The occurrence of pesticides residues in milk of ruminants is a matter of public health concern, since milk and dairy pro-ducts are widely consumed by infants, children and adults throughout the world. Similarly, the toxins in the water and feed transfer to animal products and then these contamina-ted products are eaten by humans (Ashraf et al. 2010), therefore it may have deleterious effects in the future for them. Moreover, these pollutants can enter the food chain and environment. Degradation rates of pollutants after release to the environments vary extensively between substances, with half-life from minutes to many years. One of the pesticides which is widely used is diazinon. Diazinon belongs to organophosphate pesticides. Diazinon's biological half-life in mammals equals about 12 hours and after 2 weeks only traces of this pesticide may be found in the bodies (Debski et al. 2007). Diazinon is still extensively used in sheep dip to control ectoparasites (Boucard et al. 2004; Gaworecki and Klaine, 2008; Jadhav and Rajini, 2009; Jemec et al. 2007). Organophosphate pesticides are among the leading chemicals used extensively for agricultural pests control throughout the world (Gaikwad et al. 2015). The presence of organophosphates in water and crops represents a potential hazard because of their high toxicity to mammals. Organophosphorous and carbamate insecticides affect acetylcholinesterase activity in the nervous system that lead to subsequent accumulation of toxic levels of endogenous acetylcholine in nerve tissues that affect organs in both insects and mammals (Kovac et al. 1998; Moretto and Johnson, 1987; Nistiar et al. 1984), including humans (Chambers and Levi, 1992; Lotti, 1995). Continuing emphasis on treating and preventing disease conditions caused by toxic substances is necessary because of the increasing incidence of poisoning of domestic animals resulting from the wide-spread use of pesticides in agriculture. There are three general types of antidotes for poisons and toxins. First, a mechanical antidote is one that binds a poison in the gut and prevents absorption of the poison. Second, a chemical antidote stimulates the body so that the poison is metabolized and detoxified at a faster rate. Third, a physiologic antidote counteracts the toxic effects of the poison. Several studies have used independent or in conjunction methods so as to removal organophosphate pesticide including chemical oxidation with ozone, photo degradation (Zertal et al. 2005), combined ozone and UV irradiation (Malato et al. 1999), biological degradation (Chen et al. 2009), ozonation (Hua et al. 2006), membrane filtration (Hofman et al. 1997) and adsorption (Daneshvar et al. 2007). An example of a mechanical antidote is sodium bentonite. Bentonite is the terminology used to describe the clay rock material composed mainly of montmorillonite clay (Eisenhour and Brown, 2009) with the presence of common impurities such as quartz, feldspars and other minerals. Montmorillonite is a phyllosilicate mineral belonging to the smectite group (Dixon and Schulze, 2002). Smectites have a structural layer formed by two tetrahedral sheets and one octahedral sheet that was sandwiched by the tetrahedral sheets. Montmorillonite contains predominantly Al^{3+} in the octahedral positions with some substitution by Fe^{3+} and Mg^{2+}.

The tetrahedral sheets are dominated by Si^{4+}. The unbalance charges that arise from the isomorphic substitutions give an overall negative charge to montmorillonites. The charge is compensated by exchangeable cations between the layers.

This interlayer space is highly expandable depending on hydration and cation valence. The high surface area (~800 m^2/g) and the expandable structure are important properties in smectites that allow for a wide range of applications. Kazemi et al. (2012) used increasing levels of diazinon (0, 0.7, 2.8 and 5.6 mg) as an organophosphorous pesticide and (0 and 100 mg) calcium bentonite, as a toxin binder to evaluate in vitro and in situ dry matter disappearance. They declared that effect of diazinon in different levels was significant (P<0.05) for the entire estimated parameters exception "a" fraction and dry matter degradability after 24 h incubation.

Moreover, effect of calcium bentonite on entire estimated parameters for dry matter degradability was insignificant exception dry matter degradability after 48 hours incubation. Donia et al. (2010) assessed accumulation of toxins in raw milk and found that no kind of organophosphorous pesticide were detected in cow and buffalo raw milk but organochlorine (OC) pesticides, hexachlorobenzene, lindane, aldrin, heptachlor epoxide, chlordane, endrin and dichloro diphenyl trichloroethane (DDT), were detected at a value exceeded the tolerance levels of FAO/WHO. Elucidation of the effects of sodium bentonite on blood parameters, feed digestibility and rumen fermentation parameters in sheep fed on diazinon, an organophosphate pesticide, was the aim of this study.

MATERIALS AND METHODS

The study was conducted under the supervision and approval of the animal care and use committee of Ferdowsi University of Mashhad, Iran, concerning the use of animals in research. Eight Balouchi lambs (40±2 kg, body weight (BW)) fitted with ruminal cannulas (2.5 cm i.d.) in a 2 × 2 factorial arrangement were housed in individual metabolic cages (0.5×1.2×1 m) in a temperature controlled house (approximately 22 °C). The experiment consisted of 4 periods. Each period lasted 21 days, comprising 14 days of adaptation to the experimental diet and followed by 7 days of data collection. Treatments were 1) control group, 2) control + 4% sodium bentonite, 3) control + 21 ppm diazinon and 4) control + 4% sodium bentonite + 21 ppm diazinon. The ingredients and chemical composition of the experimental diet is shown in Table 1. All diets were supplied as total mixed ration (TMR), and offered at maintenance level twice daily in equal portions at 08:00 and 20:00. Clean water was freely available ad libitum.

Table 1 Ingredients and chemical composition of experimental diets (DM basis)[a]

Items	(%)
Ingredient (%)	
Alfalfa hay	30
wheat straw	20
Barley	30
Wheat bran	12.5
Canola meal	6
Vitamin and mineral premix[b]	1
NaCl	0.2
Limestone	0.3
Total	100
Chemical composition	
Dry matter	90.4
Crude protein (CP)	11.9
Ether extracts (EE)	2.4
Neutral detergent fiber (NDF)	40.8
Acid detergent fiber (ADF)	25.2
Non fiber carbohydrates (NFC)	40.4
Ca	0.8
P	0.5

[a] Ration were formulated to supply nutrient requirement at maintenance level (NRC, 2001).
[b] Chemical composition was calculated based on tabulated composition of individual feedstuffs (Ministry of Agriculture, MOA, PRC, 2004).
[c] The mix contained (/kg of premix; DM basis): vitamin A: 330000 IU; vitamin D: 60000 IU; vitamin E: 1000 IU; Ca: 160 g; P: 85 g; Na: 63 g; Mg: 45 g; Zn: 2100 mg; Mn: 1500 mg; Cu: 535 mg; Se: 12 mg and I: 45 mg.

Ruminal fluid was collected by suction through the rumen cannula from before the morning feeding (0.0 h) to 6 hours after feeding at 15-min intervals for the determination of rumen pH on day 18 of each period. The pH of each rumen fluid sample was measured immediately with a portable pH meter (Metrohm 744, Herisau, Switzerland). Samples of ruminal fluid were taken at 0, 3 and 6 hours after feeding for the determination of ammonia nitrogen concentrations. The rumen fluid was then strained through four layers of cheesecloth and prepared for subsequent ammonia-N analyses. On the last day of each experimental period, blood samples were collected from the jugular vein (10 mL into sterile tubes containing ethylenediaminetetraacetic acid (EDTA) solution).

The samples were immediately placed on ice for processing in the laboratory. Blood samples were centrifuged (3000 g for 15 min at 5 °C). Plasma was harvested and frozen at -20 °C for later analysis. The count of white blood cells, red blood cells, hemoglobin and hematocrit were calculated using cell counter. Isolated serum was used for determining the activity of the acetylcholine esterase enzyme by titrimetric method. Feed samples were collected at each feeding and composited for later analysis. During the last 7 days of collection period, feces and refused feed were also collected and frozen. Feed, refusals and fecal samples thawed, mixed and were analyzed for DM (48 h at 60 °C)

and ash (4 h at 550 °C).

Statistical analysis

All data obtained from the experiment were subjected to ANOVA for a completely randomized design with 2 × 2 factorial arrangement of treatments using the general linear models (GLM) procedures of the statistical analysis system Institute (SAS) (SAS, 1999). Treatment means were compared by Duncan's new multiple range test (Steel and Torrie, 1960).

The statistical model was:

$$X_{idbjn} = \mu + \alpha_d + \beta_b + (\alpha\beta)_{db} + P_j + T_n + \varepsilon_{idbjn}$$

Where:
X_{idbjn}: observation idbjn.
μ: overall mean.
α_d: effect of diazinon.
β_b: effect of sodium bentonite.
$(\alpha\beta)_{db}$: interaction of two factors.
P_j: effect of period.
T_n: effect of animal.
ε_{idbjn}: experimental error.

Laboratory analyses

The dry matter content of feed ingredients was determined by oven-drying at 60 °C for 24 hours then analyzed for concentrations of DM, CP, ether extract (AOAC, 2002), NDF and ADF (Van Soest et al. 1991). For ammonia nitrogen determination, 5 mL of rumen fluid from each collection point was acidified with 5 mL of 0.2 NHCl and then analyzed for ammonia-N concentration using the distillation method (Kjeltec Auto Analyzer, Model 1030, Tecator Co. Sweden). Samples of feeds offered and individual refusals were also retained for dry matter determination after drying in a forced drought oven at 80 °C for 48 h. The stored subsamples of feeds offered and refusals dried at 50 °C then ground to pass through a 1mm screen and stored until analyzed.

RESULTS AND DISCUSSION

Rumen pH

Rumen pH changes during 6 hours after feeding is presented in Figure 1 and effects of experimental treatment in 0, 3 and 6 hours after feeding are presented in Table 2. For all treatments rumen pH decreased 3 hours after feeding and again increased gradually till 6 hours after feeding. Diazinon pesticide had no significant effect on rumen fluid pH (P>0.05) while sodium bentonite significantly affected rumen pH of different experimental groups.

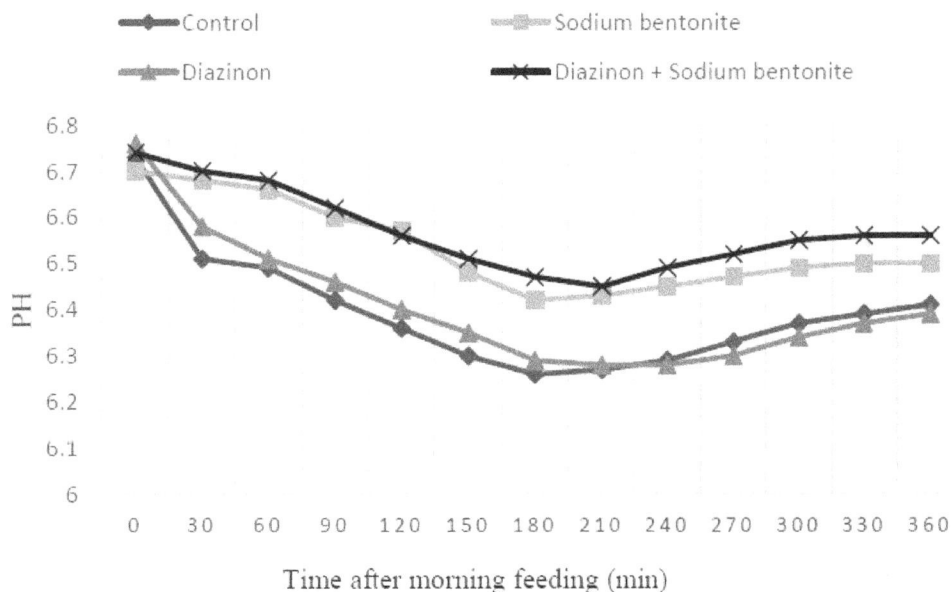

Figure 1 Effect of adding diazinon pesticide and sodium bentonite on ruminal pH

Table 2 Effect of adding diazinon pesticide and sodium bentonite on ruminal pH

Items	- Diazinon Sodium bentonite		+ Diazinon Sodium bentonite		P-value			SEM
	-	+	-	+	D	S	D × S	
Before feeding	6.73	6.70	6.76	6.74	0.24	0.39	0.95	0.12
3 h after feeding	6.26^a	6.42^a	6.29^a	6.47^b	0.56	0.04	0.88	0.05
6 h after feeding	6.41^a	6.51^b	6.39^a	6.56^b	0.79	0.04	0.50	0.05

(-) and (+) signs show respectively absence and presence of experimental treatment.
D: effect of diazinon; S: effect of sodium bentonite and D × S: interaction effect.
The means within the same row with at least one common letter, do not have significant difference (P>0.05).
SEM: standard error of the means.

Ammonia nitrogen of rumen fluid

Before feeding, ruminal ammonia nitrogen level was similar among different experimental treatments (P>0.05). Three hours after feeding ammonia nitrogen rose in all treatments and this increase was significantly higher in groups containing diazinon pesticide rather than groups containing sodium bentonite (P<0.05). Six hours after feeding ammonia nitrogen showed trend toward reduction in all groups (Table 3). Adding sodium bentonite caused a mild decrease in ammonia nitrogen content of rumen but this effect was not statistically significant which was in agreement with Helal and Abdel-Rahman (2010). Decreasing pattern of after feeding rumen ammonia nitrogen level was similar among treatments without diazinon. Groups containing diazinon and without diazinon groups showed significant differences (P<0.05). Differences among the groups containing sodium bentonite plus diazinon and the groups containing just diazinon was not statistically significant (P>0.05).

Presence or absence of diazinon resulted in a significant difference of rumen ammonia nitrogen level of treatments.

Blood parameters

Blood parameters (red blood cell count, white blood cell count, hemoglobin concentration and hematocrit) are presented in Table 4. Hemoglobin, white blood cell count and hematocrit remained unaffected by adding diazinon and sodium bentonite (P>0.05). There were significant differences in red blood cell count among groups which receive diazinon pesticide and control group (P<0.05). Adding sodium bentonite as a binder was effective to protect RBC in a same count of control group. Presence of diazinon in Balouchi sheep diet significantly reduced acetylcholine esterase activity (P<0.05).

Apparent nutrients digestibility

Means of apparent nutrients digestibility (%) including dry matter (DM), organic matter (OM), crud protein (CP) neutral detergent fiber (NDF) and acid detergent fiber (ADF) are presented in Table 5. In the present study, apparent digestibility of DM, OM, CP, NDF and ADF did not significantly differ among treatments (Table 5).

Rumen pH

Sodium bentonite because of its high capacity of cation exchange, prevented rumen pH steep reduction during first 3 hours after feeding and sharp increase till 6 hours after feeding (next 3 hours). Diets with bentonite because of exchangeable cations, Na and K, maintain rumen pH in a stable level (Stephenson et al. 1992). Rumsey et al. (1975) stated that adding organophosphate pesticide and binder have no effect on rumen parameters. Hershberger et al. (1971) reported that supplying diet of lambs which was exposed to pesticide with sodium bentonite had no significant effect on rumen pH. Schwartz et al. (1973) in a 12-day in vivo study and an in vitro study found feed contaminated with pesticide did not change rumen fluid pH pattern. Aazami et al. (2013) stated that saponin as a natural antinutritional component has no effect on rumen pH and rumen ammonia nitrogen level.

Ammonia nitrogen of rumen fluid

Fluctuation of ammonia nitrogen in group containing sodium bentonite during six hours after feeding was lower. Aghashahi et al. (2005) reported adding 2% sodium bentonite had no effect on rumen ammonia nitrogen level but it reduced rumen ammonia nitrogen level variations. Meanwhile, adding fourpercent of sodium bentonite significantly reduced rumen ammonia nitrogen level especially after 4 hours. Stephenson et al. (1992) reported that adding sodium bentonite prevented rumen ammonia nitrogen raising after feeding probably due to influence of sodium bentonite on rumen microorganism metabolism. Limiting archaea movements, sodium bentonite ceases bacteria hunting (Saleh, 1994). The more bacterial population, the more ammonia nitrogen utilization in rumen. Moreover, Williams and Withers (1993) declared that protozoa population decrease will be resulted in ammonia nitrogen level alleviation. Additionally, Wallace and Newbold (1991) stated that sodium bentonite caused interference with the motion of cilia and thereby prevented motility of protozoa, particularly the holotrichs. McCullough (1974) interpreted cows which received sodium bentonite had a lower rumen ammonia level and a bigger population of micro-organisms except protozoa. As mentioned above, in this study presence or absence of diazinon resulted in a significant difference of rumen ammonia nitrogen level of treatments. One possible reason of this significant differences can be the reduction of microbial population due to presence of diazinon.

Nistiar et al. (2000) have verified negative effects of organophosphate pesticides on rumen microbial population. By reduction in microbial population, fewer nitrogen will be used for microbial production so nitrogen is cumulated in rumen (McCullough, 1974). Rumsey et al. (1975) observed no effect on ammonia nitrogen in cows fed runnel which this is inconsistent with our findings.

Blood parameters

Similar to our findings, Garillo et al. (1995) reported that binder was not effective to prevent dropping red blood cell counts in group fed with diazinon contaminated fodders. Effect of diazinon on red blood cell producing organs (liver and bone marrow), inefficiency of bone narrow stem cells, shortening life span of red blood cells and preventing heme biosynthesis could be mentioned as reasons of reducing red blood cell count in groups containing diazinon (Betrosian, et al. 1995).

Rumsey et al. (1975) reported insignificant effect of organophosphate pesticide on blood parameter namely, red blood cell count and white blood cell count. In this study the lowest red blood cell count (8.83) appertained to the group containing diazinon and without sodium bentonite. Therefore, in this case sodium bentonite was effective in order to relieve negative actions of dizinon pesticide. Reducing acetylcholine esterase activity in group fed diazinon are consistent with Khan (2001) that interpreted dry cows and replacement heifers fed with diazinon contaminated diet had lower acetylcholine esterase activity.

Moreover, he has verified binders can reduce organophosphate pesticide absorption in gastrointestinal tract and keep acetylcholine esterase activity in a normal range. Judge et al. (2016) reported that acute in vivo diazinon exposure caused a small but statistically significant inhibition of blood and brain acetylcholine esterase activity and this was dependent on dose and tissue (blood and brain regions) but not time since exposure. Enzyme acetylcholine esterase is present in synaptic regions of neurons and mediates transmission of impulse by breaking acetylcholine into acetic acid and choline.

Diazinon induced suppressed activity of Acetylcholine esterase results in accumulation of acetylcholine at neural and neuromotor regions that causes hyper excitability (Van Cong et al. 2009).

Apparent nutrients digestibility

Ivan et al. (2001) stated adding 2% sodium bentonite had no significant effects on apparent nutrients digestibility of sheep fed with palm kernel cake by-product. Galyean and Chabot (1981) reported that supllying sudium bentonite in ruminally cannulated Hereford steers had no effect on nutrient digestibility.

Table 3 Effect of adding diazinon pesticide and sodium bentonite on ruminal ammonia nitrogen

Items	- Diazinon		+ Diazinon		P-value			SEM
	Sodium bentonite		Sodium bentonite					
	-	+	-	+	D	S	D × S	
Before feeding	17.04	17.63	18.11	19.94	0.26	0.42	0.68	1.46
3 h after feeding	21.37[a]	21.25[a]	27.87[b]	27.98[b]	0.02	0.99	0.96	1.67
6 h after feeding	20.58[a]	19.05[a]	26.53[b]	25.65[b]	0.04	0.34	0.92	1.77

(-) and (+) signs show respectively absence and presence of experimental treatment.
D: effect of diazinon; S: effect of sodium bentonite and D × S: interaction effect.
The means within the same row with at least one common letter, do not have significant difference (P>0.05).
SEM: standard error of the means.

Table 4 Effect of adding diazinon pesticide and sodium bentonite on blood parameteres

Items	- Diazinon		+ Diazinon		P-value			SEM
	Sodium bentonite		Sodium bentonite					
	-	+	-	+	D	S	D × S	
Hemoglobin (g/dL)	10.40	9.63	8.80	9.43	0.09	0.80	0.17	0.57
Hematocrit (%)	30.47	29.67	25.40	26.70	0.06	0.89	0.58	1.84
RBC count	11.02[a]	10.13[ab]	8.83[b]	9.27[ab]	0.04	0.73	0.33	0.64
WBC count	7.50	7.56	7.43	7.13	0.52	0.76	0.63	0.37
AChE activity	18.67[a]	18.33[a]	12.03[b]	20.00[a]	0.03	0.08	0.06	1.91

(-) and (+) signs show respectively absence and presence of experimental treatment.
D: effect of diazinon; S: effect of sodium bentonite and D × S: interaction effect.
RBC: red blood cell count; WBC: white blood cell count and AChE: acetylcholine esterase activity.
The means within the same row with at least one common letter, do not have significant difference (P>0.05).
SEM: standard error of the means.

Table 5 Effect of adding diazinon pesticide and sodium bentonite on apparent nutrients digestibility

Items	- Diazinon		+ Diazinon		P-value			SEM
	Sodium bentonite		Sodium bentonite					
	-	+	-	+	D	S	D × S	
Dry matter	54.33	55.10	52.66	55.47	0.57	0.14	0.22	1.41
Organic matter	58.60	58.00	56.00	60.66	0.69	0.06	0.1	1.54
Crud protein	58.17	57.92	57.87	58.22	0.09	0.82	0.44	0.91
Neutral detergent fiber	51.97	53.01	52.05	53.13	0.49	0.72	0.69	1.02
Acid detergent fiber	44.06	45.01	43.08	44.26	0.05	0.07	0.09	1.11

(-) and (+) signs show respectively absence and presence of experimental treatment.
D: effect of diazinon; S: effect of sodium bentonite and D × S: interaction effect.
SEM: standard error of the means.

Although the same result have been published by some other authors (Aguilera-Soto *et al.* 2009; Ha *et al.* 1983), but Fisher and MacKay (1983) reported that adding 0.6 to 1.6% of dry matter, sodium bentonite to diet decreased nutrients digestibility. Schwartz *et al.* (1973) examined pesticide effect on rumen function and *in vitro* nutrients digestion. They found that bordeaux mixture, toxaphene, Mema RM, dichloro diphenyl trichloroethane (DDT), O-Ethyl O-(4-nitrophenyl) phenylphosphonothioate (EPN), Zectran, dieldrin, parathion, mobam and aldrin significantly (P>0.05) reduced *in vitro* dry matter and cell wall constituent digestion at pesticide concentrations of 1000 ppm. No effect was noted with Baygon, Black-Leaf 40, malathion and 2, 4-D acid on either cell-wall constituent digestion or dry matter digestion *in vitro*. Kazemi *et al.* (2012) used different levels (0, 0.7, 2.8 and 5.6 mg) of diazinon with different levels of calcium bentonite (0 and 100 mg) as a toxin binder in order to test their toxicological effects on *in vitro* dry matter disappearance.

Authors reported that effect of diazinon with adding the different levels was significant for the entire estimated parameters exception "a" fraction and dry matter degradability after 24 h incubation. Moreover, effect of calcium bentonite on entire estimated parameters for dry matter degradability was insignificant except dry matter degradability after 48 h incubation. Chegeni *et al.* (2013) stated adding 2.5% of diet sodium bentonite had no significant effect on total tract digestibility of organic matter, NDF and ADF in Han × Dorper crossbreed sheep.

CONCLUSION

Different researches in sphere of diazinon and somewhat other organophosphate pesticide, emphatically evaluated the effects of diazinon on acetylcholineesterase activity especially in dipped sheep. Results of present study showed that using sodium bentonite as a binder was effective to reduce negative effects of diazinon on pH variation and acetylcholineesterase activity and it has no effect on hemoglobin, white blood cells, hematocrit and nutrient digestibility. So sodium bentonite can be used as an effective diazinon binder.

ACKNOWLEDGEMENT

The author would like to thank the Ferdowsi University of Mashhad and Excellence Centre for Animal Science for financial support of this study.

REFERENCES

Aazami M.H., Tahmasbi A.M., Ghaffari M., Naserian A.A., Valizadeh R. and Ghaffari A. (2013). Effects of saponins on rumen fermentation, nutrients digestibility, performance, and plasma metabolites in sheep and goat kids. *Annu. Rev. Res. Biol.* **3(4),** 596-607.

Aghashahi A.R., Mirhadi S.A. and Moradishahre Babak M. (2005). Effects of natural bentonite (Montmorillonite), processed bentonite and clinopetilolite-rich tuff on the fermentation parameters, rumen microbial population and feedlot performance in male calves. *Irnian J. Agric. Sci.* **36,** 3-10.

Aguilera-Soto J., Ramirez R., Arechiga C., Mendez-Llorente F., Lopez-Carlos M., Pina-Flores J., Medina-Flores C., Rodriguez-Frausto H., Rodriguez-Tenorio D. and Gutierrez-Banuelos H. (2009). Effect of feed additives on digestibility and milk yield of Holstein cows fed wet brewer grains. *J. Appl. Anim. Res.* **36(2),** 227-230.

AOAC. (2002). Official Methods of Analysis. Vol. I. 18th Ed. Association of Official Analytical Chemists, Arlington, VA, USA.

Ashraf M.A., Maah M.J., Yusoff I. and Mehmood K. (2010). Effects of polluted water irrigation on environment and health of people in Jamber, District Kasur, Pakistan. *Int. J. Basic Appl. Sci.* **10(3),** 37-57.

Betrosian A., Balla M., Kafiri G., Kofinas G., Makri R. and Kakouri A. (1995). Multiple systems organ failure from organophosphate poisoning. *J. Clin. Toxicol.* **33(3),** 257-260.

Boucard T.K., Parry J., Jones K. and Semple K.T. (2004). Effects of organophosphate and synthetic pyrethroid sheep dip formulations on protozoan survival and bacterial survival and growth. *FEMS Microbiol. Ecol.* **47,** 121-127.

Chamber J.E. and Levi P.A. (1992). Organophosphates: chemistry, fate and effects. Academic Press, San Diego, California.

Chegeni A., Li Y., Deng K., Jiang C. and Diao Q. (2013). Effect of dietary polymer-coated urea and sodium bentonite on digestibility, rumen fermentation, and microbial protein yield in sheep fed high levels of corn stalk. *Livest. Sci.* **157(1),** 141-150.

Chen H., He X., Rong X., Chen W., Cai P., Liang W., Li S. and Huang Q. (2009). Adsorption and biodegradation of carbaryl on montmorillonite, kaolinite and goethite. *Appl. Clay Sci.* **46,** 102-108.

Daneshvar N., Aber S., Khani A. and Rasoulifard M.H. (2007).

Investigation of adsorption kinetics and isotherms of imidacloprid as a pollutant from aqueous solution by adsorption onto industrial granular activated carbon. *J. Food Agric. Environ.* **5,** 425-429.

Debski B., Kania B.F. and Kuryl T. (2007). Transformations of diazinon, an organophosphate compound in the environment and poisoning by this compound. *Ekológia.* **26(1),** 68.

Dixon J.B. and Schulze D.G. (2002). Soil Mineralogy with Environmental Applications: Soil Science Society of America, Inc. Madison, Wisconsin, USA.

Donia M.A., Abou-Arab A., Enb A., El-Senaity M. and Abd-Rabou N. (2010). Chemical composition of raw milk and the accumulation of pesticide residues in milk products. *Glob. Vet.* **4(1),** 06-14.

Eisenhour D.D. and Brown R.K. (2009). Bentonite and its impact on modern life. *Elements.* **5(2),** 83-88.

Fisher L. and Mackay V. (1983). The effect of sodium bicarbonate, sodium bicarbonate plus magnesium oxide or bentonite on the intake of corn silage by lactating cows. *J. Anim. Sci.* **63(1),** 141-148.

Gaikwad A.S., Karunamoorthy P., Kondhalkar S.J., Ambikapathy M. and Beerappa R. (2015). Assessment of hematological, biochemical effects and genotoxicity among pesticide sprayers in grape garden. *Int. J. Occup. Med. Toxicol.* **10,** 1-11.

Galyean M. and Chabot R. (1981). Effects of sodium bentonite, buffer salts, cement kiln dust and clinoptilolite on rumen characteristics of beef steers fed a high roughage diet. *J. Anim. Sci.* **52,** 1197-1204.

Garillo E., Pradhan R. and Tobioka H. (1995). Effects of activated carbon on growth, ruminal characteristics, blood profiles and feed digestibility in sheep. *J. Anim. Sci.* **8(1),** 43-50.

Gaworecki K.M. and Klaine S.J. (2008). Behavioral and biochemical responses of hybrid striped bass during and after fluoxetine exposure. *Aquat Toxicol.* **88,** 207-213.

Ha J., Emerick R. and Embry L. (1983). *In vitro* effect of pH variations on rumen fermentation, and *in vivo* effects of buffers in lambs before and after adaptation to high concentrate diets. *J. anim. Sci.* **56(3),** 698-706.

Helal F. and Abdel-Rahman K. (2010). Productive performance of lactating ewes fed diets supplementing with dry yeast and / or bentonite as feed additives. *J. Agric. Sci.* **6(5),** 489-498.

Hershberger T., Wilson L., Chase L., Rugh M. and Varela-Alvarez H. (1971). Effect of activated carbon on lamb performance and rumen parameters1. *J. Dairy Sci.* **54(5),** 693-695.

Hofman J., Beerendonk E., Folmer H. and Kruithof J. (1997). Removal of pesticides and other micropollutants with cellulose-acetate, polyamide and ultra-low pressure reverse osmosis membranes. *Desalinat.* **113,** 209-214.

Hua W., Bennett E.R. and Letcher R.J. (2006). Ozone treatment and the depletion of detectable pharmaceuticals and atrazine herbicide in drinking water sourced from the upper Detroit River, Ontario, Canada. *Water Res.* **40,** 2259-2266.

Ivan M., Neill L., Alimon R. and Jalaludin S. (2001). Effects of bentonite on rumen fermentation and duodenal flow of dietary components in sheep fed palm kernel cake by-product. *J. Anim. Feed Sci. Technol.* **92(1),** 127-135.

Jadhav K.B. and Rajini P.S. (2009). Evaluation of sublethal effects of dichlorvos upon Caenorhabditis elegans based on a set of end points of toxicity. *J. Biochem. Mol. Toxic.* **23,** 9-17.

Jemec A., Drobne D., Tišler T., Trebše P., Roš M. and Sepčić K. (2007). The applicability of acetylcholinesterase and glutathione S-transferase in Daphnia magna toxicity test. *Comp. Biochem. Physiol. Part C: Toxicol. Pharmacol.* **144,** 303-309.

Judge S.J., Savy C.Y., Campbell M., Dodds R., Gomes L.K., Laws G., Watson A., Blain P.G., Morris C.M. and Gartside S.E. (2016). Mechanism for the acute effects of organophosphate pesticides on the adult 5-HT system. *Chem. Biol. Interact.* **245,** 82-89.

Kazemi M., Tahmasbi A.M., Valizadeh R. and Naserian A.A. (2012). Toxic influence of diazinon as an organophosphate pesticide on parameters of dry matter degradability according to *in situ* technique. *Int. J. Basic. Appl. Sci.* **12(06),** 229-233.

Khan O. (2001). Organophosphate poisoning in a group of replacement heifers and dry cows. *Canadian Vet. J.* **42(7),** 561-567.

Kovac G., Reichel P., Seidel H. and Mudron P. (1998). Effects of sorbents during organophosphate intoxication in sheep. *Czech J. Anim. Sci.* **43,** 3-7.

Lotti M. (1995). Cholinesterase inhibition: complexities in interpretation. *Clin. Chem.* **41(12),** 1814-1818.

Malato S., Blanco J., Richter C., Milow B. and Maldonado M. (1999). Solar photocatalytic mineralization of commercial pesticides: methamidophos. *Chemosphere.* **38,** 1145-1156.

McCullough M. (1974). Improving urea utilization in ruminants. *Georgia Agric. Res.* **15,** 3-10.

Ministry of Agriculture, MOA, PRC (2004). Feeding Standard of Meatproducing Sheep and Goats (NY/T 816-2004). China Agricultural Press, Beijing, China.

Moretto A. and Johnson M. (1987). Toxicology of organophosphates and carbamates. Pp. 33-48. in Toxicology of Pesticides. L.G. Costa, C.L. Galli and S.D. Murphy, Eds, Springer, Berlin, Heidelberg.

Nistiar F., Hrusovský J., Mojzis J. and Mizik P. (1984). Distribution of dichlorvos in the rat and the effect of clinoptilolite on poisoning. *Vet. Meal.* **29,** 689-698.

Nistiar F., Mojžiš J., Kovac G., Seidel H. and Racz O. (2000). Influence of intoxication with organophosphates on rumen bacteria and rumen protozoa and protective effect of clinoptilolite-rich zeolite on bacterial and protozoan concentration in rumen. *Folia Microbiol.* **6,** 567-571.

Prasad K. and Chhabra A. (2001). Organo-chlorine pesticide residues in animal feeds and fodders. *Indian J. Anim. Sci.* **71(12),** 1178-1180.

Raikwar M. and Nag S. (2003). Organochlorine pesticide residues in animal feeds. Pp. 54-57 in Proc. 40th Annu. Conven. Chem. Indian Chemical Society, India.

Rumsey T., Williams E. and Evans A. (1975). Tissue residues, performance and ruminal and blood characteristics of steers fed ronnel and activated carbon. *J. Anim. Sci.* **40,** 743-749.

Saleh M. (1994). Using of feed additives for feeding farm animals. Ph D. Thesis. Tanta University, Tanta, Egypt.

SAS Institute. (1999). SAS®/STAT Software, Release 8. SAS Institute, Inc., Cary, NC. USA.

Schwartz C.C., Nagy J. and Streeter C. (1973). Pesticide effect on rumen microbial function. *J. Anim. Sci.* **37(3),** 821-826.

Steel R.G. and Torrie J.H. (1960). Principles and procedures of statistics. McGraw-Hill Book Company, London, New York.

Stephenson R., Huff J., Krebs G. and Howitt C. (1992). Effect of molasses, sodium bentonite and zeolite on urea toxicity. *Australian J. Agric. Res.* **43,** 301-310.

Van Cong N., Phuong N.T. and Bayley M. (2009). Effects of repeated exposure of diazinon on cholinesterase activity and growth in snakehead fish (*Channa striata*). *Ecotoxicol. Environ. Saf.* **72(3),** 699-703.

Van Soest P.V., Robertson J. and Lewis B. (1991). Methods for dietary fiber, neutral detergent fiber and nonstarch polysaccharides in relation to animal nutrition. *J. Dairy Sci.* **74,** 3583-3597.

Wallace R. and Newbold C. (1991). Effects of bentonite on fermentation in the rumen simulation technique (Rusitec) and on rumen ciliate protozoa. *J. Agric. Sci. Camb.* **116,** 163-168.

Williams A.G. Withers S.E. (1993). Changes in the rumen microbial population and its activities during the refaunation period after the reintroduction of ciliate protozoa into the rumen of defaunated sheep. *Canadian J. Microbiol.* **31,** 61-69.

Zertal A., Jacquet M., Lavédrine B. and Sehili T. (2005). Photodegradation of chlorinated pesticides dispersed on sand. *Chemosphere.* **58,** 1431-1437.

Estimation of Economic Values for Fertility, Stillbirth and Milk Production Traits in Iranian Holstein Dairy Cows

H. Ghiasi[1*], A. Pakdel[2], A. Nejati-Javaremi[3], O. González-Recio[4], M.J. Carabaño[4], R. Alenda[5] and A. Sadeghi-Sefidmazgi[2]

[1] Department of Animal Science, Faculty of Agriculture, Payame Noor University, Tehran, Iran
[2] Department of Animal Science, College of Agriculture, Isfahan University of Technology, Isfahan, Iran
[3] Department of Animal Science, Faculty of Agriculture and Natural Resources, University of Tehran, Karaj, Iran
[4] Departamento de Mejora Genética, Instituto Nacional de Investigación y Tecnología Agraria y Alimentaria, Madrid, Spain
[5] Departamento de Producción Animal, Universidad Politécnica de Madrid, Madrid, Spain

*Correspondence E-mail: ghiasi@ut.ac.ir

ABSTRACT

The objective of present study was to derive the economic values for number of inseminations to conception, calving interval, milk yield and stillbirth, using economic data of 10 Iranian Holstein herds. The economic values were derived by using the profit function methods and differentiating a profit equation with respect to the traits of interest. The cow fertility costs herd amortization or replacement cost and cow feed cost were included in the profit function. The average of feed cost per cow per day was 8.65 USD. The total feed cost comprised 61 percent of milk production, 23 percent of maintenance, 12 percent of pregnancy and 4 percent of growth. In calculation of cow feed cost, the estimated cost for each Mcal and per g protein of feed were 0.0006 and 0.165 USD, respectively. The replacement cost of each heifer per cow herd was 1719 USD. The average cost of each insemination was 30 USD. The estimated economic values for the number of insemination, calving interval, milk yield and one percent unit of stillbirth, were -82, -2.08, 0.193 and -1.27 USD per cow/per year, respectively. The results of the current study suggested that improving the number of inseminations, calving interval, milk yield and stillbirth will have a positive effect on the profitability of Iranian Holstein cows.

KEY WORDS calving interval, cow feed cost, cow fertility cost, fertility, number of insemination, selection index.

INTRODUCTION

According to Fewson (1993), the breeding objective is defined as "developing vital animals. Which ensure that the profit will be as high as possible under future commercial conditions of production. On the other hand, the aim of breeding programs in the dairy cattle industry is to produce cows that maximize profitability for the farmer (Kearney, 2007). The profitability of dairy herds depends on the productive life-time which depends on a number of traits including level of production and functional traits. The functional traits are animal characteristics that increase the effi-

ciency via reduction of production costs (Forabosco, 2005). Fertility is one of the most important functional traits, so that improvement of this trait will increase the net returns via reduction of calving interval, involuntary culling rate, replacement cost and also by increasing milk production (Bagnato and Oltenacu, 1994). Moreover, the level of milk production is the major source of income in dairy farm. However selection on milk production alone could increase the production cost, through negative genetic correlations that exist between milk production and functional traits (Young, 1970). In the past decades and in many breeding goals, the most emphasis had focused on milk production except in Scandinavian and North American countries. However, in the beginning of 2000, the most of developing countries have shifted their emphasis from milk production to functional traits because of quota-based milk marketing systems, price constraints, Labor costs and deterioration of the functional traits. The average of relative emphasis for production and reproduction traits in selection indices in 2003 was 59.5 and 12.6%, respectively. In comparison among countries, Danish S-Index focused the most emphasis (37%) on health and reproduction performance (Miglior et al. 2005).

The first and the most important steps to define a breeding program is definition of breeding objectives. The economic values and the genetic parameters are two components that should be estimated in defining breeding objective and to develop selection indices (Dekkers, 2003). In other words, the economic values are the key point in defining the breeding objectives and a criterion for evaluating the livestock improvement programs (Groen et al. 1997). Moreover, estimation of economic value for the important traits was required to establish an economic total merit index.

In animal breeding, the economic value of the trait was defined as the value of a unit change in the mean of the trait while keeping constant the other traits in the aggregate genotype (Berry et al. 2005). Therefore, the breeding objectives have to define in economic terms and the important production and functional traits should be included in the breeding goal, based on their economic importance. Knowing the economic value for key traits in the breeding goal enables the establishment of an economic total merit index (TMI). The TMI then permits the assessment of these traits on sustainable breeding programs that will benefit future farm productivity. The objective of the present study was to derive the economic values for number of insemination, calving interval, milk yield and stillbirth in Iranian Holstein cows.

MATERIALS AND METHODS

Estimation of economic values for fertility traits
Economic data

The economic information from 10 large Iranian herds was used to estimate economic values. The economic data including production cost of each kg milk, the price of each kg milk, the price of a 3 months of age calves (male and female), the cost of raising a 3 months of age calve, the price of heifer, the cost of raising a heifer, the average of salvage value, the average cost of each doses of semen, the cost of hormonal treatment for reproduction purpose, the average veterinary fee per insemination and the average cost of genetic counselor (Table 1).

Table 1 Economic and performance parameters used to estimate economic value

Parameter	Value 2008
M (kg)	9200
P (USD)	0.51
CMP (USD)	0.36
CP (USD)	555
CRC (USD)	405
CI (day)	415
INS	3
CM (%)	0.035
SB (%)	0.05
SP (USD)	20
EHTC (USD)	11
CIN (USD)	10
L	3
SL	1000
LFH	3.33
CRH (USD)	2777
FXC (USD)	500
COM (%)	3

M: milk production for the average calving interval; P: average price of milk per kg including bonus; CMP: average cost of milk production per kg including lactation feed cost and other variable costs related to milk production; CP: average price of 3 month female and male calf; CRC: average cost of rearing 3 months female or male calf; CI: average calving interval; INS: number of insemination per conception; CM: mortality of calves from birth to selling; SB: average stillbirth; SP: cost of doses of semen; EHTC: cost of per hormonal treatment; CIN: veterinary fee per insemination; L: average herd life in number of lactation; SL: salvage value; LFH: Labor of the farmer per hour ($); CRH: cost of replacement heifer; FXC: fixed costs per cow per year and COM: average cow mortality.

Profit equation

The profit function was constructed as follows:

$$P = R - C$$
$$R = [((M_{AVCI}+(a \times CI)) \times P) + ((1-CM) \times (1-SB-NH)) \times CP] \times 365 / CI$$
$$C = FXC + CFERC + HA + FCP + [(((1-CM) \times (1-SB-NH)) \times CRC) + ((M_{AVCI}+(a \times CI)) \times CMP) + (CI \times FCMA)] \times 365 / CI$$

Where:

R: average revenue per cow per year.

C: average cost per cow per year.

M_{AVCI}: milk production for the average calving interval. CI: average calving interval.

a: regression coefficient for the linear regression (kg milk/day CI).

P: average price of milk per kg including bonus.

CM: mortality of calves from birth to selling.

SB: average stillbirth.

NH: portion of female calves determined for replacement (as proportion of all born calves, including stillbirth).

CP: average price of 3 month female and male calf.

FXC: fixed costs per cow per year.

CFERC: cow fertility costs.

HA: herd amortization or replacement cost.

FCP: cow feed cost for pregnancy.

CRC: average cost of rearing 3 months female or male calf.

CMP: average cost of milk production per kg including lactation feed cost and other variable costs related to milk production.

FCMA: feed costs for maintenance.

It should be noted that in order to estimate the economic value for calving interval, milk production and feed cost were expressed as a function of calving interval. Feed cost for pregnancy was independent of calving interval and were assumed that the cow was mature and, therefore, feed costs for growth were not included. Feed cost for lactation was added to other variable costs for milk production and related to the calving interval. Cow feed cost for maintenance was calculated per cow per day and then multiplied by calving interval.

Cow fertility cost (CFERC)

Cow fertility costs were obtained from the costs of insemination and hormonal treatments.

Average insemination cost (AIC)

To calculate AIC, veterinary fees per insemination, cost of each dose of semen and cost of genetic counselor were included.

$$AIC = INS \times (SP+CIN) + GC$$

Where:

INS: number of insemination per conception.

SP: average cost of doses of semen.

CIN: veterinary fee per insemination.

GC: cost of genetic counselor.

In these herds that were used in the current study and before of each insemination, a cow genetic counselor determines which cow should be inseminated with which of the available sperm. Therefore the cost of genetic counselor was added to the average of insemination cost.

Cost of hormonal treatment for fertility problems (THC)

This cost is only for the cows that need more than one insemination per conception and was calculated as follows:

$$THC = (INS-1) \times (EHTC)$$

Where:

EHTC: cost of one hormonal treatment including veterinary cost and treatment for removing corpus luteum, synchronization and other related costs.

Total cow fertility cost was then obtained by summing of THC and AIC:

$$CFERC = AIC + THC$$

Herd amortization or replacement cost (HA)

To calculate HA, the following equation was used:

$$HA = 1 / (LH \times (CI/365)) \times [(1/(1-HM) \times THC) - ((1-CM) \times SV)]$$

Where:

LH: average number of lactations.

CI: average calving interval.

THC: total cost of rearing a heifer.

HM: average heifer mortality.

COM: average cow mortality.

SV: average salvage value.

Derivation of economic values

The economic value of trait X was estimated by partially deriving the profit function with respect to the trait X. As we know, by each day increase in calving interval, the revenue and the cost of milk will be changed, but milk production and calving interval were included in the aggregate genotype. Therefore, in order to estimate economic value for calving interval to use in index we should assume that milk production will be constant. Therefore, we use M instead of $(MAVCI+(a \times CI))$ in the profit function and for economic evaluation of calving interval the profit function with $(MAVCI+(a \times CI))$ were used.

RESULTS AND DISCUSSION

Cow feed cost

The cost of each Mcal and each g/protein plus cow feed cost for each cow with 650 kg body weight and 28 kg daily milk production are presented in Table 2. The cost of each Mcal and g/protein was 0.0006 and 0.152 USD, respectively. From the total cow feed cost, 61 percent was the cost of milk production, 23 percent was the cost of maintenance, 12 percent was the cost of pregnancy and 4 percent was the cost of growth.

Replacement or herd amortization cost

The cost of replacement of one heifer instead of calling cow was 3094.5 USD.

The result showed that if a farmer decided to produce replacement heifer in owner farm of 36 percent of all born female calf should be kept as a replacement.

Table 2 Daily cow feed cost and cost of each Mcal and g/protein

Variable	USD
Each M/cal	0.165
Each g/protein	0.0006
Cow feed cost for maintenance	1.95
Cow feed cost for milk production	5.21
Cow feed cost for growth	0.31
Cow feed cost for pregnancy	1.02
Total feed cost	8.65

Cow fertility cost

The cost of each insemination was 30 USD. If the cow doesn't pregnant in the first insemination, she needs hormonal treatment that average of each hormonal treatment was 10 USD.

Economic values for milk yield and fertility traits

The most important trait was INS, milk yield, calving interval and stillbirth were 0.01, 0.06 and 7.8 percent as important as number of inseminations per conception, respectively. Estimated economic values for milk production, CI, INS and SB are shown in Table 3.

Table 3 Estimated absolute and relative (ratio to INS) economic values (USD) for milk production (M), number of inseminations per conception (INS), calving interval (CI) and one percent unit of stillbirth (SB)

Traits	Absolute economic value economic value ($/cow per year)	Relative economic value
M (kg)	0.193	0.0001
INS	-82	1.0000
CI (day)	-2.08	-0.0006
SB (%)	-1.27	-0.0779

A positive economic value was estimated for milk yield. Therefore selection to increase milk production should be considered in selection programs in Iranian Holstein population. An increase of 1 kg/yr in milk yield will cause 0.193 USD more profit per cow. Negative economic values were obtained for INS, CI and SB; meaning that an increase in either INS, CI or SB will decrease profit. An increase of CI by one day over the average calving interval will decrease profitability by 2.08 USD per cow/year. Estimated economic value for calving interval by González-Recio et al. (2004); Pérez-Cabal and Alenda (2003) and Plaizier et al. (1997) were -4.9 USD (US), -0.37 € and -4.7 USD per cow per year, respectively.

The economic loss associated to enlarging CI results an increase in food cost. To estimate economic values for calving interval we assumed that the dry period was constant. Therefore, a change in calving interval is the result of changes in lactation length. In this case, longer calving intervals imply that cows enlarge the last part of lactation (lower daily yields), which can be reducing the average daily milk yield per lactation.

Then, we would expect that lower yielding cows loose more income per day from enlarging calving interval than higher yielding cows. The negative economic value estimated for INS indicated that for each unit increase in INS, the profitability will decrease by 82 USD per cow/year, which is slightly larger than the 67.52 USD reported by González-Recio et al. (2004). In the present study, the economic values for two fertility traits (CI and INS) were estimated.

Economic value of INS is higher than the economic value of CI. Therefore if the aim is to improve fertility performance, INS is an important fertility trait that should be considered in the breeding goal. One unit increase in INS, will increase fertility cost (doses of semen, veterinary fee and hormonal treatment) and the cost associated with increasing calving interval.

One unit increase in INS causes an increase in CI interval of at least 21 days. It is obvious that cows with low conception rates need more INS. Oltenacu et al. (1981) and McMahon et al. (1985) reported an increase in net income of 3.50 USD and 2 USD per each unit increase in sire conception rate, respectively.

Economic value for stillbirth

Economic loss due to one percent stillbirth rate was -1.27 USD. Given that there is no specialized beef breeds in Iran, many farmers buy a male calf from dairy herds for fattening. This large economic value of stillbirth indicates that in Iranian dairy farms selling calves is the most important revenue for farmer after milk sales. In this study only direct stillbirth was considered. Therefore, indirect costs associated with stillbirth (decrease in milk culling risk) were not considered for calculation of economic loss due to stillbirth.

CONCLUSION

The results of present study suggest that improving number of insemination and milk yield, while reducing calving interval and the incidence of, stillbirth will have positive effect on the profitability of Iranian Holstein cows. The positive economic value was estimated for milk yield which indicates in Iranian economic condition selection based on milk production can be reasonable. Stillbirth should be included in the breeding goal because of its large economic value. Due to unfavorable genetic correlation reported between milk production and fertility trait and because of economically importance of milk production and fertility, it is better to include both milk yield and INS in the breeding goal.

ACKNOWLEDGEMENT

The author grateful to University of Tehran for founding this project.

REFERENCES

Bagnato A. and Oltenacu P. (1994). Phenotypic evaluation of fertility traits and their association with milk production of Italian Friesian cattle. *J. Dairy Sci.* **77,** 874-882.

Berry D.P., Shalloo L., Cromie A.R., Olori V.E. and Amer P. (2005). Economic Breeding Index for Dairy Cattle in Ireland. Dairy Production Department, Teagasc, Moorepark Production Research Center, Fermoy, Co. Cork, Ireland.

Dekkers J.C.M. (2003). Design and Economics of Animal Breeding Strategies. Notes for summer short course, Iowa State University, USA.

Fewson D. (1993). Definition of Breeding Objective. Design of Livestock Breeding Programs. Animal Genetics and Breeding Unit, University of New England, Armidale, NSW, Australia.

Forabosco F. (2005). Breeding for longevity in Italian Chianina cattle. MS Thesis. Wageningen Univ., Netherlands.

González-Recio O., Pérez-Cabal M. and Alenda R. (2004). Economic value of female fertility and its relationship with profit in Spanish dairy cattle. *J. Dairy Sci.* **87,** 3053-3061.

Kearney F. (2007). Improving dairy herd fertility through genetic selection. *Irish Vet. J.* **60,** 376-379.

McMahon R.T., Blake R.W., Shumway C.R., Leatham D.J., Tomaszewski M.A. and Butcher K.R. (1985). Effects of planning horizon and conception rate on profit-maximizing selection of artificial insemination sires. *J. Dairy Sci.* **68,** 2295-2302.

Miglior F., Muir B. and Van Doormaal B. (2005). Selection indices in Holstein cattle of various countries. *J. Dairy Sci.* **88,** 1255-1263.

Oltenacu P., Rounsaville T., Milligan R. and Foote R. (1981). Systems analysis for designing reproductive management programs to increase production and profit in dairy herds. *J. Dairy Sci.* **64,** 2096-2104.

Pérez-Cabal M. and Alenda R. (2003). Lifetime profit as an individual trait and prediction of its breeding values in Spanish Holstein cows. *J. Dairy Sci.* **86,** 4115-4122.

Plaizier J., King G., Dekkers J. and Lissemore K. (1997). Estimation of economic values of indices for reproductive performance in dairy herds using computer simulation. *J. Dairy Sci.* **80,** 2775-2783.

Young C. (1970). What additional traits need to be considered in measuring dairy cattle utility in the future? *J. Dairy Sci.* **53,** 847-851.

Anti-oxidative Effects of Ethanol Extract of *Origanum vulgare* on Kinetics, Microscopic and Oxidative Parameters of Cryopreserved Holstein Bull Spermatozoa

H. Daghigh Kia[1*], R. Farhadi[1], I. Ashrafi[1] and M. Mehdipour[1]

[1] Department of Animal Science, Faculty of Agriculture, University of Tabriz, Tabriz, Iran

*Correspondence E-mail: daghighkia@tabrizu.ac.ir

ABSTRACT

Origanum vulgare contains high levels of phenolic compounds such as gallic acid and polyphenols such as rosmarinic acid and quercetin. The purpose of this study was to investigate the effects of *Origanum vulgare* extract, as a natural antioxidant, on freezing-thawing semen quality in Holstein bulls. Three Holstein bulls (5-6 years old, mean live weight 800 kg) were used for semen collection twice a week for two months. Ethanol extract of *Origanum vulgare* (2, 4, 8, 12, 16 and 20 mL/dL extender) was added to a citrate-yolk-base extender. After freezing-thawing sperm motility parameters, viability and membrane integrity were determined using a CASA system, eosin-nigrosin staining, and hypo-osmotic swelling test, respectively; malondialdehyde concentration and activity of superoxide dismutase and catalase were also measured. The percentages of motility were higher ($P<0.05$) in the freezing extender containing 4 mL/dL *Origanum vulgare* extract (72.34 ± 7.98). Addition of 2 and 4 mL/dL extract of *Origanum vulgare* significantly improved the motility, viability and plasma membrane integrity of the spermatozoa following freezing-thawing process compared to the control group. Addition of 4 mL/dL extract significantly reduced the concentration of malondialdehyde compared to the control group ($P<0.05$). The activity of superoxide dismutase and catalase increased significantly by inclusion of 4 and 8 mL/dL extract of *Origanum vulgare* to the extender (1.86 ± 0.18 and 1.92 ± 0.28 U/mg protein; 4.54 ± 0.13 and 4.28 ± 0.28 U/mg protein, respectively). The activity of superoxide dismutase and catalase was significantly increased by inclusion of 4 and 8 mL/dL extract to the extender. In conclusion, addition of 2 and 4 mL/dL extract of *Origanum vulgare* to the semen extender improved the post-thawed quality of semen, which may be due to increasing in antioxidant enzyme activity and reduction in lipid peroxidation.

KEY WORDS antioxidant, bull sperm, cryopreservation, *Origanum vulgare*.

INTRODUCTION

The cryopreservation of spermatozoa has provided special opportunities for the preservation of genetic resources and improving breed programs by the artificial insemination technique (Holt, 1996). Nowadays, semen cryopreservation has many applications such as solving problems of infertility, some diseases, conservation semen and DNA conserva-tion of some important species (Barbas and Mascarenhas, 2009). However, sperm cryopreservation stimulates intra-cellular ice crystals formation, increasing osmotic and chilling injury that causes sperm cell damage (Isachenko *et al.* 2003). Freezing and thawing processes impose physical and chemical insults on the sperm membrane that decrease sperm viability and fertilizing ability (Alvarez and Storey, 1992). Both damages are associated with excessive generation of reactive oxygen species (ROS) and peroxidation of the phospholipids in the membrane (Wang *et al.* 1997; Lasso *et al.* 1994). The imbalance between ROS production and biological systems that control free radicals is the main cause of damage to the membrane and structural composition of sperm (Bilodeau *et al.* 2000).

Mammalian semen contains antioxidant compounds including superoxide dismutase (SOD), catalase (CAT), reduced glutathione (GSH), glutathione peroxidase (GSH-PX), taurine and hypotaurine that protect the sperm against peroxidative damage (Bilodeau *et al.* 2000; Zini *et al.* 1993). However, the endogenous anti-oxidative capacity may be insufficient to prevent lipid peroxidation (LPO) during cooling storage of sperm; therefore, the use of antioxidants during the freeze-thaw process may be beneficial (Aurich *et al.* 1997). Using natural antioxidants is increasing because of the toxicity problems of synthetic antioxidants such as butylated hydroxyl toluene (BHT), propylgalate (PG), butylatedhyd roxylanisole (BHA) and tertiary butyl hydroquinone (TBHQ) that are commonly used in lipid-containing foods. However, there are limited data regarding the effects of natural antioxidants on spermatozoa (Yanishlieva and Marinova, 1996).

Origanum vulgare, (Lamiaceae) is an aromatic plant with a wide distribution throughout Asia and especially in Iran. *Origanum vulgare* contains some aqueous compounds such as rosmarinic acid, eriocitrin, luteolin-7-oglucoside, apigenin-7-o-glucoside (Kulišić *et al.* 2007), origanol A and B (Matsuura *et al.* 2003) and ursolic acid (Heo *et al.* 2002). Rosmarinic acid and origanol A and B, the most important components of the aqueous extract of *Origanum vulgare* have anti-oxidative activities (Matsuura *et al.* 2003; Kulišić *et al.* 2007).

No study has reported the effects of *Origanum vulgare* extract in semen extenders against cryo-damage to bull sperm. The present study was conducted to determine the effect of an ethanol extract of *Origanum vulgare* in the semen extender on sperm motility parameters, viability, plasma membrane integrity, total antioxidant capacity (TAC) and LPO, as well as antioxidant activities in terms of superoxide dismutase (SOD) and catalase (CAT), in post-thawed bull semen.

MATERIALS AND METHODS

Preparation of *Origanum vulgare* extract
Collected *Origanum vulgare* plants were dried for 10 days at room temperature and powdered by an electric mill. Fifty grams of the plant powder was soaked in 60% ethanol for 24 hours and the mixture was filtered. Soxhlet apparatus was used to evaporate organic solvent and the concentrated material was obtained as extracts and maintained at 4 °C

until used. Different concentrations of *Origanum vulgare* extract were prepared in accordance with the protocol of Dehghan *et al.* (2007).

Semen source and preparation
This experiment was performed at the Animal Breeding Center located in Tabriz city, northwest of Iran. Ejaculates were collected from three Holstein bulls (5-6 years old, mean live weight of 800±50 kg), regularly used for breeding purpose, based on their fertility estimation through *in vitro* tests, including viability and motility evaluation and field fertility tests. Semen samples were collected twice a week for two months by an artificial vagina (45 °C). The ejaculates were immediately transferred to the laboratory and submerged in a water bath (34 °C), until semen evaluation was done. The volume of the ejaculate was estimated in a conical tube graduated at 0.1 mL intervals. Sperm concentration was determined by means of an Accucell photometer (IMV Technologies, L'Aigle, France). The ejaculates meeting the following criteria were used: volume between 5 and 10 mL; sperm concentration $\geq 1 \times 10^9$ sperm mL^{-1}; percentage of motile sperm $\geq 70\%$; and $\leq 10\%$ abnormal sperm. The ejaculates were pooled to compensate for within-individual variations in seminal quality. A citrate-egg yolk extender containing sodium citrate dihydrate (2.9 g dL^{-1}), penicillin (1000 IU mL^{-1}), streptomycin (1000 µg mL^{-1}), 25% hen egg yolk, 7% glycerol and double-distilled water (to 100 mL) was used (Ashrafi *et al.* 2011).

The extender was divided into two parts (A and B); 3% and 11% (v/v) glycerol were added to part A and part B, respectively. Part A was warmed to 37 °C while part B cooled to 5 °C. *Origanum vulgare* extract was added to part A at 0, 2, 4, 8, 12, 16 and 20 mL/dL. The pooled ejaculate was divided into seven equal aliquots and diluted in the semen extender containing the extract, at a final concentration of 1×10^8 sperm per mL. This was then cooled slowly to 5 °C, mixed with part B, packaged in 0.5 mL straws, sealed and frozen on liquid nitrogen vapor at approximately −15 °C min^{-1} from +5 to −150 °C. Then straws were stored in a liquid nitrogen tank. Frozen straws were thawed at 37 °C for 30 s in a water bath for microscopic evaluation (Ashrafi *et al.* 2011).

Semen evaluation
Viability evaluation
Sperm viability was assessed by means of the nigrosin–eosin staining method. The final composition of the stain was: eosin-Y 1.67 g, nigrosin 10 g and sodium citrate 2.9 g, dissolved in 100 mL of distilled water. A sub-sample (5 µL) of frozen-thawed semen was placed on a pre-warmed slide and mixed with 5 µL nigrosin-eosin stain and spread into a uniform smear (Mahmood and Ijaz, 2006). After air-

drying, the smear was observed under a microscope (Nikon, Japan) at 1000 × magnification, using immersion oil to count the live (unstained) and dead (stained/partial stained) sperm.

Evaluation of motility parameters

For evaluating the motility parameters, sperm samples were incubated after thawing for about 5 min at 37 °C. A computer-assisted sperm motility analysis (CASA, Hoshmand Fanavaran, Version 6, Amirkabir Medical Engineering Co., Tehran, Iran) was used to analyze sperm motion characteristics. A 5 µL of diluted semen was transferred onto a pre-warmed microscope slide (37 °C) and covered by a cover slip. The following variables were obtained: total motility (TM, %), progressive motility (PM, %), average path velocity (VAP, µm/s), straight linear velocity (VSL, µm/s), curvilinear velocity (VCL, µm/s), amplitude of lateral head displacement (ALH, µm), linearity (LIN, %). At least 200 spermatozoa were assessed in each CASA analysis.

Assessment of sperm membrane integrity

The hypoosmotic swelling test (HOST) was used to evaluate the functional integrity of the sperm membrane, based on coiled and swollen tails, where 50 µL of thawed semen was mixed with 500 µL of a 100 mOsM hypo-osmotic solution (9 g fructose+4.9 g sodium citrate per liter distilled water) and was incubated at 37 °C for 60 min. Then, 5 µL of the mixture was spread with a cover slip on a warm slide at 37 °C. A total of 200 sperm were evaluated using brightfield microscopy (400×magnification) and the sperm with swollen or coiled tails were recorded (Revell and Mrode, 1994).

Biochemical assay

Frozen straws were thawed for 60 seconds at 37 °C for biochemical measurements. Five-hundred µL of semen sample was centrifuged at 800 × g for 10 min, the supernatant was removed and the pellet washed with phosphate buffered saline (PBS) (pH 7.4). This procedure was repeated for three times. After the last centrifugation, 1mL of deionized water was added to the pellet and was snap-frozen and stored at -70 °C until further analysis (Roca et al. 2004).

Concentration of MDA (MDA), as an indicator of lipid peroxidation in the semen sample, was measured using the thiobarbituric acid reaction as described by Placer et al. (1996).

The absorbance of the supernatant was read using a spectrophotometer (UV-1200, Shimadzu, Japan) at 532 nm. Superoxide dismutase (SOD) was measured according to the method described by Ukeda et al. (1997). The reaction is dependent on the presence of superoxide anions that cause pyrogallol oxidation. The inhibition of pyrogallol

oxidation by SOD was monitored using a spectrophotometer at 420 nm and the amount of enzyme producing 50% inhibition was defined as one unit of enzyme activity (U/mg protein). Catalase (CAT) was measured by monitoring the decomposition of hydrogenperoxide (Aebi, 1984).

Statistical analysis

Each treatment was replicated 5 times. The experiment was done by completely randomized design. Data were analyzed by SAS (2001) software using the GLM procedure. The Tukey-Kramer's test was used for mean comparisons. Data were expressed as mean ± SD. The probability values less than 0.05 were considered as statistically significant.

RESULTS AND DISCUSSION

The effects of Origanum vulgare extract in the freezing extender on post-thaw motility parameters, viability and membrane integrity in bull sperm are presented in Table 1. The percentages of TM were higher (P<0.05) in the freezing extender containing 4 mL/dL Origanum vulgare extract (72.34±7.98).

The percentages of PM, LIN and VAP were higher (P<0.05) in the extender containing 4 mL/dL extract (64.91±5.62; 62.25±1.19 and 37.75±3.39, respectively). For parameters of VCL the highest performance (P<0.05) was observed at 2, 4 and 8 mL/dL of extract. Addition of 2 and 4 mL/dL extract of Origanum vulgare to the extender significantly improved VSL parameter compared to the control group (37.98±5.77 and 40.11±2.68, respectively). The ALH values were not different (P<0.05) between treatments.

Addition of 2 and 4 mL/dL extract of Origanum vulgare to the extender improved (P<0.05) the viability parameter (73.91±4.85 and 79.13±6.55) and addition of 4 mL/dL extract to the extender improved (P<0.05) the sperm plasma membrane integrity compared to the control group. The effects of Origanum vulgare extract on oxidative parameters in thawed bull semen are summarized in Table 2. Inclusion of 4 mL/dL extract of Origanum vulgare significantly resulted in lower concentration of MDAMDA compared to the control group (15.69±0.76 versus 18.16±1.08 nmol/dL). On the other hand, with increasing extract levels, MDAMDA concentration increased, therefore adding of 20 mL/dL of extract (P<0.05) increased MDAMDA concentrations compared to the control group (20.19±1.61 vs. 18.16±1.08 nmol/dL). The activity of SOD and CAT increased significantly by inclusion of 4 and 8 mL/dL extract of Origanum vulgare to the extender (1.86±0.18 and 1.92±0.28 U/mg protein; 4.54±0.13 and 4.28±0.28 U/mg protein, respectively). Sperm cooling causes some functional and structural damages to sperm in a time-dependent process in most species.

Table 1 Mean (±SEM) percentages of sperm motion parameters, viability and membrane integrity of frozen–thawed Holstein bull sperm at different levels of *Origanum vulgare* extract

Parameters	Levels of *Origanum vulgare* extract (mL/dL)						
	Control	2	4	8	12	16	20
TM (%)	56.56[bc]±6.51	68.79[ab]±4.53	72.34[a]±7.98	59.03[ab]±11.62	49.92[bcd]±11.71	40.69[cd]±8.66	33.55[d]±5.51
PM (%)	49.18[bc]±4.25	57.12[ab]±5.45	64.91[a]±5.62	49.78[bc]±5.78	41.15[cd]±10.61	33.53[de]±9.14	24.05[e]±6.49
VCL (μm/s)	42.34[b]±0.99	52.32[a]±3.61	52.63[a]±4.97	50.02[a]±7.11	42.83[b]±9.07	36.97[bc]±6.39	30.42[c]±3.59
VSL (μm/s)	29.11[bc]±1.47	37.98[a]±5.77	40.11[a]±2.68	34.28[ab]±3.74	28.28[bc]±6.47	23.22[cd]±4.24	17.81[d]±3.15
VAP (μm/s)	27.77[bcd]±3.09	35.39[ab]±6.81	37.75[a]±3.39	31.68[abc]±3.52	25.92[cde]±3.52	22.91[de]±5.81	18.04[e]±2.08
ALH (μm)	1.94[ab]±0.07	2.34[a]±0.18	2.31[a]±0.23	2.15[ab]±0.38	2.11[ab]±0.32	1.92[ab]±0.16	1.86[b]±0.21
LIN (%)	54.58[bc]±3.75	58.11[ab]±6.44	62.25[a]±1.19	52.48[bcd]±1.77	49.01[cd]±4.65	46.68[d]±0.81	38.94[e]±2.51
Viability (%)	66.31[cd]±1.95	73.91[ab]±4.85	79.13[a]±6.55	69.13[bc]±6.65	60.82[d]±7.74	50.99[e]±5.09	46.06[e]±2.66
HOST (%)	63.45[bc]±3.59	71.92[ab]±4.15	76.42[a]±6.65	65.33[abc]±7.37	57.03[cd]±5.58	48.25[de]±6.25	41.97[e]±4.66

TM: total motility (%); PM: progressive motility (%); LIN: linearity (%); ALH: amplitude of lateral head displacement (μm); VCL: curvilinear velocity (μm/s); VAP: average path velocity (μm/s) and HOST: hypo-osmotic swelling test (%).
The means within the same row with at least one common letter, do not have significant difference (P>0.05).
SEM: standard error of the means.

Table 2 Mean (±SEM) of the oxidative parameters in frozen-thawed bull semen at different levels of *Origanum vulgare* extract

Parameters	Levels of *Origanum vulgare* extract (mL/dL)						
	Control	2	4	8	12	16	20
MDA (nmol/dL)	18.16[ab]±1.08	17.31[bc]±1.69	15.69[c]±0.76	18.57[ab]±1.79	18.82[ab]±2.03	20.04[a]±1.69	20.19[a]±1.61
SOD (U/mg protein)	1.46[bc]±0.13	1.62[abc]±0.26	1.86[a]±0.18	1.92[a]±0.28	1.32[c]±0.18	1.25[c]±0.19	1.23[c]±0.21
CAT (KU/mg protein)	0.12±3.81[b]	3.88[b]±0.13	4.54[a]±0.13	4.28[a]±0.28	3.72[b]±0.24	3.61[b]±0.25	3.53[b]±0.24

MDA: malondialdehyde; SOD: activity level of superoxide dismutase and CAT: catalase.
The means within the same row with at least one common letter, do not have significant difference (P>0.05).
SEM: standard error of the means.

For preventing these damages, the semen needs to be diluted with appropriate extenders for freezing-storage (Buyukleblebici *et al.* 2014). *In vitro* studies have shown that antioxidants remove free radicals such as superoxide ion, hydrogen peroxide, hydroxyl radicals and peroxyl (Ashrafi *et al.* 2013). Due to the ability of phenolic compounds to act as chelating in binding the hydroxyl group and metal ions, they inhibit the activity of free radicals (Osawa, 1994). Antioxidant activity of *Origanum vulgare* extract against lipid peroxidation has been demonstrated by several researchers (Vekiari *et al.* 1993). Several studies have shown that *Origanum vulgare* contains phenolic compounds and flavonoids, especially rosmarinic acid, quercetin, kaempferol, apigenin, rutinand origanol which can cleanse free radicals such as superoxide anion and hydroxyl (Kikuzaki and Nakatani, 1989; Cervato *et al.* 2000). In the current study, sperm motion parameters as well as sperm viability increased with inclusion of 2 and 4 mL/dL *Origanum vulgare* extract in the semen extender. The peroxidation of sperm lipids damages the structure of lipid matrix in the spermatozoa membranes and it is related to quick loss of intracellular ATP leading to axonemal damage and reduced sperm viability (Sanocka and Kurpisz, 2004).

There is a strong correlation between ROS production and decreased sperm motility (Armstrong *et al.* 1999), so it has been determined that peroxide (H_2O_2) radical can be distributed across the sperm membranes and there is an inhibited activity of key enzymes such as glucose-6-phosphate dehydrogenase (Aitken *et al.* 1997).

This enzyme controls glucose concentration through diverting path of hexose monophosphate and NADPH activity which have major role in ATP production and sperm motility (Aitken *et al.* 1997). The results of the present study showed that inclusion of 4 mL/dL *Origanum vulgare* extract in the freezing media increased plasma membrane integrity of frozen-thawed bull sperm. Previous work (Cervato *et al.* 2000) have shown that *Origanum vulgare* extract is effective in prevention of the peroxidative procedure by counteracting different type of free radicals in the first step, then stopping peroxidation catalysis by iron-chelating and iron-oxidizing properties and at last pausing lipid-radical chain reactions. Another advantageous activity was glycosylation of lipoproteins which is directly related to their peroxidation (Cervato *et al.* 2000). Several studies reported a negative correlation between the MDA production and sperm viability (Guthrie and Welch, 2012). The MDA production is usually used to determine LPO in various cell types including sperm cells (Sikka, 1996).

The semen antioxidant system having both enzymatic and non-enzymatic antioxidants prevents or limits peroxides formation. Inadequate amounts of antioxidants, or the prevention of antioxidant enzymes, increases oxidative stress, damaging spermatozoa. One of the byproducts of lipid peroxides decomposition is MDA, which is commonly used in biochemical assays to show the degree of peroxidative damage produced by spermatozoa (Najafi *et al.* 2014).

The results of present study showed that in the freezing extender containing 4 mL/dL *Origanum vulgare* extract

MDA concentration was significantly lower than that of the control group. Phenolic compounds (especially flavonoids) can change peroxidation kinetics by modifying the lipid packing order. They also stabilize membranes by reducing membrane fluidity and inhibit free radicals diffusion and restrict peroxidative reaction (Arora et al. 2000; Blokhina et al. 2003). Our data are in agreement with the finding Zhao et al. (2009), who found a significant correlation between Rhodiola sacra aqueous extract concentrations and MAD in frozen-thawed boar semen.

Membrane permeability increases after cooling and this may be due to the increased membrane leakiness and specific protein channels. Calcium regulation is influenced by cooling and this has harmful effects on cell function, consequently cell death. Calcium absorption during cooling affects capacitation and fusion events between plasma membrane and acrosomal membrane (Purdy, 2006). Peroxidation of polyunsaturated fatty acids in sperm cell membranes is an autocatalytic, self-propagating reaction, which can increase cell dysfunction associated with loss of membrane function and integrity.

The HOST test assessed the resistance of the sperm plasma membrane to damage induced by the loss in permeability under the stress of swelling driven by the hypoosmotic treatment. Thus, this provided a form of a membrane stress-test, which is particularly useful when testing the membrane-stabilizing action of antioxidants (Sariozkan et al. 2015).

Flavonoids increase membranes integrity by preventing the access of deleterious molecules to the hydrophobic region of the bilayer, including those that can affect membrane stability and those that induce oxidative damage to the membrane components (Michalak, 2006).

SOD is an enzymatic biological antioxidant, which scavenges ROS, therefore controls oxidative stress in mammalian sperm. Furthermore, GSH is able to directly react with many ROS and is a co-factor for GSH-Px, catalysing toxic H2O2 and hydroperoxides reduction (Bilodeau et al. 2001).

In our study the activity of CAT and SOD increased by inclusion of 4 and 8 mL/dL of extract in the extender. It seems that phenolic compounds stimulate the activities of antioxidant enzymes such as SOD and CAT which reduce the number of free radicals and may also increase the production of molecules protecting sperm cells against oxidative stress (Cervato et al. 2000). In the present study, the most effective concentration of Origanum vulgare extract in microscopic evaluations of bull sperm freezing extender was 4 mL/dL. On the other hand, almost for all measured parameters, inclusion of 16 and 20 mL/dL extract in the freezing extender can counteract the ROS-induced oxidative stress. So it may impede the ROS-associated functions of sperm (Roca et al. 2004) and it can be due to the osmotic

changes, pH and disturbing the balance extender compounds.

Previous studies reported that addition of rosemary extract in goat (Zanganeh et al. 2013) and boar (Malo et al. 2010; Malo et al. 2011) semen freezing extender had beneficial effects. So freezing extender improved the post-thaw sperm quality, showing a negative significant correlation between rosemary concentration and MDA concentration and also the number of free radicals, ROS. On the other hand, the use of Rosmarinus officinalis in semen extender increased the sperm motility parameters, sperms viability and production of molecules protecting sperm cells against oxidative stress. In addition to the negative effects on motility, viability and LP, ROS can also damage sperm mitochondria and consequently sperm motility. A negative correlation between sperm LPO and sperm motility in cryopreserved semen has also been observed (Amini et al. 2015).

In another study, addition of rosemary extract (10 g L^{-1}) to bull semen freezing extender significantly improved the post-thaw quality of bull semen by means of increasing the sperm motility, viability and reducing the amount of LPO (Daghigh Kia et al. 2014).

The results of this study showed that rosemary extract significantly increased intracellular defense systems and cell membrane compounds against the ROS production following cryopreservation (Daghigh Kia et al. 2014). Effects of the antioxidant properties of rosemary are related to some compounds such as carnosic acid, carnosol, rosmarinic acid and 3,4-dihydroxyphenyllactate (phenolic depside) (Bai et al. 2010; Mulinacci et al. 2011).

In similar studies, antioxidant properties of Rodhiola sacra aqueous, agenus of Chinese herb have been investigated and a correlation between concentrations of this antioxidant and MDA after thawing was observed (Zhao et al. 2009).

Recently, the beneficial effect of Ilex paraguayensis was exhibited in semen cryopreservation, presenting higher percentages of total and progressive motility (Malo et al. 2010).

The excessive production of ROS not only causes LPO and DNA fragmentation, but also influences ATP, NADPH production and proteins phosphorylation in the spermatozoa (Cocchia et al. 2011). Adenosine triphosphate is necessary for sperm motility. Also, phosphorylation of proteins plays a main role in capacitation, the acrosome reaction and sperm penetration of the zonapellucida (Cocchia et al. 2011). Therefore, increased ROS levels may be reduced fertilizing ability of the spermatozoa (Wishart, 1982).

Our results are in agreement with other studies showing that botanical extracts could benefit mammal semen post-thaw quality. However, this is the first study reporting the use of Origanum vulgare as an anti-oxidative supplement in semen extender. Also, it should be noticed that the animal species, the type of freezing extender and freezing methods of the present study were different from the mentioned studies. There is a need to research which of the different plants evaluated have a better performance, or if the use of mixes of botanical extracts can show a better performance.

CONCLUSION

Addition of *Origanum vulgare* extract to the semen extender at 2 and 4 mL/dL improved the quality of frozen-thawed bull semen, probably due to polyphenolic compounds having antioxidant activity.

REFERENCES

Aebi H. (1984). Catalase *in vitro. Method. Enzymol.* **105**, 121-126.

Aitken R.J., Fisher H.M., Fulton N., Gomez E., Knox W., Lewis B. and Irvine S. (1997). Reactive oxygen species generation by human spermatozoa is induced by exogenous NADPH and inhibited by the flavoprotein inhibitors diphenylene iodonium and quinacrine. *Mol. Reprod. Dev.* **47**, 468-482.

Alvarez J.G. and Storey B.T. (1992). Evidence for increased lipid peroxidative damage and loss of superoxide dismutase activity as a mode of sublethal cryodamage to human sperm during cryopreservation. *J. Anthol.* **13**, 232-241.

Amini M.R., Kohram H., Zare-Shahaneh A., Zhandi M., Sharideh H. and Nabi M.M. (2015). The effects of different levels of catalase and superoxide dismutase in modified Beltsville extender on rooster post-thawed sperm quality. *Cryobiology.* **70**, 226-232.

Armstrong J.S., Rajasekaran M., Chamulitrat W., Gatti P., Hellstrom W.J. and Sikka S.C. (1999). Characterization of reactive oxygen species induced effects on human spermatozoa movement and energy metabolism. *Free Radic. Biol. Med.* **26**, 869-880.

Arora A., Byrem T.M., Nair M.G. and Strasburg G.M. (2000). Modulation of liposomal membrane fluidity by flavonoids and isoflavonoids. *Arch. Biochem. Biophys.* **373**, 102-109.

Ashrafi I., Kohram H. and Ardabili F.F. (2013). Anti-oxidative effects of melatonin on kinetics, microscopic and oxidative parameters of cryopreserved bull spermatozoa. *Anim. Reprod. Sci.* **139**, 25-30.

Ashrafi I., Kohram H., Naijian H., Bahreini M. and Mirzakhani H. (2011). Effect of controlled and uncontrolled cooling rate on motility parameters of cryopreserved ram spermatozoa. *Afican J. Biol.* **10**, 8965-8969.

Aurich J., Schönherr U., Hoppe H. and Aurich C. (1997). Effects of antioxidants on motility and membrane integrity of chilled-stored stallion semen. *Theriogenology.* **48**, 185-192.

Barbas J.P. and Mascarenhas R.D. (2009). Cryopreservation of domestic animal sperm cells. *Cell. Tissue. Bank.* **10**, 49-62.

Bilodeau J.F., Blanchette S., Gagnon C. and Sirard M.A. (2001). Thiols prevent H_2O 2-mediated loss of sperm motility in cryopreserved bull semen. *Theriogenology.* **56**, 275-286.

Bilodeau J.F., Chatterjee S., Sirard M.A. and Gagnon C. (2000). Levels of antioxidant defenses are decreased in bovine spermatozoa after a cycle of freezing and thawing. *Mol. Reprod. Dev.* **55**, 282-288.

Blokhina O., Virolainen E. and Fagerstedt K.V. (2003). Antioxidants, oxidative damage and oxygen deprivation stress: a review. *Ann. Bot.* **91**, 179-194.

Buyukleblebici S., Tuncer P.B., Bucak M.N., Tasdemir U., Eken A., Buyukleblebici O., Durmaz E., Sariozkan S. and Endirlik B.U. (2014). Comparing ethylene glycol with glycerol and with or without dithiothreitol and sucrose for cryopreservation of bull semen in egg-yolk containing extenders. *Cryobiology.* **69**, 74-78.

Cervato G., Carabelli M., Gervasio S., Cittera A., Cazzola R. and Cestaro B. (2000). Antioxidant properties of oregano (*Origanum vulgare*) leaf extracts. *J. Food Biochem.* **24**, 453-465.

Cocchia N., Pasolini M., Mancini R., Petrazzuolo O., Cristofaro I., Rosapane I., Sica A., Tortora G., Lorizio R. and Paraggio G. (2011). Effect of sod (superoxide dismutase) protein supplementation in semen extenders on motility, viability, acrosome status and ERK (extracellular signal-regulated kinase) protein phosphorylation of chilled stallion spermatozoa. *Theriogenology.* **75**, 1201-1210.

Daghigh-Kia H., Olfati-Karaji R., Hoseinkhani A. and Ashrafi I. (2014). Effect of rosemary (*Rosmarinus officinalis*) extracts and glutathione antioxidants on bull semen quality after cryopreservation. *Spanish J. Agric. Res.* **12**, 98-105.

Dehghan G., Shafiee A., Ghahremani M.H., Ardestani S.K. and Abdollahi M. (2007). Antioxidant potential of various extracts from Ferula szovitsiana. in relation to their phenolic content. *Pharm. Biol.* **45**, 691-699.

Guthrie H. and Welch G. (2012). Effects of reactive oxygen species on sperm function. *Theriogenology.* **78**, 1700-1708.

Heo H.J., Cho H.Y., Hong B., Kim H.K., Heo T.R., Kim E.K., Kim S.K., Kim C.J. and Shin D.H. (2002). Ursolic acid of *Origanum majorana* reduces amyloid beta-induced oxidative injury. *Mol. cells.* **13**, 5-11.

Holt W. (1996). Alternative strategies for the long-term preservation of spermatozoa. *Reprod. Fertil. Dev.* **9**, 309-319.

Isachenko E., Isachenko V., Katkov I.I., Dessole S. and Nawroth F. (2003). Vitrification of mammalian spermatozoa in the absence of cryoprotectants: from past practical difficulties to present success. *Reprod. Biomed. Online.* **6**, 191-200.

Kikuzaki H. and Nakatani N. (1989). Structure of a new anti-oxidative phenolic acid from oregano (*Origanum vulgare*). *Agric. Biol. Chem.* **53**, 519-524.

Kulišić T., Kriško A., Dragović-Uzelac V., Miloš M. and Pifat G. (2007). The effects of essential oils and aqueous tea infusions of oregano (*Origanum vulgare*), thyme (*Thymus vulgaris*) and wild thyme (*Thymus serpyllum*) on the copper-induced oxidation of human low-density lipoproteins. *Int. J. Food Sci. Nutr.* **58**, 87-93.

Lasso J.L., Noiles E.E., Alvarez J.G. and Storey B.T. (1994). Mechanism of superoxide dismutase loss from human sperm cells during cryopreservation. *J. Androl.* **15**, 255-255.

Mahmood S.A. and Ijaz A. (2006). Effect of cold shock on frozen-thawed spermatozoa of buffalo and cow. Pp. 701-708 in Proc. 5th Asian Buffalo. Congr. Naning, China.

Malo C., Gil L., Cano R., Martínez F. and Galé I. (2011). Antioxidant effect of rosemary (*Rosmarinus officinalis*) on boar epididymal spermatozoa during cryopreservation. *Theriogenology*. **75**, 1735-1741.

Malo C., Gil L., Gonzalez N., Martinez F., Cano R., De Blas I. and Espinosa E. (2010). Anti-oxidant supplementation improves boar sperm characteristics and fertility after cryopreservation: comparison between cysteine and rosemary (*Rosmarinus officinalis*). *Cryobiology*. **61**, 142-147.

Matsuura H., Chiji H., Asakawa C., Amano M., Yoshihara T. and Mizutani J. (2003). DPPH radical scavengers from dried leaves of oregano (*Origanum vulgare*). *Biosci. Biotechnol. Biochem*. **67**, 2311-2316.

Michalak A. (2006). Phenolic compounds and their antioxidant activity in plants growing under heavy metal stress. *Polish J. Environ. Stud*. **15**, 523-601.

Najafi A., Daghigh Kia H., Mohammadi H., Najafi M.H., Zanganeh Z., Sharafi M., Martinez-Pastor F. and Adeldust H. (2014). Different concentrations of cysteamine and ergothioneine improve microscopic and oxidative parameters in ram semen frozen with a soybean lecithin extender. *Cryobiology*. **69**, 68-73.

Osawa T. (1994). Novel natural antioxidants for utilization in food and biological systems. Pp. 241-251 in Postharvest Biochemistry of Plant Food Materials in the Tropics. I. Uritani, V.V. Garcia and E.M. Mendoza, Eds. Tokyo, Japan.

Placer Z.A., Cushman L.L. and Johnson B.C. (1966). Estimation of product of lipid peroxidation (malonyl dialdehyde) in biochemical systems. *Anal. Biochem*. **16**, 359-364.

Purdy P. (2006). A review on goat sperm cryopreservation. *Small Rumin. Res*. **63**, 215-225.

Revell S. and Mrode R. (1994). An osmotic resistance test for bovine semen. *Anim. Reprod. Sci*. **36**, 77-86.

Roca J., Gil M.A., Hernandez M., Parrilla I., Vazquez J.M. and Martinez E.A. (2004). Survival and fertility of boar spermatozoa after freeze thawing in extender supplemented with butylated hydroxytoluene. *J. Androl*. **25**, 397-405.

Sanocka D. and Kurpisz M. (2004). Reactive oxygen species and sperm cells. *Reprod. Biol. Endocrinol*. **2**, 1-7.

Sariozkan S., Bucak M.N., Tuncer P.B., Buyukleblebici S., Eken A. and Akay C. (2015). Influence of fetuin and hyaluronan on the post-thaw quality and fertilizing ability of Holstein bull semen. *Cryobiology*. **71**, 119-124.

SAS Institute. (2001). SAS®/STAT Software, Release 9.1. SAS Institute, Inc., Cary, NC. USA.

Sikka S.C. (1996). Oxidative stress and role of antioxidants in normal and abnormal sperm function. *Front. Biosci*. **1**, 78-86.

Ukeda H., Maeda S., Ishii T. and Sawamura M. (1997). Spectrophotometric assay for superoxide dismutase based on tetrazolium salt 3'-{1-[(phenylamino)-carbonyl]-3, 4-tetrazolium}-bis (4-methoxy-6-nitro) benzenesulfonic acid hydrate reduction by xanthine-xanthine oxidase. *Anal. Biochem*. **251**, 206-209.

Vekiari S., Oreopoulou V., Tzia C. and Thomopoulos C. (1993). Oregano flavonoids as lipid antioxidants. *J. Am. Oil Chem. Soc*. **70**, 483-487.

Wang Y., Sharma R. and Agarwal A. (1997). Effect of cryopreservation and sperm concentration on lipid peroxidation in human semen. *Urology*. **50**, 409-413.

Wishart G. (1982). Maintenance of ATP concentrations in and of fertilizing ability of fowl and turkey spermatozoa *in vitro*. *J. Reprod. Fertil*. **66**, 457-462.

Yanishlieva N.V. and Marinova E.M. (1996). Anti-oxidative effectiveness of some natural antioxidants in sunflower oil. *Z. Lebensm. Unters. For*. **203**, 220-223.

Zanganeh Z., Zhandi M., Zare-Shahneh A., Najafi A., Nabi M.M. and Mohammadi-Sangcheshmeh A. (2013). Does rosemary aqueous extract improve buck semen cryopreservation? *Small Rumin. Res*. **114**, 120-125.

Zhao H.W., Li Q.W., Ning G.Z., Han Z.S., Jiang Z.L. and Duan Y.F. (2009). Rhodiola sacra aqueous extract (RSAE) improves biochemical and sperm characteristics in cryopreserved boar semen. *Theriogenology*. **71**, 849-857.

Zini A., Lamirande E. and Gagnon C. (1993). Reactive oxygen species in semen of infertile patients: levels of superoxide dismutase and catalase like activities in seminal plasma and spermatozoa. *Int. J. Androl*. **16**, 183-188.

Effect of Total and Differential Somatic Cell Counts, Lactation Stage and Lactation Number on Lipolysis and Physicochemical Composition in Camel (*Camelus dromedaries*) and Cow Milk

H. Hamed[1*], A.F. El Feki[1] and A. Gargouri[1]

[1] Laboratory of Animal Ecophysiology, Faculty of Science, Sfax University, Sfax, Tunisia

*Correspondence E-mail: houdahamed1@yahoo.fr

ABSTRACT

The present study was carried out to investigate the effects of somatic cell counts (SCC), differential SCC (macrophage (MAC), lymphocyte (LYM) and polymorphonuclear leukocytes (PMN)), number and stage lactation on milk composition in camel and cow milk. Camel milk appeared to contain significantly (P<0.05) a higher content of minerals. Lipolysis level is similar in camel milk compared to cow milk. Lipolysis level increased as MAC level increased in camel's milk but not in cow's milk. Our results suggest that MAC play a role in the degradation of dromedary milk fat. Mineral compositions were significantly affected by the SCC in camel milk. The milk composition was not affected by lactation number in both species. Total solid, Ca and Na content in camel's milk were gradually decreased through lactation.

KEY WORDS camel milk, lipolysis, physicalchemical composition, somatic cell count.

INTRODUCTION

Milk normally contains some level of somatic cells: neutrophils (PMN cells), lymphocytes (LYM) and macrophages. Macrophages (MAC) comprise the major cell type in milk from healthy udders (Dosogne *et al.* 2003; Lindmark-Månsson *et al.* 2006). When there is bacterial infection, tissue damage, or other inflammation processes affecting the mammary tissue, the SCC in milk dramatically increases (Sharma *et al.* 2011; Katherine *et al.* 2013). This increase in SCC results from the transfer of white blood cells from the blood to the mammary gland (Kelly *et al.* 2000; Sládek *et al.* 2006). In addition, the relative proportions of cell types present in milk change significantly, with an increased in PMN level (up to 90%) to protect the udder from bacterial challenge (Alhussien *et al.* 2016; Kehrli and Shuster, 1994; Zecconi and Smith, 2000). The increase in

SCC milk causes the change in the components of cow's milk. The variation of the many components of cow milk with SCC was observed by many authors (Aysan *et al.* 2011; Brandt *et al.* 2010; Kelly *et al.* 2000; Somers *et al.* 2003; Bansal *et al.* 2005; Lindmark-Månsson *et al.* 2006) have reported changes in the composition of milk obtained from cows with infection, but little is known about such changes in camel milk. The relationship between SCC and lipolysis was investigated. It has been suggested that milk cells contribute to the lipolysis of milk fat to provide flavor defects (Azzara and Dimick, 1985; Ma *et al.* 2000; Santos *et al.* 2003; Gargouri *et al.* 2008). On the contrary, other studies indicated no relationship between milk SCC and lipolysis level (Lee *et al.* 1980; Cartier and Chilliard, 1990). Milk composition varies according to factors such as breed, age, mammary gland health, lactation stage, nutritional management and season (Dobranié *et al.* 2008).

Similar, the variation in the constituents of camel milk may be attributed to factors such as breed, age, the number of calving, nutrition, management, the stage of lactation, and the sampling technique used (Alshaikh and Salah, 1994). The purposes of this study were, firstly, to investigate the relationship among SCC, lipolysis and chemical composition and, secondly, to evaluate the influence of lactation stage on these variables in lactating camels and cows.

MATERIALS AND METHODS

Sampling

The study was carried out using individual milk samples from 36 dromedary animals (*Camelus dromedarius*) of Maghrabi breed from the south and the center of Tunisia. A total of 52 lactating dairy cows housed either in a free stall barn were used. Samples were obtained from each cow at days < 100 (n=15), between 100 and 240 days (n=25) and > 240 (n=12) after parturition. Of the 36 dromedaries, 11 were at early lactation (100 days in lactation), 18 at mid lactation (between 100 and 240 days lactation) and 7 at late lactation (between 100 and 240 days lactation). The camels were fed exclusively on natural browse. For cows, their nutrition is based on forage and concentrates. The milk was collected during the routine morning milking. Bovine samples were obtained by automated milking systems, but dromedary samples were obtained by manual milking. All the animals were free from clinical mastitis during the sampling period. Milk samples were taken to the laboratory immediately after collection and 250 mL were kept at 4 °C until the SCC. The rest was stored at -18 °C up to the rest of analysis.

Somatic cell counts

Somatic cells were counted using a Fossomatic 5000 (FossElectric, Hillerod, Denmark) according to International Dairy Federation Standard (IDF, 1995).

Milk analyses

Milk was analyzed for pH, titratable acidity (AOAC, 1995), total solids by drying at 102 °C (IDF, 1987), milk fat by gerber method (IDF, 1981). The extent of lipolysis in milk was measured using the bureau of dairy industries (BDI) method (IDF, 1991) and was expressed as acid degree value in meq FFA/100 g of fat. The mineral content was estimated using an Automate Synchron CX9 (Beckman coulter®). All analyses were performed in duplicate

Stastics

Statistical evaluations were performed using SPSS software (SPSS, 2011). The effect of lactation stage and lactation number on the different data was analyzed by one-way analysis of variance (ANOVA) and group means were compared by the Tukey's least significant difference test. Secondly, pearson's correlation coefficients (r) were also established to determine the relationships between the various parameters studied. The results were considered significant if the associated P-value was < 0.05.

RESULTS AND DISCUSSION

The overall results of physicalchemical parameters of dromedary and cow milk are resumed in Table 1. The Pearson correlation coefficients between total and differential SCC and physicalchemical parameters of dromedary and cow milks are presented in Tables 2 and 3. The pearson correlation coefficients between stage and number of lactation and physicalchemical parameters of dromedary and cow milk are presented in Table 4.

Milk characteristics

The data obtained showed a wide range of variation in some parameters studied between different individual camel and cow milk samples. There were no significant difference between pH values, titratable acidity and ash content of fresh camel and cow milk (Table 1).

Table 1 Composition of the camel's and cow's milk (Mean±SE)

Component	Dromedary milk	Cow milk	P-value
pH	6.38±0.22	6.7±0.19	0.44
Acidity (%)	16.8±0.36	17.09±0.22	0.08
Ash (%)	0.63±0.04	0.67±0.04	0.53
Total solid (%)	11.32±0.38	10.21±0.22	0.01
Fat (%)	3.71±0.23	3.43±0.13	0.30
Lipolysis (meq/100 mg fat)	2.96±0.39	2.29±0.24	0.13
Mg (mmol/L)	3.45±0.35	4.32±0.40	0.09
Cl (mmol/L)	61.4±3.02	37.58±1.08	0.000
K (mmol/L)	58.72±2.29	41.53±1.09	0.000
Na (mmol/L)	31.86±0.68	21.73±0.94	0.000
Ca (mmol/L)	10.32±0.44	6.37±0.30	0.000
Na^+/K^+	0.573±0.66	0.520±0.019	0.118

SE: standard error.

The obtained result of titrable acidity of camel milk was lower in comparison to the acidity of camel milk reported by Khaskheli et al. (2005). These results were similar to those reported by Aljumaah et al. (2012) and Hammadi et al. (2010). The value of titrable acidity found in camel milk is similar to that of cow. The ash content (0.63%) of camel milk found in this study was similar to that reported by Mehaia et al. (1995). The average of total solids content in camel milk was significantly higher (P<0.01) in comparison to cow's milk. The total solids content of camel milk was higher to that reported by others authors (Farag and Kebary, 1992; Ahmed, 1990; FAO, 1982).

According to Mehaia *et al.* (1995), the total solids content ranged from 10.0 to 14.4% in camel milk. The average value of fat in camel milk was 3.7%, which is similar to the content of fat in cow's milk (Table 1). The findings of this study agrees with Konuspayeva *et al.* (2009), who as a result of a meta-analysis of the literature data reported 3.82% an average fat content of camel milk. From the results shown in Table 1, it appears that lipolysis level in camel milk was not significantly different from lipolysis level found in cow's milk. The results of lipolysis of cows' milk were comparable to of the results reported by Andrews (1983). It was estimated from previous research (Bodyfelt *et al.* 1988) that the sensory threshold for detection of off-flavor would be about 1.0 meq/100 g of fat.

Table 2 Correlation coefficients (r) between total and differential SCC and physicochemical parameters of camel's milk. (values noted in bold are significant at P<0.05)

Parameter	SCC[1]	MAC[1]	LYM[1]	PMN[1]
pH	0.256	0.156	0.320	0.165
Ash	0.058	0.247	0.148	0.022
D°	-0.118	-0.217	-0.217	-0.077
TS	0.126	0.283	0.294	0.066
Fat	0.091	0.083	0.142	0.010
Lipolysis	0.213	**0.350**	-0.303	0.168
Mg	**0.404**	0.315	0.293	**0.385**
Cl	0.163	-0.078	0.022	0.193
K	0.015	-0.320	-0.319	0.071
Na	**0.635**	0.093	0.216	0.312
Ca	0.301	0.017	**0.366**	0.304

SCC: somatic cells count; MAC: macrophage; LYM: lymphocyte; PMN: polymorphonuclear leukocytes; D°: acidity (°Doronic) and TS: total solids.

Results from this study are in general higher than this threshold and the average lipolysis neared 2 meq/100 g of fat. The higher lipolysis level has been described as the most important factor that contributed to the lower sensory quality and shorter shelf life of milk (Azzara and Dimick, 1985; Ma *et al.* 2000). The levels of K, Cl, Na and Ca were significantly higher (P<0.05) in dromedary (Table 1), which is in agreement with others studies (Mehaia *et al.* 1995; Sawaya *et al.* 1984). An average Ca concentration in camel milk was 10.32 mmol/L, a little lower than that reported by Faye *et al.* (2008). The K, Na and Cl contents of camel milk were higher than the value reported by Kamoun (1990). Magnesium content of camel milk mean value was higher than the value reported by Ahmed (1990). High variability was observed in some studies regarding the mineral content of camel milk (Dukwal *et al.* 2007; Haddadin *et al.* 2008; Ayadi *et al.* 2009) and it could be attributed to the breed difference, intervals between milking, feeding, analytic procedures and water intake (Haddadin *et al.* 2008; Mehaia *et al.* 1995). Mehaia *et al.* (1995) considered that genetic factors could significantly affect the milk composition, especially under non controlled environmental conditions, as is mostly the case locally. The calculated Na:K

ratio in camel milk was higher than that reported by Aljumaah *et al.* (2012). The variation in concentration of minerals and the increments in Na:K ratio were studied in dairy goats (Boutinaud *et al.* 2003) and dairy cows (Stelwagen *et al.* 1999; Delamaire and Guinard-Flament, 2006). Alterations in the Na:K ration could interfere with a number of intracellular processes. Increased Na:K ratio reduce mammary protein system in dairy goats (Stelwagen *et al.* 1999). In dairy camels, the regulatory mechanism seems not to operate (Ayadi *et al.* 2009). Instead, this difference might be related to the adaptation of the camels to the desert conditions.

Effect of SCC

The results show that the SCC and differential cell count did not have any significant correlation with pH, titratable acidity, ash and total solid values in both species. Table 3 shows that there was no significant relationship between lipolysis and total and differential SCC count in cow milk, which in agreement with Chazal and chillard (1986); Lee *et al.* (1980) and Cartier and Chilliard (1990). However, a positive correlation (P<0.05) between lipolysis and MAC was found in dromedary milk (Table 2). This suggests that the macrophages secreted lipolytic enzymes into the gradient while fractions containing polymorphonuclear leukocytes and lymphocytes did not possess lipolytic activity. These results confirm those of Russell *et al.* (1977) and Azzara and Dimick (1985), who found that lipolytic enzymes produced by monocytes and macrophages are believed to play a role in the degradation of cow milk fat ingested by those cells. A positive correlation (P<0.05) between fat, SCC and PMN count was found in cow milk but not in camel milk, which is in accordance with others studies (Aysan *et al.* 2011; Paura *et al.* 2002; Sawa and Piwczynski. 2002).

Table 3 Correlation coefficients (r) between total and differential SCC and physicochemical parameters of cow's milk. (values noted in bold are significant at P<0.05)

Parameter	SCC	MAC	LYM	PMN
pH	0.189	0.105	0.300	0.165
Ash	0.058	0.247	0.148	0.022
D°	0.193	0.144	0.136	0.299
TS	0.129	0.121	-0.014	0.134
Fat	**0.428**	0.195	0.195	**0.408**
Lipolysis	0.097	0.025	-0.191	0.165
Mg	**-0.076**	0.113	0.150	-0.002
Cl	-0.019	0.099	-0.195	-0.026
K	-0.111	-0.027	**-0.373**	-0.031
Na	0.019	0.066	-0.170	0.063
Ca	0.015	0.152	-0.240	0.023

SCC: somatic cells count; MAC: macrophage; LYM: lymphocyte; PMN: polymorphonuclear leukocytes; D°: acidity (°Doronic) and TS: total solids.

This positive correlation reported in this study and others (Barbano *et al.* 1989; Pereira *et al.* 1999; Ma *et al.* 2000) may be ascribed to the strong reduction in milk production consequently to mammary epithelium damages (Akers and Thompson, 1987). Mineral composition was significantly affected by the SCC in camel milk. Table 2 illustrates the positive correlation between SCC, PMN and Mg content. A high positive correlation was also observed between the SCC and the Na content, which is in agreement with Bruchmaier *et al*. (2004).

However, mineral compositions was not significantly correlated with SCC in cow's milk, except in the case of K content which is in a negative correlation with LYM. Potassium declines because of paracellular passage out of the alveolar lumen between damaged epithelial cells (Harmon, 1994). The ion concentrations in milk may be due to increased blood capillary permeability, the destruction of tight junctions, and the destruction of the active ion-pumping systems.

Effect of lactation

The pH in camel milk was significantly (P<0.05) affected by the stage of lactation (Table 4), in agreement with Aljumaah *et al*. (2012).

Table 4 Correlation coefficients (r) between stage of lactation, number of lactation and physicochemical parameters of camel's and cow's milk (values noted in bold are significant at P<0.05)

Parameter	Camel milk		Cow milk	
	SL	NL	SL	NL
pH	**0.396**	0.006	0.256	0.004
Ash	0.044	0.145	0.133	0.014
D°	**0.338**	0.007	0.075	0.018
TS	**-0.357**	0.003	0.103	-0.040
Fat	0.102	0.009	0.017	0.080
Lipolysis	0.185	0.094	0.087	**-0.308**
Mg	-0.278	-0.257	0.097	-0.079
Cl	-0.005	0.239	0.251	-0.136
K	0.248	0.214	0.031	0.022
Na	**-0.363**	-0.160	0.102	0.288
Ca	**-0.491**	0.201	0.038	0.067

SL: stage of lactation and NL: number of lactation.

Fat content was not affected by the stage of lactation (SL) in both species, which also observed Abeni *et al*. (2005). The ash content in camel milk was higher in the late stage compared to the initial stage of lactation. These results confirmed those of El-Hatmi *et al*. (2004) and Raziq *et al*. (2011), who reported that the ash content increased during lactation. The higher ash contents during late lactation stage suggest that camel milk can provide a satisfactory level of minerals (Mal *et al*. 2007). There was a negative relationship between total solid and lactation stage in dromedary. This decrease may be due to the increase in the milk water content during the last stage of lactation. These results confirmed those of Zeleke (2007), who demonstrate that total solid of camel milk decreased from 11.7% in the

first stage of lactation to 10.1% by the end of lactation. In this study, Ca and Na content in camel milk showed a significant (P<0.05) decrease throughout the lactation, as observed by Aljumaah *et al*. (2012).

The variations in the major mineral contents of camel milk could be due to breed, feeding, stage of lactation, drought conditions, or analytical procedures (Haddadin *et al*. 2008; Farah, 1993; Mehaia *et al*. 1995). There was a negative, but not significant, relationship between lipolysis and lactation number (NL) in cow milk (r=-0.308; P>0.05) and camel milk (r=-0.09; P>0.05). This suggests that lipolysis seems to be higher in primaparous cows than in multiparous. These results showed no effect of lactation number on camel's milk composition.

CONCLUSION

In view of the observed results on the camel milk, it could be concluded that physicochemical properties was comparable to that of cow's milk. However, in present study, cow milk was found to contained lower mineral content compared to camel milk. The higher level of lipolysis was observed in camel's milk that contained a high percentage of MAC. This may also indicate that MAC in milk could play an important part in determining the lipolysis level in camel's milk. Negative relationship between lactation number and lipolysis level was found in cow's milk. For this species, the lactation stage not affected the physicalchemical compostion. The present study emphasizes that the variations in the camel's milk composition could be attributed to SCC and lactation stage.

ACKNOWLEDGEMENT

The authors would like to thank the "Ministère de l'enseignement Supérieur et de la Recherche Scientique, Tunisie'' for the support of this research work.

REFERENCES

Abeni F., Degano L., Calza F., Giangiacomo R. and Pirlo G. (2005). Milk quality and automatic milking: fat globule size, natural creaming and lipolysis. *J. Dairy Sci*. **88**, 35189-3529.

Ahmed M.M. (1990). The analysis and quality of camel milk. MS Thesis. Universities of Great Britain, UK.

Akers R.M. and Thompson W. (1987). Effect of induced leucocyte migration on mammary cell morphology and milk component biosynthesis. *J. Dairy Sci*. **70**, 1685-1695.

Alhussien M., Manjari P., Sheikh A.A., Mohammed Seman S., Reddi S., Mohanty A.K., Mukherjee J. and Dang A.K. (2016). Immunological attributes of blood and milk neutrophils isolated from crossbred cows during different physiological conditions. *Czech J. Anim. Sci*. **61**, 223-231.

Aljumaah R.S., Almutairi F.F., Ismail E., Alshaikh M.A., Sami A. and Ayadi M. (2012). Effects of production system, breed, parity and stage of lactation on mil composition of dromedary camels in Saudie Arabia. *J. Anim. Vet. Adv.* **11**, 141-147.

Alshaikh M.A. and Salah M.S. (1994). Effect of milking interval on secretion rate and composition of camel milk in late lactation. *J. Dairy Res.* **61**, 451-456.

Andrews A.T. (1983). Breakdown of caseins by proteinases in bovine milks with high somatic cell counts arising from mastitis or infusion with bacterial endotoxin. *J. Dairy Res.* **50**, 57-66.

AOAC. (1995). Official Methods of Analysis. Vol. I. 16th Ed. Association of Official Analytical Chemists, Arlington, VA, USA.

Ayadi M., Hammadi M., Khorchani T., Barmat A., Atigui M. and Caja G. (2009). Effect of milking interval and cisternal udder evaluation in Tunisian Maghrebi Dairy Dromedaries (*Camelus dromedarius*). *J. Dairy Sci.* **92**, 1452-1459.

Aysan T., Hizli H., Yazgan E., Kara U. and Gok K. (2011). The effect of somatic cell count on milk urea nitrogen and milk composition. *Kafkas Univ. Vet. Fac. J.* **17**, 659-662.

Azzara C.D. and Dimick P.S. (1985). Lipoprotein lipase activity of milk from cows with prolonged subclinical mastitis. *J. Dairy Sci.* **68**, 3171-3175.

Bansal B.K., Hamann J., Grabowski N.T. and Singh K.B. (2005). Variation in the composition of selected milk fraction samples from healthy and mastitic quarters, and its significance for mastitis diagnosis. *J. Dairy Res.* **72**, 144-152.

Barbano D.M., Rudan M.A. and Rasmussen R.R. (1989). Influence of milk composition and somatic cell and psychrotrophic bacteria counts on ultrafiltration flux. *J. Dairy Sci.* **72**, 1118-1123.

Bodyfelt F.W., Tobias J. and Trout G.M. (1988). The Sensory Evaluation of Dairy Products. AVI Westport, Connecticut, USA.

Boutinaud M.C., Rousseau D.H., Keisler H. and Jammes J. (2003). Grouth hormone and milking frequency act differently on goat mammary gland in late lacatation. *J. Dairy Sci.* **86**, 509-520.

Brandt M., Haeussermann A. and Hartung E. (2010). Invited review: technical solutions for analysis of milk constituents and abnormal milk. *J. Dairy Sci.* **93**, 427-436.

Cartier P. and Chilliard Y. (1990). Spontaneous lipolysis in bovine milk : combined effects of nine characteristics in native milk. *J. Dairy Sci.* **73**, 1178-1186.

Chazal P. and Chilliard Y. (1986). Effect of stage of lactation, stage of pregnancy, milk yield and herd management on seasonal variation in spontaneous lipolysis in bovine milk. *J. Dairy Res.* **53**, 529-538.

Delamaire E. and Guinard- Flament J. (2006). Longer milking intervals alter mammary epithelial permeability and the udder's ability to extract nutrients. *J. Dairy Sci.* **89**, 2007-2016.

Dobranié V., Njari B., Samardžija M., Mioković B. and Resanović R. (2008). The influence of the season on the chemical composition and the somatic cell count of bulk tank cow's milk. Vet. Arhiv. **78**, 235-242.

Dosogne H., Vangroenweghe F., Mehrzad J., Massart-leen A.M. and Burvenich C. (2003). Differential leucocyte count method for bovine low somatic cell count milk. *J. Dairy Sci.* **86**, 828-834.

Dukwal V., Modi S. and Singh M. (2007). A comparative study of nutritional composition of camel and cow's milk. Pp. 91-92 Int. Camel Conf. Bikaner, India.

El-Hatmi H., Hammadi M., Khorchani T., Abdennebi M. and Attia H. (2004). Effects of diet supplementation on production and composition of camels milk during lactation under Tunisian arid range conditions. *J. Camel Pract. Res.* **11**, 147-152.

FAO. (1982). Camels and camel milk. Animal Production and Health Papers. Rome, Italy.

Farag S.I. and Kabary K.M. (1992). Chemical composition and physical properties of camel's milk and milk fat. Pp. 325-326 in Proc. 5th Egyptian Conf. Dairy Sci. Technol. Cairo, Egypt.

Farah Z. (1993). Composition and characteristics of camel milk. *J. Dairy Res.* **60**, 603-626.

Faye B., Konuspayeva G., Messad S. and Loiseau G. (2008). Discriminant milk components of Bactrian camel (*Camelus bactrianus*), dromedary (*Camelus dromedarius*) and hybrids. *Dairy Sci. Technol.* **88**, 607-617.

Gargouri A., Hamed H. and ElFeki A. (2008). Total and differential bulk cow milk somatic cell counts and their relation with lipolysis. *Livest. Sci.* **113**, 274-279.

Haddadin M.S.Y., Gammoh S.I. and Robinson R.K. (2008). Seasonal variations in the chemical composition of camel milk in Jordan. *J. Dairy Res.* **75**, 8-12.

Hammadi M., Atigui M., Ayadi M., Barmat A., Belgacem G., Khalid G. and Khorchani T. (2010). Training period and short time effects of machine milking on milk composition in Tunisian maghrebi camels. *J. Camels Pract. Res.* **17**, 1-7.

Harmon R.J. (1994). Symposium: mastitis and genetic evaluation for somatic cell count-physiology of mastitis and factors affecting somatic cell counts. *J. Dairy Sci.* **77**, 2103-2112.

IDF. (1981). Milk-Determination of Fat Content. Gerber Butyrometers. IDF Standard, vol. 105. International Dairy Federation, Brussels, Belgium.

IDF. (1987). Milk, Cream and Evaporated Milk. Determination of Total Solids Content. IDF Standard, vol. 21B. International Dairy Federation, Brussels, Belgium.

IDF. (1991). Determination of Free Fatty Acids in Milk and Milk Products. Bulletin, vol. 265. International Dairy Federation, Brussels, Belgium.

IDF. (1995). Enumeration of Somatic Cells, FIL-IDF Standard No. 148A. International Dairy Federation, Brussels, Belgium.

Kamoun M. (1990). La production de fromage à partir du lait de Dromadaire. Les petits ruminants et leurs productions laitières dans la région Méditerranéenne. *Opt. Méditerranéennes.* **12**, 119-124.

Katherine M.H., Janet E.W., Bahman S., Martha K.H., Rebecca B., Robert T., Michelle K.M. and Mark A.M.C. (2013). Mastitis is associated with increased free fatty acids, somatic cell count, and interleukin-8 concentrations in human milk. *Breastfeed. Med.* **8**, 105-110.

Kehrli Jr M.E. and Shuster D.E. (1994). Factors affecting milk somatic cells and their role in health of the bovine mammary gland. *J. Dairy Sci.* **77**, 619-627.

Kelly A.L., Tiernan D.,'Sullivan C.O. and Joyce P. (2000). Correlation between bovine milk somatic cell count and polymorphonuclear leukocyte level for samples of bulk milk and milk from individual cows. *J. Dairy Sci.* **83**, 300-304.

Khaskheli M., Arain M.A., Chaudhry S., Soomro A.H. and Qureshi T.A. (2005). Physico-chemical quality of camel milk. *J. Agric. Soc. Sci.* **2**, 164-166.

Konuspayeva G., Faye B. and Loiseau G. (2009). The composition of camel milk: a meta-analysis of the literature data. *J. Food Comp. Anal.* **22**, 95-101.

Lee C.S., Wooding F.B. and Kemp P. (1980). Identification, properties and differential counts of cell populations using electron microscopy of dry cows secretions, colostrum and milk from normal cows. *J. Dairy Res.* **47**, 39-50.

Lindmark-Mansson H., Branning C., Alden G. and Paulsson M. (2006). Relationship between somatic cell count, individual leukocyte populations and milk components in bovine udder quarter milk. *Int. Dairy J.* **16**, 717-727.

Ma Y., Ryan C., Barbano D.M., Galton D.M., Rudan M. and Boor K. (2000). Effects of somatic cell count on quality and shelf-life of pasteurized fluid milk. *J. Dairy Sci.* **83**, 264-274.

Mal G., Suchitra Sena D. and Sahani M.S. (2007). Changes in chemical and macro-minerals content of dromedary milk during lactation. *J. Camel Pract. Res.* **14(2)**, 195-197.

Mehaia M.A., Hablas M.A., Abdel-Rahman K.M. and El-Mougy S.A. (1995). Milk composition of Majaheim, Wadah and Hamra camels in Saudi Arabia. *Food Chem.* **52**, 115-122.

Paura L., Kairisha D. and Jonkus D. (2002). Repeatability of milk productivity traits. *Vet. Zootec.* **19**, 90-93.

Pereira A.R., Prada e Silva L.F., Molon L.K., Machado P.F. and Barancelli G. (1999). Efeito do nível de células somáticas sobre os constituintes do leite. I-gordura e proteína. *Brazilian J. Vet. Res. Anim. Sci.* **36**, 1413-9596.

Raziq A., Tareen A.M. and De Verdier K. (2011). Characterization and significance of Raigi camel, a livestock breed of the Pashtoon pastoral people in Afghanistan and Pakistan. *J. Livest. Sci.* **2**, 11-19.

Russell D.G., Lievers K.W. and Lovering J. (1977). Effects of change in rate of harvest and selected management variables in timothy silage yields, quality, and net returns at Charlottetown: I. Direct-cut silage. *Canadian Agric. Engin.* **19**, 29-36.

Santos M.V., Ma Y., Caplan Z. and Barbano D.M. (2003). Sensory threshold of off-flavors caused by proteolysis and lipolysis in milk. *J. Dairy Sci.* **86**, 1601-1607.

Sawa A. and Piwczynski D. (2002). Somatic cell count and milk yield and composition in Black and White × Holstein-Friezian cows. *Med. Vet.* **58**, 636-640.

Sawaya W.N., Khalil J.K., Shalhat A.A.L. and Mohammad H.A.L. (1984). Chemical composition and nutritional quality of camel milk. *J. Food Sci.* **49**, 744-749.

Sharma N., Singh N.K. and Bhadwal M.S. (2011). Relationship of somatic cell count and mastitis: an overview. *Asian-Australas J. Anim. Sci.* **24**, 429-438.

Sladek Z., Rysanek D. and Rysanek D. (2006). Macrophages of the bovine heifer mammary gland: morphological features during initiation and resolution of the inflammatory response. *Anat. Histol. Embryol.* **35**, 116-124.

Somers J.G.C.J., Frankena K., Noordhuizen-Stassen E.N. and Metz J.H.M. (2003). Prevalence of claw disorders in Dutch dairy cows exposed to several floor systems. *J. Dairy Sci.* **86**, 2082-2093.

SPSS Inc. (2011). Statistical Package for Social Sciences Study. SPSS for Windows, Version 20. Chicago SPSS Inc.

Stelwagen K., Farr V.C. and McFadden H.A. (1999). Alteration of the sodium to potassium ratio in milk and the effect on milksecretion in goats. *J. Dairy Sci.* **82**, 52-59.

Zecconi A. and Smith K.L. (2000). International dairy federation standard position paper on ruminant mammary gland immunity. Pp. 1-120 in Proc. Symp. Immunol. Rumin. Mammary Gland. Stresa, Italy.

Zeleke M.Z. (2007). Major non-genetic factors affecting milk yield and milk composition of traditionally managed camels (*Camelus dromedarius*) in eastern Ethiopia. Pp. 89-90 in Proc. Int. Camel Conf. Bikaner, India.

Permissions

All chapters in this book were first published in IJAS, by Central Tehran Branch. IAU; hereby published with permission under the Creative Commons Attribution License or equivalent. Every chapter published in this book has been scrutinized by our experts. Their significance has been extensively debated. The topics covered herein carry significant findings which will fuel the growth of the discipline. They may even be implemented as practical applications or may be referred to as a beginning point for another development.

The contributors of this book come from diverse backgrounds, making this book a truly international effort. This book will bring forth new frontiers with its revolutionizing research information and detailed analysis of the nascent developments around the world.

We would like to thank all the contributing authors for lending their expertise to make the book truly unique. They have played a crucial role in the development of this book. Without their invaluable contributions this book wouldn't have been possible. They have made vital efforts to compile up to date information on the varied aspects of this subject to make this book a valuable addition to the collection of many professionals and students.

This book was conceptualized with the vision of imparting up-to-date information and advanced data in this field. To ensure the same, a matchless editorial board was set up. Every individual on the board went through rigorous rounds of assessment to prove their worth. After which they invested a large part of their time researching and compiling the most relevant data for our readers.

The editorial board has been involved in producing this book since its inception. They have spent rigorous hours researching and exploring the diverse topics which have resulted in the successful publishing of this book. They have passed on their knowledge of decades through this book. To expedite this challenging task, the publisher supported the team at every step. A small team of assistant editors was also appointed to further simplify the editing procedure and attain best results for the readers.

Apart from the editorial board, the designing team has also invested a significant amount of their time in understanding the subject and creating the most relevant covers. They scrutinized every image to scout for the most suitable representation of the subject and create an appropriate cover for the book.

The publishing team has been an ardent support to the editorial, designing and production team. Their endless efforts to recruit the best for this project, has resulted in the accomplishment of this book. They are a veteran in the field of academics and their pool of knowledge is as vast as their experience in printing. Their expertise and guidance has proved useful at every step. Their uncompromising quality standards have made this book an exceptional effort. Their encouragement from time to time has been an inspiration for everyone.

The publisher and the editorial board hope that this book will prove to be a valuable piece of knowledge for researchers, students, practitioners and scholars across the globe.

List of Contributors

M. Jafarpour Boroujeni, M. Danesh Mesgaran, A.R. Vakili and A.A. Naserian
Department of Animal Science, Faculty of Agriculture, Ferdowsi University of Mashhad, Mashhad, Iran

H. Kharrati Koopaee
Institute of Biotechnology, Shiraz University, Shiraz, Iran

M. Pasandideh
Department of Genetics and Animal Breeding, Faculty of Animal Science and Fishery, Sari Agricultural Sciences and Natural Resources University, Sari, Iran

M. Dadpasand
Department of Animal Science, Faculty of Agriculture, Shiraz University, Shiraz, Iran

A. Esmailizadeh Koshkoiyeh and M.R. Mohammad Abadi
Department of Animal Science, Faculty of Agriculture, Shahid Bahonar University of Kerman, Kerman, Iran

D. W. Mengistu
Department of Animal Science, Selale University, Selale, Ethiopia

K. A. Wondimagegn and M. H. Demisash
Department of Animal Production and Technology, Bahirdar University, Bahirdar, Ethiopia

M. Momen, A. Ayatollahi Mehrgardi, A. K. Esmailizadeh and M. Assadi Foozi
Department of Animal Science, Faculty of Agriculture, Shahid Bahonar University of Kerman, Kerman, Iran

A. Sheikhy
Department of Animal Science, Faculty of Mathematics and Computer Science, Shahid Bahonar University of Kerman, Kerman, Iran

M. M. Momin, M.K.I. Khan and O. F. Miazi
Department of Genetics and Animal Breeding, Chittagong Veterinary and Animal Science University, Khulshi, Chittagong - 4225, Bangladesh

A. Mirshamsollahi
Department of Animal and Science, Markazi Agricultural and Natural Resources Research and Education Center, Arak, Iran

F. Khalilavi, M. Mamouei, S. Tabatabaei and M. Chaji
Department of Animal Science, Faculty of Animal and Food Science, Ramin Agricultural and Natural Resources University, Mollasani, Ahvaz, Iran

A. Ahmadi, R. Talebi and A. Farahavar
Department of Animal Science, Faculty of Agriculture, Bu Ali Sina University, Hamedan, Iran

F. Afraz and S.M.F. Vahidi
Department of Genomics and Animal, Agricultural Biotechnology Research Institute (ABRI), North branch, Rasht, Iran

K. Forutan, M. Amin Afshar, M. Chamani and N. Emam Jome Kashan
Department of Animal Science, Science and Research Branch, Islamic Azad University, Tehran, Iran

K. Zargari
Department of Agronomy, Varamin - Pishva Branch, Islamic Azad University, Varamin, Iran

R. Hajializadeh Valilou, S.A. Rafat and M. Ebrahimi
Department of Animal Science, Faculty of Agriculture, University of Tabriz, Tabriz, Iran

M. Firouzamandi
Department of Pathobiology, Faculty of Veterinary Medicine, University of Tabriz, Tabriz, Iran

Z. Patiabadi and S. Varkoohi
Department of Animal Science, Faculty of Agriculture, Razi University, Kermanshah, Iran

S. Savar-Sofla
Animal Science Research Institute of Iran, Agricultural Research, Education and Extension Organization (AREEO), Karaj, Iran

M. A. Norouzian
Department of Animal Science, College of Abouraihan, University of Tehran, Tehran, Iran

M. Vakili Alavijeh
Department of Mathematics, Faculty of Mathematical Science, Shahid Beheshti University, Tehran,   Iran

M. Bouyeh
Department of Animal Science, Rasht Branch, Islamic Azad University, Rasht, Iran

O. K. Gevorgyan
Department of Veterinary and Animal science, Armenia National Agrarian University, Yerevan, Armenia

T. Farahvash, R. Vaez Torshizi and A.A. Masoudi
Department of Animal Science, Faculty of Agriculture, Tarbiat Modares University, Tehran, Iran

H.R. Rezaei
Department of Environmental Science, Faculty of Fisheries and Environmental Science, Gorgan University of Agricultural Science and Natural Recourses, Gorgan, Iran

M. Tavallaei
Human Genetic Research Center, Baqiyatallah University of Medical Science, Tehran, Iran

M. Sharifi
Department of Animal Science, Faculty of Agriculture, University of Tabriz, Tabriz, Iran

R. Pahlavan
Department of Animal Science, Faculty of Agriculture and Natural Resources, University of Tehran, Karaj, Iran

A. Aghaei
Department of Management and Accounting, Farabi Compus University of Tehran, Qom, Iran

Z. Davari Varanlou, S. Hassani, M.Ahani Azari and F. Samadi
Department of Animal Science, Gorgan University of Agricultural Science and Natural Resources, Golestan, Iran

S. Zakizadeh
Department of Animal Science and Veterinary, Khorasan Razavi Agricultural and Natural Resources Research and Education Center, AREEO, Mashhad, Iran

A. R. Khan Ahmadi
Department of Animal Science, Faculty of Agricultural and Natural Resources, Gonbad University, Gonbad, Iran

M. Muhaghegh-Dolatabady and H. Hossainy-Dolatabady
Department of Animal Science, Faculty of Agriculture, Yasouj University, Yasouj, Iran

E. Heidari Arjlo and R. Mahmoudi
Cellular and Molecular Research Center, Yasouj University of Medical Science, Yasouj, Iran

M. Moridi, S. H. Hosseini Moghaddam and S. Z. Mirhoseini
Department of Animal Science, Faculty of Agricultural Science, University of Guilan, Rasht, Iran

M. Hosseinpour Mashhadi
Department of Animal Science, Mashhad Branch, Islamic Azad University, Mashhad, Iran

S. Khamoshi, F. Kafilzadeh and V. Naseri
Department of Animal Science, Faculty of Agriculture, Razi University, Kermanshah, Iran

H. Jahani-Azizabadi
Department of Animal Science, Faculty of Agriculture, University of Kurdistan, Sanandaj, Iran

S. R. Ebrahimi-Mahmoudabad
Department of Animal Science, Shahr - e - Qods Branch, Islamic Azad University, Tehran, Iran

M. Taghinejad-Roudbaneh
Department of Animal Science, Tabriz Branch, Islamic Azad University, Tabriz, Iran

S. Savar Sofla, H .R. S eyedabadi a nd A . Javanrouh Aliabad
Animal Science Research Institute of Iran (ASRI), Agricultural Research Education and Extension Organization (AREEO), Karaj, Iran

R. Seyed Sharifi
Department of Animal Science, Faculty of Agricultural and Natural Resources, University of Mohaghegh Ardabili, Ardabil, Iran

A. Teimouri Yansari
Department of Animal Science, Faculty of Agricultural Science, Sari University of Agricultural Science and Natural Resources, Sari, Iran

S. Kozat and N. Yüksek
Department of Internal Medicine, Faculty of Veterinary, Yuzuncu Yıl University, Zeve Campus, Van, Turkey

Z. Nikzad, M. Chaji, K. Mirzadeh, T. Mohammadabadi and M. Sari
Department of Animal Science, Ramin Agricultural and Natural Resources University, Mollasani, Ahvaz, Iran

M.H. Aazami, A.M. Tahmasbi, V. Forouhar and A.A. Naserian
Department of Animal Science, Faculty of Agriculture, Ferdowsi University of Mashhad, Mashhad, Iran

H. Ghiasi
Department of Animal Science, Faculty of Agriculture, Payame Noor University, Tehran, Iran

A. Pakdel and A. Sadeghi-Sefidmazgi
Department of Animal Science, College of Agriculture, Isfahan University of Technology, Isfahan, Iran

A. Nejati-Javaremi
Department of Animal Science, Faculty of Agriculture and Natural Resources, University of Tehran, Karaj, Iran

O. González-Recio and M.J. Carabaño
Departamento de Mejora Genética, Instituto Nacional de Investigación y Tecnología Agraria y Alimentaria, Madrid, Spain

R. Alenda
Departamento de Producción Animal, Universidad Politécnica de Madrid, Madrid, Spain

H. Daghigh Kia, R. Farhadi, I. Ashrafi and M. Mehdipour
Department of Animal Science, Faculty of Agriculture, University of Tabriz, Tabriz, Iran

H. Hamed, A. F. El Feki and A. Gargouri
Laboratory of Animal Ecophysiology, Faculty of Science, Sfax University, Sfax, Tunisia

Index

A

Acidogenic Value, 1-4, 6
Amelogenin, 98-103
Ann, 13, 31, 63, 85-89, 96, 154, 198

B

Baluchi, 60-62, 64-65, 67-69, 80, 82
Barley Grain, 105, 144-148, 171, 174, 176, 178-179
Bathan Farming System, 33, 35-36, 39-40
Bestkeeper, 119-120, 122, 124
Bmp15, 54-56, 58, 60-62
Breeding, 8-9, 13, 20, 23-29, 31-34, 39, 41-44, 46-52, 54, 64, 68, 71, 73, 75, 77-79, 81-84, 86, 91-92, 96, 103, 110, 113-114, 117, 125, 129, 135-136, 150, 188-189, 194
Buffering Capacity, 1-4, 6
Bull Sperm, 193, 195-197

C

Calving Interval, 15, 20-22, 24, 33, 35-36, 40, 188-191
Camel Milk, 200-205
Carcass, 40, 68, 86, 89-97, 103, 144-149
Cellulolytic Bacteria, 6, 138-140, 142, 177
Cidr, 42-44, 46, 49, 52
Cow Feed Cost, 188, 190-191
Cow Fertility Cost, 188, 190-191
Crossbred, 15-16, 18-21, 23-24, 33-36, 39, 41, 52, 75, 84, 125-126, 203
Cryopreservation, 193, 197-199
Cyp19 Gene, 113-116

D

Dairy Cows, 1-2, 6-7, 13, 15-16, 18, 20-21, 23-24, 41, 104-105, 111, 119-120, 134, 136-137, 143, 148, 155, 164-165, 167-170, 176, 178-179, 188, 201-202, 205
Dairy Heifers, 18, 119-120, 122, 178
Dgat1, 8-11, 13-14, 130, 136-137
Diazinon, 180-187

E

Ecg, 42-44, 47-52

F

Farahani, 42-44, 60
Fermentation, 1, 3-7, 111, 138-140, 142-146, 148-149, 155, 171-172, 175, 177-181, 186-187

G

Gastrointestinal Nematodes, 70-72, 75-76

Gdf9, 54-63
Gene Expression, 64-68, 73, 114, 120, 123-124, 126-128, 130
Genetic Diversity, 9-10, 40, 78, 84, 99, 150, 152, 154
Genomic Prediction, 25-27, 29-31, 76
Ghezel Sheep, 61, 70, 73, 75-76
Growth Traits, 23, 78, 82, 84, 113, 116-117
Guilan Native Cows, 125-126

H

Herd Amortization, 188, 190
Holstein Friesian, 15-16, 18-21, 23-24, 136

I

Inbreeding Depression, 78-79, 82-84
Iranian Sheep, 62, 80, 150-154
Ivdmd, 1, 3-6, 139

K

Ketosis, 164-170

L

Lipolysis, 168, 200-204
Livestock, 15, 23-24, 68, 71, 73, 77-78, 104, 111, 126, 130, 140, 150, 179-180, 189, 192, 205
Lysine, 9, 14, 91-94, 96-97

M

Malva Sylvestris, 138-140, 142
Methionine, 91-94, 96-97
Microsatellite Polymorphisms, 70-74
Milk Composition, 8, 13, 104, 108, 130, 136, 178, 200, 203-205
Milk Somatic Cell, 119, 123, 204
Missense, 13, 54-56, 58, 61, 63
Mstn, 64-68
Mtnr1a Gene, 113-114, 116-117
Myog, 64-68

N

Number of Insemination, 189-190
Nutrient Digestibility, 171, 177, 180, 184

O

Oestrus Synchronization, 42-44, 46-47, 51-52
Olr1, 13-14, 132-137
Opn, 8-11, 13, 130
Organic System, 104-111
Organophosphate Pesticide, 180-181, 184-185

P

Partitioning Factor, 138-140

Pcr, 8-11, 55, 62, 64-66, 68, 70-73, 76, 99, 103, 110, 113-120, 122-124, 126-127, 129-130, 132-135, 150-151, 178

Phylogenetic, 58, 60, 98-99, 101-102, 150-154

Pmn, 200, 202-203

Polymorphism, 10, 13-14, 25, 32, 54-55, 60, 62, 70-71, 74-77, 113-118, 130, 132, 134, 136-137, 152, 154

Ppargc1a, 8-11, 13-14, 130, 136-137

Protozoa, 6, 138-140, 142-143, 172, 177, 184, 187

R

Reference Gene, 119-120, 122-124

Reproductive Performance, 15-16, 23-24, 35, 39, 41-42, 46-47, 51-52

Reproductive Traits, 15, 23, 35-36, 41, 54, 82, 89, 113-115, 136

Ruminal Ammonia, 144-147, 183

S

Scd1, 13, 132-137

Semi-intensive Farming, 36, 38, 40

Sex Determination, 98-100, 102-103

Snp, 10, 13-14, 25-27, 31-32, 61, 71, 76, 103, 130, 132-133, 135-136

Somatic Cell Count, 104, 106, 108, 111, 120, 200, 204-205

W

Wrm, 171-175, 177-178

www.ingramcontent.com/pod-product-compliance
Lightning Source LLC
Chambersburg PA
CBHW080648200326
41458CB00013B/4777